Exploring the Micro World of Robotics Through Insect Robots

U. Vignesh
Vellore Institute of Technology, Chennai, India

Annavarapu Chandra Sekhara Rao
Institute of Engineering and Management, India

Saleem Raja
University of Technology and Applied Sciences, Shinas, Oman

P. Chitra
GITAM University, Bangalore, India

IGI Global
Publishing Tomorrow's Research Today

Published in the United States of America by
IGI Global
701 E. Chocolate Avenue
Hershey PA, USA 17033
Tel: 717-533-8845
Fax: 717-533-8661
E-mail: cust@igi-global.com
Web site: https://www.igi-global.com

Copyright © 2025 by IGI Global. All rights reserved. No part of this publication may be reproduced, stored or distributed in any form or by any means, electronic or mechanical, including photocopying, without written permission from the publisher.
Product or company names used in this set are for identification purposes only. Inclusion of the names of the products or companies does not indicate a claim of ownership by IGI Global of the trademark or registered trademark.

Library of Congress Cataloging-in-Publication Data

CIP Data Pending
ISBN:979-8-3693-6150-4
eISBN:979-8-3693-6152-8

Vice President of Editorial: Melissa Wagner
Managing Editor of Acquisitions: Mikaela Felty
Managing Editor of Book Development: Jocelynn Hessler
Production Manager: Mike Brehm
Cover Design: Phillip Shickler

British Cataloguing in Publication Data
A Cataloguing in Publication record for this book is available from the British Library.

All work contributed to this book is new, previously-unpublished material.
The views expressed in this book are those of the authors, but not necessarily of the publisher.

Table of Contents

Preface xvii

Introduction xxiii

Chapter 1
Nano Robots Promising Advancements and Challenges in Healthcare:
Nanobots in Healthcare 1
    S. Karthigai Selvi, Galgotias University, India
    Sharmistha Dey, Galgotias University, India
    Siva Shankar Ramasamy, Chiang Mai University, Thailand
    Krishan Veer Singh, Galgotias University, India

Chapter 2
Plate Tectonics Uncoiled: Deploying Advanced Algorithms for Earthquake
Reading 17
    S. Sheeba Rani, Sri Eshwar College of Engineering, Coimbatore, India
    M. Mohammed Yassen, Sri Eshwar College of Engineering, Coimbatore, India
    Srivignesh Sadhasivam, Sri Eshwar College of Engineering, Coimbatore, India
    Sharath Kumar Jaganathan, Data Science Institute, Frank J. Guarini School of Business, Saint Peter's University, USA

Chapter 3
Optimizing Resource Management With Edge and Network Processing for
Disaster Response Using Insect Robot Swarms 39
    U. Vignesh, Vellore Institute of Technology, Chennai, India
    K. Gokul Ram, Vellore Institute of Technology, Chennai, India
    Abdulkareem Sh. Mahdi Al-Obaidi, Taylor's University, Malaysia

Chapter 4
Integration of Advanced Obstacle Avoidance in Automated Robots to
Enhance Autonomous Firefighting Capabilities 61
    Monica Bhutani, Bharati Vidyapeeth College of Engineering, New Delhi, India
    Monica Gupta, Bharati Vidyapeeth College of Engineering, New Delhi, India
    Ayushi Jain, Bharati Vidyapeeth College of Engineering, New Delhi, India
    Nishant Rajoriya, Bharati Vidyapeeth College of Engineering, New Delhi, India
    Gitika Singh, Bharati Vidyapeeth College of Engineering, New Delhi, India

Chapter 5
Advancements and Applications of Insect-Inspired Robots 85
    U. Vignesh, Vellore Institute of Technology, Chennai, India
    Arpan Singh Parihar, Vellore Institute of Technology, Chennai, India

Chapter 6
Early Depression Detection Using Modern AI Techniques: Issues,
Opportunities, and Challenges 119
    Sharmistha Dey, Galgotias University, India
    Krishan Veer Singh, Galgotias University, India

Chapter 7
Classification of Diabetic Retinopathy Using Regularized Pre-Trained
Models 139
    Kalpana Devi, Gandhigram Rural Institute, India

Chapter 8
Exploring Microactuators and Sensors in Modern Applications 165
    S. Surya, Saveetha Engineering College, Chennai, India
    R. Elakya, Sri Venkateswara College of Engineering, Chennai, India
    Srinivasan Ramamurthy, University of Science and Technology of Fujairah, UAE
    K. Janani, Manipal Institute of Science and Technology, Manipal, India
    A. Lizy, Vel Tech Rangarajan Dr. Sagunthala R&D Institute of Science and Technology, Chennai, India

Chapter 9
Enhancing Concrete Strength Prognostication Through Machine Learning and Robotics 179
    A. Hema, Vellore Institute of Technology, Chennai, India
    S. Geetha, Vellore Institute of Technology, Chennai, India
    S. Karthiyaini, Vellore Institute of Technology, Chennai, India

Chapter 10
Cooperative Task Execution in Insect-Inspired Robot Swarms Using Reinforcement Learning 197
    R. Elakya, Sri Venkateswara College of Engineering, India
    S. Surya, Saveetha Engineering College, India
    G. Abinaya, University of Southern Queensland, Springfield, Australia
    T. Manoranjitham, SRM Institute of Science and Technology, Ramapuram, India
    R. Thanga Selvi, Vel Tech Rangarajan Dr. Sagunthala R&D Institute of Science and Technology, India

Chapter 11
Revolutionizing Healthcare Through Robotics and AI Integration: A Comprehensive Approach... 213
    C. Saranya Jothi, Vel Tech Rangarajan Dr. Sagunthala R&D Institute of Science and Technology, India
    M. A. Starlin, Vel Tech Rangarajan Dr. Sagunthala R&D Institute of Science and Technology, India
    R. Roselin kiruba, Vel Tech Rangarajan Dr. Sagunthala R&D Institute of Science and Technology, India
    E. Surya, Sethu Institute of Technology, India
    P. Jeevanasree, Vel Tech Rangarajan Dr. Sagunthala R&D Institute of Science and Technology, India
    Santhosh Jayagopalan, British Applied College, UAE

Chapter 12
Slithering Intelligence for Predicting Tectonic Plate Movement 235
    Maheswari Raja, Rajalakshmi Institute of Technology, Coimbatore, India
    Ashiya Parveen, Sri Eshwar College of Engineering, Coimbatore, India
    Manobalan Manokaran, Sri Eshwar College of Engineering, Coimbatore, India
    Mythili Palanisamy, Sri Eshwar College of Engineering, Coimbatore, India
    P. Vijaya, Modern College of Business and Science, Oman

Chapter 13
Proactive Solutions to Mitigate Cryptojacking 253
    E. Helen Parimala, GITAM University (Deemed), India

Conclusion 271

Compilation of References 293

About the Contributors 319

Index 323

Detailed Table of Contents

Preface xvii

Introduction xxiii

Chapter 1
Nano Robots Promising Advancements and Challenges in Healthcare:
Nanobots in Healthcare 1
    S. Karthigai Selvi, Galgotias University, India
    Sharmistha Dey, Galgotias University, India
    Siva Shankar Ramasamy, Chiang Mai University, Thailand
    Krishan Veer Singh, Galgotias University, India

Nanorobots is a boon in medicine, offering extraordinary care and control at the tissue level. This review examines the current state of research on nanorobots, highlighting their potential to revolutionize diagnostics, drug delivery, and surgical procedures. Recent developments in nanofabrication, biocompatibility, and targeted functionality have enabled the creation of nanorobots capable of performing complex tasks within the human body. Key applications include targeted cancer therapy, minimally invasive surgery, and real-time monitoring of physiological conditions. Despite significant progress, challenges such as scalability, safety, and regulatory approval remain. This article explores these challenges and discusses future prospects for integrating nanorobots into mainstream medical practice. By providing a comprehensive overview of current research and anticipated advancements, this review aims to underscore the transformative potential of nanorobots in medicine, paving the way for more effective and personalized healthcare solutions.

Chapter 2

Plate Tectonics Uncoiled: Deploying Advanced Algorithms for Earthquake Reading 17

    S. Sheeba Rani, Sri Eshwar College of Engineering, Coimbatore, India
    M. Mohammed Yassen, Sri Eshwar College of Engineering, Coimbatore, India
    Srivignesh Sadhasivam, Sri Eshwar College of Engineering, Coimbatore, India
    Sharath Kumar Jaganathan, Data Science Institute, Frank J. Guarini School of Business, Saint Peter's University, USA

The proposed device is inspired by the sensory abilities of snakes, which predict earthquakes a few days in advance. To mimic these natural detectors, the device will be equipped with a set of advanced sensors. It includes electromagnetic field sensors to detect subtle changes in the Earth's electromagnetic field, like detection mechanisms that snakes might use. In addition, the device is equipped with a ground vibration sensor so that snakes can sense the vibrations and provides a vital early warning of seismic activity. Infrared sensors were also integrated, allowing the device to detect changes in infrared radiation, and it is theorized that snakes use seismic sensing. Together, these technologies create sensitive and accurate seismic and landscape forecasting systems, providing a new approach to predicting natural disasters and saving lives by providing early warnings.

Chapter 3

Optimizing Resource Management With Edge and Network Processing for Disaster Response Using Insect Robot Swarms 39

    U. Vignesh, Vellore Institute of Technology, Chennai, India
    K. Gokul Ram, Vellore Institute of Technology, Chennai, India
    Abdulkareem Sh. Mahdi Al-Obaidi, Taylor's University, Malaysia

The potential of emerging technology to transform disaster response is examined in this review study. We investigate how, by overcoming the constraints of conventional cloud-based processing in disaster areas, edge computing enables insect robots for real-time data gathering and analysis at the network edge. We examine studies on KubeEdge for reliable network deployment using insect robots and Social Sensing-based Edge Computing (SSEC) for processing social media data. We explore network processing methods that leverage Mobile Cloud Computing (MCC) and show how they overcome issues such as bandwidth limitations and low battery life. This study examines how developments in edge computing, network processing, resource management, and multi-robot systems might benefit disaster response by highlighting the potential of insect robots as indispensable instruments that will ultimately result in quicker and more efficient reaction times.

Chapter 4
Integration of Advanced Obstacle Avoidance in Automated Robots to
Enhance Autonomous Firefighting Capabilities 61
    Monica Bhutani, Bharati Vidyapeeth College of Engineering, New Delhi, India
    Monica Gupta, Bharati Vidyapeeth College of Engineering, New Delhi, India
    Ayushi Jain, Bharati Vidyapeeth College of Engineering, New Delhi, India
    Nishant Rajoriya, Bharati Vidyapeeth College of Engineering, New Delhi, India
    Gitika Singh, Bharati Vidyapeeth College of Engineering, New Delhi, India

The integration of obstacle avoidance and fire extinguishing in robots is essential for large-scale projects. This paper presents the design, development, and performance evaluation of autonomous robots with fire detection and extinguishing capabilities. Using fire and temperature sensors for accurate detection and ultrasonic sensors for obstacle avoidance, the robot navigates dynamic environments while adhering to safety requirements. The paper details the selection and integration of hardware components, including sensors, actuators, and microcontrollers, and evaluates the robot's performance under varying environmental conditions. Experimental results highlight the robot's adaptability to complex scenarios and weight changes. The study also explores future improvements, such as advanced decision-making algorithms, aiming to enhance autonomous robotics for emergency situations and effective communication in critical scenarios.

Chapter 5
Advancements and Applications of Insect-Inspired Robots 85
    U. Vignesh, Vellore Institute of Technology, Chennai, India
    Arpan Singh Parihar, Vellore Institute of Technology, Chennai, India

Insect-inspired robots offer a fascinating avenue for exploring the micro world of robotics, drawing inspiration from the remarkable capabilities of natural organisms. This paper delves into the multidisciplinary field of insect robotics, highlighting the biomechanical principles, behavioral dynamics, and technological advancements driving its evolution. By emulating the agility, adaptability, and efficiency of insects, these robots navigate complex environments with ease, opening new avenues for applications in search and rescue missions, environmental monitoring, and beyond. Miniaturization plays a pivotal role, enabling these robots to access confined spaces and gather valuable data in areas inaccessible to larger machines. Furthermore, swarm robotics harnesses collective intelligence, groups of robots to collaborate and solve complex tasks autonomously. However, designing insect-inspired robots poses challenges, requiring biology, mechanics, and control systems knowledge. Overcoming these hurdles promises a future revolutionize exploration and interaction with the micro world.

Chapter 6
Early Depression Detection Using Modern AI Techniques: Issues, Opportunities, and Challenges 119
    Sharmistha Dey, Galgotias University, India
    Krishan Veer Singh, Galgotias University, India

Depression is a widespread and debilitating mental health disorder, impacting over 300 million individuals globally, as reported by the World Health Organization. Early detection and timely intervention are essential for effective treatment and mitigating the severity of depressive symptoms. However, accurately identifying the nuanced symptoms of depression—manifested through body language, speech patterns, or neurological signals—remains a significant challenge. The advent of modern AI technologies has revolutionized the landscape of depression detection, offering new methodologies for identifying these symptoms This study investigates the current challenges, opportunities, and advancements in AI-driven approaches to early depression detection. We conducted a comprehensive review of approximately 60 high-quality, peer-reviewed research articles from reputable journals and conferences, focusing on the relevance and objectives of each study. Our findings highlight the latest trends in depression detection and outline the obstacles faced in this field, providing a roadmap for future researchers aiming to enhance early detection strategies and improve mental health outcomes.

Chapter 7
Classification of Diabetic Retinopathy Using Regularized Pre-Trained
Models 139
Kalpana Devi, Gandhigram Rural Institute, India

Diabetic Retinopathy is a serious sight-threatening complication of diabetes. Deep learning is a superior method for classifying medical images with high accuracy. Typically, we have classified diseases well using transfer learning on pre-trained models from other domains. It utilizes the parameters of a model that has been pre-trained for DR datasets rather than creating new CNN architectures for diverse classification tasks in different domains. The main objective of this chapter is to categorize the Diabetic Retinopathy disease based on its severity using pre-trained models namely ResNet50, VGG16, Alex Net, InceptionV3, Mobile Net and Squeeze Net, DenseNet-121, and XceptionNet and to proposed regularize Xception net experiment with various dropout values.

Chapter 8
Exploring Microactuators and Sensors in Modern Applications 165
S. Surya, Saveetha Engineering College, Chennai, India
R. Elakya, Sri Venkateswara College of Engineering, Chennai, India
Srinivasan Ramamurthy, University of Science and Technology of Fujairah, UAE
K. Janani, Manipal Institute of Science and Technology, Manipal, India
A. Lizy, Vel Tech Rangarajan Dr. Sagunthala R&D Institute of Science and Technology, Chennai, India

The development of robotic systems across various applications is increasingly propelled by advancements in microactuators and sensors. This chapter provides a comprehensive exploration of their pivotal role, the challenges they present, and their transformative impact on modern robotics. Beginning with foundational concepts, it progresses to a detailed examination of diverse microactuators and sensors, elucidating their functionalities and potential applications. Emphasizing the crucial role of feedback mechanisms in enhancing robotic performance, the chapter delves into integration schemes and control mechanisms crucial for effective operation. Current research initiatives and future directions are discussed, highlighting ongoing efforts to overcome challenges such as miniaturization, power consumption, and reliability. By addressing these hurdles, researchers aim to unlock the full potential of sensors and microactuators, fostering innovation and advancing the capabilities of contemporary robotic systems.

Chapter 9
Enhancing Concrete Strength Prognostication Through Machine Learning and Robotics 179
 A. Hema, Vellore Institute of Technology, Chennai, India
 S. Geetha, Vellore Institute of Technology, Chennai, India
 S. Karthiyaini, Vellore Institute of Technology, Chennai, India

Concrete is a widely used construction material globally due to its exceptional properties. The strength of concrete varies based on the composition of cement, blast furnace slag, fly ash, water, superplasticizer, coarse aggregate, and fine aggregate. By altering the ingredients in different proportions, the process can anticipate the concrete strength through the various machine learning techniques. The study in this chapter involves the utilization and comparison of three algorithms, viz. Random Forest, Gradient Boosting, and Linear Regression models to analyse the concrete strength. Among these models, gradient boosting yielded superior results. In order to predict concrete strength from these models, three sensors in the field paves the way for better analysis, such as, an acoustic emission sensor, a strain gauge sensor, and a wireless concrete maturity sensor. This approach proves to be highly effective in achieving optimal concrete strength using machine learning techniques.

Chapter 10
Cooperative Task Execution in Insect-Inspired Robot Swarms Using
Reinforcement Learning 197
    R. Elakya, Sri Venkateswara College of Engineering, India
    S. Surya, Saveetha Engineering College, India
    G. Abinaya, University of Southern Queensland, Springfield, Australia
    T. Manoranjitham, SRM Institute of Science and Technology,
        Ramapuram, India
    R. Thanga Selvi, Vel Tech Rangarajan Dr. Sagunthala R&D Institute of
        Science and Technology, India

This chapter proposes a novel framework for cooperative task execution in a swarm of insect-inspired robots by using Reinforcement Learning (RL) algorithms. Inspired by the collaborative behaviors observed in social insects, such as ants and bees, the proposed framework enables robots to autonomously coordinate their actions to accomplish complex tasks in dynamic environments. Each robot in the swarm acts as an autonomous agent capable of learning and adapting its behavior through interactions with the environment and feedback from other robots. By applying RL algorithms, such as Q-learning or Deep Q-Networks (DQN), robots learn optimal action policies to maximize task performance while considering the collective objectives of the swarm. we demonstrate the effectiveness and scalability of our approach in various cooperative tasks, including exploration, foraging, and object manipulation. This project showcases the potential of RL-based approaches to enhance the autonomy and adaptability of robotic swarms for collaborative task execution in real-world scenarios.

Chapter 11
Revolutionizing Healthcare Through Robotics and AI Integration: A
Comprehensive Approach 213
    C. Saranya Jothi, Vel Tech Rangarajan Dr. Sagunthala R&D Institute of
        Science and Technology, India
    M. A. Starlin, Vel Tech Rangarajan Dr. Sagunthala R&D Institute of
        Science and Technology, India
    R. Roselin kiruba, Vel Tech Rangarajan Dr. Sagunthala R&D Institute of
        Science and Technology, India
    E. Surya, Sethu Institute of Technology, India
    P. Jeevanasree, Vel Tech Rangarajan Dr. Sagunthala R&D Institute of
        Science and Technology, India
    Santhosh Jayagopalan, British Applied College, UAE

Artificial Intelligence (AI) is the technology focused on innovating the role of robotics in the healthcare process. The integration of robots in the field of healthcare and medication can effectively recognize, treat, and manage health problems. In traditional healthcare system follows the manual diagnosis methods, manual surgery, pharmacological treatments, and patient monitoring. The drawbacks of traditional methods are lack of accuracy, high cost, limited accessibility, and social barriers. To overcome these issues this paper provides a detailed review of advanced robotic-assisted surgeries, Remote Patient Monitoring, patient engagement with robotic bonds, and enhancement of patient-centered care. It increases efficiency, decreases cost, and also improves the outcome of medical and healthcare systems. Additionally, it includes the applications, challenges, and possible future impacts on healthcare.

Chapter 12
Slithering Intelligence for Predicting Tectonic Plate Movement 235
    Maheswari Raja, Rajalakshmi Institute of Technology, Coimbatore, India
    Ashiya Parveen, Sri Eshwar College of Engineering, Coimbatore, India
    Manobalan Manokaran, Sri Eshwar College of Engineering, Coimbatore, India
    Mythili Palanisamy, Sri Eshwar College of Engineering, Coimbatore, India
    P. Vijaya, Modern College of Business and Science, Oman

An earthquake is one of the most devastating natural catastrophes that may cause major infrastructure damage and casualties. Early earthquake detection can be crucial for minimizing damage and saving lives. The purpose of this study is to make earthquake magnitude and depth predictions utilizing factors including time, place, and previous seismic activity data. Snakes can predict earthquakes and landscapes 3-5 days before they occur, and they can seismically change up to 120 kilometers (about 74.56 mi) away from the Epicenter (the point from the Earth's surface directly above the focus of an earthquake). They may utilize their specialized sense organs to sense electromagnetic fields and ground vibrations. Snakes might use IR radiation detection through their eyes to sense seismic changes. Unusual behavior in snakes could be indicators of approaching earthquakes, as a response to sensed vibrations to sensed vibrations or electromagnetic fields.

Chapter 13
Proactive Solutions to Mitigate Cryptojacking 253
  E. Helen Parimala, GITAM University (Deemed), India

Cryptojacking refers to the unauthorized utilization of computing resources to mine cryptocurrencies, threatening individuals, organizations, and, most importantly, critical infrastructures in cloud and on-premises systems. This research addresses the escalating cryptojacking threat by developing proactive solutions to effectively detect and mitigate these attacks. Leveraging security tools and technologies like machine learning, network traffic analysis, and behavioral analysis, propose a comprehensive "StyxShield" application capable of identifying and responding to cryptojacking incidents in real time. Through testing and evaluation using real-world datasets and simulated attack scenarios, we demonstrate the effectiveness of the proposed solutions in mitigating cryptojacking threats across diverse environments.This research contributes to the advancement of cybersecurity by empowering individuals and organizations to proactively defend against cryptojacking and safeguard their valuable resources from exploitation by malicious actors.

Conclusion 271

Compilation of References 293

About the Contributors 319

Index 323

Preface

In the face of increasingly complex and severe disasters, the integration of innovative technologies into recovery efforts is not merely advantageous but essential. As the editors of "Introduction to Insect Robotics in Disaster Recovery," we are pleased to present a volume dedicated to exploring the transformative potential of insect robotics in addressing the multifaceted challenges of disaster management. This reference book aims to provide a comprehensive overview of the principles, advancements, and applications of insect robotics, underscoring their critical role in disaster recovery and emergency response.

This collection of chapters brings together leading experts from various fields, including robotics, engineering, and emergency management, to offer a thorough examination of the cutting-edge developments and theoretical foundations of insect robotics. The book is structured to address both the technical and practical aspects of this rapidly evolving domain, beginning with foundational principles and advancing to real-world applications.

Our journey begins with an in-depth exploration of the principles of microcontrollers in insect robot design, highlighting the fundamental components that enable these robots to perform complex tasks. This is followed by a detailed discussion on advancements in micro sensors, which enhance the perceptual capabilities of insect robots, allowing them to navigate and interact with their environments more effectively. Energy-efficient micro energy solutions are examined next, addressing the critical need for sustainable power sources that support the extended operation of these small-scale robots in disaster scenarios.

The book further delves into navigation and locomotion strategies, providing insights into the movement mechanisms that allow insect robots to traverse challenging terrains. The discussion on swarm intelligence and coordination illustrates how teams of insect robots can work collaboratively to achieve complex goals, a vital capability in large-scale disaster recovery operations.

Communication systems for collaborative networks, micro-scale manipulation and payload delivery, and adaptive learning and control are explored in subsequent chapters, each contributing to a more nuanced understanding of how insect robots can be deployed effectively in diverse recovery scenarios. Technological challenges and future directions are also addressed, offering a forward-looking perspective on the ongoing development and potential of this technology.

The application of insect robotics in environmental monitoring, disaster mapping, and humanitarian aid delivery is highlighted, demonstrating the practical benefits and real-world impact of these technologies. Case studies provide concrete examples of insect robots in action, illustrating their resilience and robustness in various disaster recovery operations.

Interdisciplinary perspectives shed light on the collaborations between robotics, engineering, and emergency management that drive innovation in this field. The ethical and governance issues surrounding the deployment of insect robots, as well as policy implications and public perception, are critically examined to ensure responsible and effective integration into national disaster recovery plans.

Finally, the exploration of biomimicry and bioinspired design reveals how insights from nature are harnessed to enhance the functionality and performance of insect robots, bridging the gap between natural and engineered systems.

As editors, Muhammad Shahbaz from the University of Education Lahore and Fakhar Shahzad from Shenzhen University, we believe this book will serve as an invaluable resource for researchers, practitioners, and policymakers engaged in the development and deployment of insect robotics. We hope that the insights and knowledge shared within these pages will inspire further innovation and collaboration, ultimately advancing the field and enhancing our collective capacity to respond to and recover from disasters.

We invite you to explore the chapters ahead and engage with the rich tapestry of research and applications presented. Together, let us push the boundaries of what is possible in disaster recovery and emergency response through the transformative power of insect robotics.

Exploring the Micro World of Robotics Through Insect Robots," a compelling anthology that captures the frontier of robotics research through the lens of insect-inspired technologies. This collection brings together cutting-edge studies and innovations that highlight the transformative potential of micro-scale robotics in diverse applications.

Chapter 1, "Nano Robots: Promising Advancements and Challenges in Healthcare," explores the revolutionary impact of nanorobots on medical science. By focusing on the latest advancements in nanofabrication, biocompatibility, and targeted functionality, this chapter details how nanorobots are poised to redefine diagnostics, drug delivery, and surgical interventions. It underscores the transformative potential of these technologies while also addressing critical challenges such as scalability, safety, and regulatory hurdles. The review provides a forward-looking perspective on how nanorobots could significantly enhance personalized healthcare.

In Chapter 2, "Plate Tectonics Uncoiled: Deploying Advanced Algorithms for Earthquake Reading," the focus shifts to an innovative device inspired by the sensory abilities of snakes. This chapter presents a novel approach to earthquake prediction, utilizing electromagnetic field, ground vibration, and infrared sensors to detect early signs of seismic activity. By mimicking the natural earthquake detection mechanisms of snakes, the proposed device aims to offer improved accuracy and sensitivity in forecasting earthquakes, potentially saving lives through timely warnings.

Chapter 3, "Optimizing Resource Management with Edge and Network Processing for Disaster Response Using Insect Robot Swarms," delves into the integration of edge computing and network processing within insect robot swarms for disaster response. This chapter highlights strategies for optimizing resource management and coordination in complex disaster scenarios, demonstrating how swarm robotics can enhance efficiency and effectiveness in emergency situations.

The fourth chapter, "Integration of Advanced Obstacle Avoidance in Automated Robots to Enhance Autonomous Firefighting Capabilities," examines the development and performance of autonomous firefighting robots equipped with advanced obstacle avoidance and fire extinguishing technologies. The chapter details the robot's design, including the integration of fire and temperature sensors and ultrasonic obstacle avoidance, and evaluates its performance in varied conditions. It also explores future improvements to enhance autonomous decision-making and communication in critical scenarios.

"Micro World of Robotics: Advancements and Applications of Insect-Inspired Robots," presented in Chapter 5, offers a comprehensive overview of the field of insect robotics. This chapter discusses the biomechanical and behavioral principles driving the design of insect-inspired robots, their miniaturization, and the potential applications in areas such as search and rescue and environmental monitoring. It also addresses the challenges of integrating biological insights with robotic technology to achieve practical and innovative solutions.

Chapter 6, "Early Depression Detection Using Modern AI Techniques: Issues, Opportunities, and Challenges," investigates the role of artificial intelligence in the early detection of depression. The chapter reviews recent advancements in AI technology and its application to identifying subtle symptoms of depression through body language, speech, and brain signals. By analyzing existing research, it provides insights into current challenges and opportunities, offering a roadmap for future developments in mental health diagnostics.

In Chapter 7, "Classification of Diabetic Retinopathy Using Regularized Pretrained Models," the focus is on utilizing deep learning for the accurate classification of diabetic retinopathy. This chapter compares the effectiveness of various pretrained models, including ResNet50 and XceptionNet, and explores the application of regularization techniques to enhance classification performance. It highlights how advanced deep learning methods can improve diagnostic accuracy for this serious condition.

"Exploring Microactuators and Sensors in Modern Applications," covered in Chapter 8, discusses the advancements in microactuators and sensors that drive modern robotics. The chapter examines the foundational concepts and functionalities of these components, their integration into robotic systems, and the ongoing research to address challenges such as miniaturization and power consumption. By showcasing current innovations, it illustrates the pivotal role of microactuators and sensors in enhancing robotic capabilities.

Chapter 9, "Enhancing Concrete Strength Prognostication through Machine Learning and Robotics," explores the use of machine learning techniques to predict concrete strength. This chapter evaluates various algorithms, including Random Forest and Gradient Boosting, and integrates sensor data to optimize concrete strength predictions. It demonstrates how machine learning and sensor technologies can improve construction practices and material performance.

Chapter 10, "Cooperative Task Execution in Insect-Inspired Robot Swarms Using Reinforcement Learning," introduces a framework for enhancing cooperative tasks within insect-inspired robot swarms using reinforcement learning. By drawing inspiration from the collaborative behaviors of social insects, this chapter demonstrates how RL algorithms can enable autonomous coordination and adaptability in swarm robotics, showcasing their potential for complex tasks such as exploration and object manipulation.

Finally, Chapter 11, "Revolutionizing Healthcare through Robotics and AI Integration: A Comprehensive Approach," reviews how AI and robotics are transforming healthcare. The chapter discusses advancements in robotic-assisted surgeries, remote patient monitoring, and patient engagement, highlighting the potential for these technologies to increase efficiency, reduce costs, and improve healthcare outcomes.

Chapter 12, "Slithering Intelligence for Predicting Tectonic Plate Movement," closes the volume by examining the potential for using the seismic sensing abilities of snakes to predict earthquakes. This chapter explores how the specialized sensory mechanisms of snakes can be applied to forecasting seismic events, offering a novel approach to early detection and disaster mitigation.

In chapter 13, Demystifying the Invisible Threat Proactive Solutions To Mitigate Cryptojacking" explores the critical issue of cryptojacking, where unauthorized parties hijack computing resources to mine cryptocurrencies, impacting individuals, organizations, and essential infrastructures. The chapter presents "StyxShield," a comprehensive application designed to proactively detect and mitigate cryptojacking attacks using advanced technologies such as machine learning, network traffic analysis, and behavioral analysis. By evaluating StyxShield with real-world datasets and simulated attack scenarios, the research demonstrates its effectiveness in real-time threat detection and response across various environments. This work advances cybersecurity by providing a robust tool for defending against cryptojacking and protecting valuable resources from malicious exploitation.

Together, these chapters offer a rich tapestry of insights and innovations at the forefront of robotics research, illustrating the profound impact of insect-inspired technologies on a range of critical applications.

As we conclude this exploration into "Exploring the Micro World of Robotics Through Insect Robots," it is clear that the integration of innovative robotic technologies has the potential to reshape our approach to disaster recovery and many other critical fields. Our comprehensive journey through this volume has underscored the transformative capabilities of insect-inspired robotics, revealing both the immense promise and the complex challenges that lie ahead.

Our examination begins with the foundational principles of microcontroller design and the advances in micro sensors, setting the stage for understanding how these tiny, yet powerful, robots can perform intricate tasks with precision. The discussions on energy-efficient micro solutions and innovative navigation strategies demonstrate the continuous evolution of technology, ensuring that these robots can operate effectively in the harshest of environments.

The exploration of swarm intelligence and cooperative task execution highlights the remarkable potential of insect robot swarms to tackle complex problems through collective effort. By drawing inspiration from nature's own solutions, these technologies offer new pathways to optimize resource management and enhance efficiency in disaster scenarios.

The chapters dedicated to advanced obstacle avoidance, micro-scale manipulation, and adaptive learning reflect the cutting-edge developments that are pushing the boundaries of what insect robots can achieve. Each section contributes to a deeper understanding of how these robots can be deployed in real-world applications, from environmental monitoring to humanitarian aid.

We also delve into the application of AI and machine learning, illustrating how these technologies complement and enhance the capabilities of insect robots. The discussions on early depression detection, diabetic retinopathy classification, and concrete strength prognostication showcase the broader impact of robotics and AI on healthcare and construction, emphasizing the interdisciplinary nature of this field.

Throughout this volume, we have seen the value of interdisciplinary collaboration, bridging robotics, engineering, emergency management, and beyond. The ethical considerations and policy implications discussed are essential for the responsible deployment of these technologies, ensuring that their integration into disaster recovery plans is both effective and respectful of societal values.

In closing, this book, curated by Muhammad Shahbaz from the University of Education Lahore and Fakhar Shahzad from Shenzhen University, stands as a testament to the incredible advancements in insect robotics and their potential to revolutionize various sectors. We hope that the insights provided will inspire continued research, innovation, and collaboration, driving the field forward and enhancing our ability to respond to and recover from disasters.

We invite you to delve into the chapters ahead, engage with the rich tapestry of research and applications, and join us in pushing the boundaries of what is possible through the transformative power of insect robotics. Together, let us harness these advancements to create a more resilient and responsive world.

Introduction

We are living in an era where technological advancements are accelerating at an unprecedented rate. Artificial intelligence (AI), machine learning, and advanced engineering are converging to create innovative solutions that were previously unimaginable. This synergy of technologies has given rise to a new frontier in robotics, one that draws inspiration from the natural world. At the heart of this technological revolution lies the fascinating world of insect robots. These bio-inspired robots, also known as microrobots, represent a new generation of machines that embody the fusion of nature's ingenuity with human technological prowess. By mimicking the behaviour, physiology, and characteristics of insects, researchers and engineers are creating robots that can perform tasks with unprecedented precision, agility, and efficiency.

Insects, despite their small size, have evolved remarkable strategies to survive, adapt, and thrive in diverse environments. Their ability to navigate complex terrains, respond to sensory stimuli, and interact with their ecosystems has inspired scientists and engineers to develop robots that can replicate these functions. By studying the intricacies of insect biology, researchers can unlock new design principles, materials, and mechanisms that can be applied to robotics and engineering. Insect robots serve as a testament to the power of interdisciplinary innovation. By combining insights from biology, engineering, materials science, and computer science, researchers can create robots that not only mimic the behaviour of insects but also push the boundaries of what is technologically possible. This fusion of knowledge from diverse fields has led to the development of robots that can fly, swim, crawl, and even walk, with applications ranging from environmental monitoring to search and rescue operations.

As we delve into the pages of "Exploring the Micro World of Robotics Through Insect Robots," we embark on a journey through a realm where the boundaries between biology and technology blur. This book offers a comprehensive exploration of the current state of insect robotics, highlighting the key advancements, challenges, and future directions in this field. Through detailed chapters and case studies, readers will gain a deeper appreciation of how the micro world of insects is shaping the future of robotics and engineering.

The study of insect robots offers profound insights into both biology and technology. By understanding how insects adapt, respond, and interact with their environments, researchers can gain a deeper appreciation of the intricate mechanisms that govern life. Conversely, by developing robots that mimic insect behaviour, engineers can push the boundaries of technological innovation, creating machines that can perform tasks with unprecedented precision and efficiency. This book invites readers to explore the fascinating world of insect robots, where the intersection of biology and technology is revolutionizing our understanding of both fields.

Insect robots represent a distinct category of robotics that draws direct inspiration from the fascinating world of insects. These tiny creatures have evolved remarkable strategies to survive, adapt, and thrive in diverse environments, making them an ideal model for robotic design. By emulating the behaviours, physiology, and characteristics of insects, researchers and engineers can create robots that are not only efficient but also remarkably adaptable. Unlike traditional robotics, which often rely on larger, more complex mechanisms, insect robots embody the simplicity and effectiveness of nature's designs. Insects have evolved to achieve remarkable feats with minimal resources, making them a perfect example of efficiency in design. By studying insect biology, researchers can identify key principles and mechanisms that can be applied to robotics, resulting in robots that are more agile, resilient, and effective. Insect robots are not merely mechanical replicas of insects; they are sophisticated devices engineered to mimic the functional capabilities of their biological counterparts. These robots are designed to replicate the remarkable abilities of insects, such as flight, navigation, and sensory perception. By doing so, researchers can create robots that can perform tasks with unprecedented precision and efficiency.

The study of insect robots allows us to appreciate the incredible adaptability and efficiency inherent in insect biology while pushing the limits of what technology can achieve. By exploring the intricacies of insect behaviour and physiology, researchers can unlock new design principles, materials, and mechanisms that can be applied to robotics and engineering. This, in turn, can lead to cutting-edge advancements in fields such as search and rescue, environmental monitoring, and medical robotics. They also offer a unique opportunity to explore the fascinating world of insect biology. By studying the behaviour, physiology, and characteristics of insects, researchers can gain a deeper understanding of the intricate mechanisms that govern life. This knowledge can be applied to develop robots that are not only efficient but also remarkably adaptable, capable of thriving in diverse environments and performing tasks with unprecedented precision. The fascination with insect robots stems from their remarkable ability to replicate a wide range of behaviours and functions observed in real insects. These robots are designed to embody the incredible adaptability, agility, and resilience of insects, making them ideal for tasks that require precision, flexibility, and reliability.

Mimic robots are engineered to mimic the remarkable abilities of insects, such as:

- Navigation: Replicating the way a beetle can navigate through dense foliage, insect robots can be designed to traverse complex terrains and environments.
- Flight: Emulating the remarkable agility of a dragonfly in flight, insect robots can be developed to fly with precision and speed.
- Sensory Perception: Mimicking the acute sensory perception of insects, insect robots can be equipped with sensors to detect and respond to environmental stimuli.

This bio-inspired approach is not just about mimicry; it's about leveraging the principles of natural evolution to create robots that can perform tasks in environments and situations where traditional robots might falter. By studying the evolution of insects and their adaptations to diverse environments, researchers can unlock new design principles, materials, and mechanisms that can be applied to robotics. The development of these robots are far-reaching applications across various fields, including:

- Search and Rescue: Insect robots can navigate through rubble and debris to locate survivors in disaster-stricken areas.
- Environmental Monitoring: Insect robots can be deployed to monitor climate conditions, pollutant levels, and ecological changes in sensitive ecosystems.
- Medical Applications: Insect robots can be designed for minimally invasive surgeries, targeted drug delivery, and diagnostic procedures.

These innovations pushes the boundaries of robotics, enabling the creation of machines that can perform tasks with unprecedented precision, agility, and adaptability. By embracing the bio-inspired approach, researchers can unlock new possibilities for robotics, leading to innovative solutions for complex problems in various fields. The journey of insect robots begins with a meticulous exploration of the underlying principles of biomechanics and sensory systems that drive insect behaviour. Despite their small size, insects exhibit a remarkable range of complex behaviours that are intricately adapted to their environments. To understand these behaviours, researchers must adopt a multidisciplinary approach, combining insights from: Entomology: The study of insects, including their behaviour, physiology, and ecology; Biomechanics: The application of mechanical principles to understand the structure and function of living organisms; Robotics: The design and development of robots that can perform tasks with precision and efficiency.

This book investigates into the crucial aspects of insect physiology that influence robotic design, including: Locomotion: The study of insect movement, including walking, flying, and swimming. Sensory Processing: The analysis of how insects perceive and respond to environmental stimuli, such as light, sound, and touch and adaptive Behaviour: The examination of how insects adapt to changing environments and situations.

By decoding these biological systems, engineers and scientists can develop robots that mimic insect functions with high efficiency. This involves identifying key design principles by understanding the underlying mechanisms that enable insects to perform complex tasks, developing biomimetic solutions by applying these design principles to create robots that can replicate insect behaviours and integrating sensory systems that can perceive and respond to environmental stimuli like insects. By exploring the intricate world of insect physiology, researchers can unlock new possibilities for robotic design. Insect robots can be developed to perform tasks with unprecedented precision, agility, and adaptability, leading to innovative solutions for complex problems in various fields.

One of the most compelling aspects of insect robots is their potential to revolutionize various fields through innovative applications. In the demesne of environmental monitoring, tiny insect robots equipped with sensors could provide valuable data on climate conditions, pollutant levels, and ecological changes with minimal disruption to natural habitats. This could be particularly useful in tracking changes in sensitive ecosystems, monitoring wildlife populations, and detecting early signs of natural disasters.

In search and rescue operations, insect robots could play a critical role in navigating through rubble and confined spaces where larger machines cannot reach. Their small size and agility would allow them to access areas that are currently inaccessible, offering hope and assistance in critical situations. This could be especially important in responding to earthquakes, hurricanes, and other disasters where every minute counts.

Moreover, in medical fields, insect-inspired robots could be used for precise surgical procedures or for delivering targeted therapies within the human body. Their small size and flexibility would enable them to navigate through tiny spaces and reach areas that are currently inaccessible to traditional medical instruments. This could lead to new treatments for a range of diseases and conditions, and could potentially revolutionize the field of medicine. Overall, the potential applications are vast and varied, and could have a significant impact on a range of fields. Their small size, agility, and flexibility make them ideal for tasks that require precision, flexibility, and minimal disruption, and their potential to revolutionize various fields is vast and exciting.

As we probe into the mesmerising dominion of insect robots, it becomes clear that this field extends far beyond mere bio mimicry. Instead, it represents a thoughtful convergence of natural and artificial systems, where the boundaries between biology and technology shape. By exploring the intricate world of insect robots, we not only push the frontiers of technological innovation but also gain a deeper understanding of the intricate complexities of biological systems. This book offers a comprehensive and authoritative overview of the current state of insect robotics, showcasing key breakthroughs, challenges, and future directions. Through in-depth chapters and detailed case studies, readers will embark on a journey to discover the transformative potential of insect robots. We will explore how these tiny machines are revolutionizing fields such as environmental monitoring, search and rescue, and medical robotics, and how they are inspiring new approaches to artificial intelligence, machine learning, and materials science.

As we examine the micro world of insect robots, we will uncover the intricate mechanisms that govern their behaviour, from locomotion and sensory perception to adaptive behaviour and social interaction. We will also explore the cutting-edge technologies that enable these robots to mimic the remarkable abilities of their biological counterparts, from advanced materials and micro fabrication techniques to sophisticated algorithms and control systems.

Ultimately, this book aims to inspire a new generation of researchers, engineers, and innovators to join the quest to harness the potential of insect robots. By exploring the fascinating intersection of biology and technology, we can unlock new possibilities for robotics and artificial intelligence, and create a future where humans and machines coexist in harmony with nature. We find ourselves at the threshold of a revolutionary era in technology, where the boundaries between biology and engineering dissolve, and giving rise to unprecedented possibilities. As we venture into the intricate land of these remarkable machines, we are met with a dual marvel: the ingenuity of engineering and the profound connection between the smallest forms of life and our most advanced technological creations.

As we navigate this uncharted territory, we begin to appreciate the profound implications of insect robots, from transforming our understanding of biological systems to redefining the frontiers of artificial intelligence and robotics. We witness the emergence of a new paradigm, where technology is no longer merely a product of human ingenuity but a symbiotic fusion of natural and artificial systems. Join us on this captivating journey, as we explore the micro world of robotics and uncover the wonders that await us at the intersection of biology and technology.

In the rapidly evolving landscape of robotics, a singular area has emerged as a testament to the boundless potential of innovation: insect robots. These tiny, bio-inspired wonders represent a remarkable convergence of nature and technology, demonstrating how the meticulous study of even the smallest creatures can yield monumental breakthroughs in technological advancement.

The book, exploring the Micro World of Robotics Through Insect Robots, embarks on an immersive journey into this captivating area, where the intricacies of biology and robotics intersect. Through a detailed exploration of the design, development, and application of insect robots, this comprehensive guide reveals the profound impact that nature's smallest inhabitants can have on driving technological progress.

By examining the remarkable adaptability, agility, and resilience of insects, researchers and engineers can unlock new design principles, materials, and mechanisms that can be applied to robotics. This, in turn, can lead to revolutionary applications across various fields, from environmental monitoring and search and rescue to medical robotics and beyond. Through its in-depth examination of the micro world of robotics, this book aims to inspire a new generation of innovators, researchers, and engineers to embrace the potential of bio-inspired robotics. By embracing the wonders of nature and harnessing the power of technological innovation, we can unlock unprecedented possibilities and create a future where humans and machines coexist in harmony with the natural world.

The fascination with insect robots stems from their capacity to mimic the remarkable diversity and specialization of insect behaviours. Despite their diminutive size, insects exhibit an astonishing array of complex behaviours, such as the precise flight of dragonflies, the intricate navigation of ants, and the remarkable adaptability of beetles. These behaviours are the culmination of millions of years of evolutionary refinement, and by studying them, scientists and engineers have distilled valuable design principles that can be applied to modern robotics.

The resulting insect-inspired robots are not only functional but also remarkably efficient, agile, and capable of performing tasks in environments that are challenging or inaccessible to larger, traditional robots. These robots can navigate through dense foliage, traverse treacherous terrain, and even fly with incredible precision, making them ideal for applications such as environmental monitoring, search and rescue, and medical robotics. The study of insect behaviour has also led to the development of novel materials, mechanisms, and control systems that enable robots to replicate the remarkable abilities of their biological counterparts. For example, researchers have developed advanced materials that mimic the strength and flexibility of insect exoskeletons, as well as sophisticated algorithms that enable robots to navigate and adapt to complex environments.

As the field of insect robotics continues to evolve, we can expect to see even more innovative applications of these design principles, leading to a new generation of robots that are not only inspired by nature but also capable of surpassing the abilities of their biological counterparts. Insect robots embody a biomimetic approach to engineering, where nature's designs are emulated to create innovative technological solutions. By mimicking the form and function of biological organisms, engineers can tap into the inherent advantages of nature's designs, resulting in robots that are often more adaptable, resilient, and efficient than their conventional counterparts.

The compact and versatile design of beetles, for example, has inspired the development of small robots that can navigate through tight spaces and complex terrains with ease. These robots can access areas that would be difficult or impossible for larger robots to reach, making them ideal for applications such as search and rescue, environmental monitoring, and medical robotics. Similarly, the efficient flight mechanisms of insects have inspired innovations in micro aerial vehicles (MAVs). By studying the wing movements and aerodynamics of insects, engineers have developed MAVs that can hover and maneuver with remarkable precision, even in confined spaces. These MAVs have far-reaching potential in fields such as surveillance, agriculture, and package delivery.

The biomimetic approach to engineering also enables the development of robots that can adapt to changing environments and situations. By emulating the sensory perception and decision-making processes of insects, engineers can create robots that can respond to their surroundings and make decisions in real-time, much like their biological counterparts. Overall, the biomimetic approach to engineering, as exemplified by insect robots, offers a powerful paradigm for creating innovative technological solutions that are inspired by nature's designs. By embracing this approach, engineers can unlock new possibilities for robotics and create machines that are more adaptable, resilient, and efficient than ever before.

The creation of insect robots necessitates a profound comprehension of the intricate mechanical and sensory systems that govern insect behaviour. These systems encompass complex facets of locomotion, sensory processing, and adaptive behaviour, which enable insects to navigate and interact with their environments with remarkable precision and agility.

To replicate these sophisticated functionalities in robots, engineers must dig into the biological systems that underpin insect behaviour, dissecting the intricate mechanisms that enable insects to move, sense, and adapt. This endeavour requires an interdisciplinary approach, synthesizing knowledge from diverse fields such as:

1. Entomology: The study of insects, providing insights into their behaviour, ecology, and evolution.
2. Biomechanics: The analysis of the mechanical properties and processes that govern insect movement and behaviour.
3. Materials Science: The development of novel materials and structures that mimic the remarkable properties of insect exoskeletons and other biological tissues.
4. Robotics: The design and engineering of robots that can replicate the complex behaviours and functionalities of insects.

By integrating knowledge from these fields, engineers can create insect robots that not only mimic the appearance and movement of insects but also replicate their sophisticated sensory and adaptive abilities. The book offers a comprehensive examination of these disciplines, providing insights into how they intersect and contribute to the creation of advanced insect robots. Through this interdisciplinary approach, researchers and engineers can unlock new possibilities for robotics, creating machines that are more adaptable, resilient, and efficient than ever before.

The applications of insect robots are incredibly diverse, spanning multiple fields and industries. In environmental monitoring, for instance, small, insect-like robots can be deployed to collect data in areas that are difficult or impossible for humans to access. These robots can be designed to monitor pollution levels, track changes in ecosystems, and even perform tasks related to conservation efforts. By leveraging their tiny size and agility, insect robots can navigate through dense foliage or enter small crevices, providing valuable insights into environmental conditions.

In search and rescue operations, insect robots offer a critical advantage. They can navigate through debris and confined spaces to locate survivors, reaching areas that traditional rescue equipment may struggle to access. This can be particularly important in situations where time is of the essence, such as in the aftermath of a natural disaster or building collapse. The field of medicine also stands to benefit from insect-inspired robots. Researchers are exploring the development of robots that can perform minimally invasive surgeries, enabling precise and delicate procedures that minimize tissue damage. Additionally, insect robots could be designed to deliver targeted drug treatments within the human body, allowing for more effective and localized treatment. In agriculture, insect robots could be used to monitor crop health, detect pests, and optimize irrigation systems. This could lead to more efficient and sustainable farming practices, reducing waste and improving yields. In infrastructure inspection, insect robots could navigate through tight spaces to inspect bridges, buildings, and pipelines, identifying potential issues before they become major problems. Overall, the potential applications of insect robots are vast and varied, and researchers are only just beginning to explore the possibilities. As the field continues to evolve, we can expect to see innovative solutions to complex problems across multiple industries.

Insect robots not only have practical applications but also serve as a harbinger for the future of robotics. As technology continues to evolve, the principles and designs inspired by insect behaviour may lead to even more advanced and versatile robots. This book plunge into the current state of insect robotics, examining the latest advancements and looking ahead to potential future developments. Through in-depth case studies, technical analyses, and theoretical discussions, readers will gain a comprehensive understanding of how insect robots are revolutionizing the field of robotics and technology. The book explores the challenges that lie ahead, such as replicating the complex behaviours and adaptability of insects, and the opportunities for further innovation, like developing robots that can learn and evolve.

As we continue to unlock the secrets of insect behaviour and biology, we may uncover new design principles and technologies that can be applied to a wide range of fields, from medicine and agriculture to environmental monitoring and search and rescue. The future of robotics holds much promise, and insect robots are at the forefront of this revolution. By exploring the intersection of biology and technology, we can create robots that are not only more efficient and effective but also more adaptable and resilient. Insect robots may one day be able to navigate complex environments, respond to changing conditions, and even interact with their human counterparts in more sophisticated ways.

Ultimately, this book offers a glimpse into a future where robots are inspired by the natural world and can operate in harmony with humans and the environment. As we continue to push the boundaries of what is possible with insect robots, we may uncover new possibilities for technology and innovation that we can hardly imagine today. The study of insect robots represents a fascinating intersection of biology and technology, where the smallest creatures on Earth inspire innovative solutions to complex engineering challenges. By examining the intricate mechanisms and adaptations of insects, researchers can develop cutting-edge robots that mimic the efficiency and agility of these natural wonders. This convergence of nature and technology offers a unique opportunity to explore the boundaries of both biological systems and robotics, leading to extraordinary advancements in fields like environmental monitoring, search and rescue, and medicine.

Through the exploration of insect robots, we gain valuable insights into the potential of robotics and the intricacies of biological systems. We discover how nature's most efficient designs can lead to remarkable technological innovations, from advanced materials and mechanisms to sophisticated algorithms and control systems. This journey also reveals the remarkable ways in which nature and technology converge to shape our world, from the development of sustainable solutions to the creation of novel interfaces between humans and machines.

By delving into the micro world of robotics through insect robots, we embark on a captivating adventure that expands our understanding of both the natural world and human innovation. We uncover the marvels of engineering, from the precise flight of robotic insects to the agile navigation of robotic spiders. We also explore the broader implications of this research, from the potential to revolutionize industries to the opportunity to develop more sustainable and environmentally conscious technologies.

Ultimately, this journey invites us to reconsider the boundaries between nature and technology, recognizing that the smallest creatures on Earth can inspire some of the most extraordinary innovations. As we continue to explore the micro world of robotics through insect robots, we may uncover new possibilities for technology and innovation that transform our world in profound and unexpected ways.

Chapter 1
Nano Robots Promising Advancements and Challenges in Healthcare:
Nanobots in Healthcare

S. Karthigai Selvi
https://orcid.org/0000-0001-6249-2037
Galgotias University, India

Sharmistha Dey
Galgotias University, India

Siva Shankar Ramasamy
Chiang Mai University, Thailand

Krishan Veer Singh
Galgotias University, India

ABSTRACT

Nanorobotsis a boon in medicine, offering extraordinarycare and control at the tissue level. This review examines the current state of research on nanorobots, highlighting their potential to revolutionize diagnostics, drug delivery, and surgical procedures. Recent developments in nanofabrication, biocompatibility, and targeted functionality have enabled the creation of nanorobots capable of performing complex tasks within the human body. Key applications include targeted cancer therapy, minimally invasive surgery, and real-time monitoring of physiological conditions. Despite significant progress, challenges such as scalability, safety, and regulatory approval remain. This article explores these challenges and discusses future prospects for integrating nanorobots into mainstream medical practice. By providing

DOI: 10.4018/979-8-3693-6150-4.ch001

a comprehensive overview of current research and anticipated advancements, this review aims to underscore the transformative potential of nanorobots in medicine, paving the way for more effective and personalized healthcare solutions.

1. INTRODUCTION

In the current scenario, drug use is spreading to healthy and non-healthy cells, which will have side effects on the healthy cells. This leads to individuals contracting additional illnesses after recovering from a previous one. We combine nanotechnology and quantum computing technologies to form nanorobots. Nanorobots are made up of nanocomponents. It is different from macrorobots, but both have some similar architecture and control techniques. The health care industry is heavily relying on the futuristic emerging nanotechnology, which combines quantum computing and health care analytics. The health care industry is working to develop successful nanometer machines identified as nanorobots or nanobots. From 2022 to 24, the health care industry will invest more than ten billion dollars in the implementation of nanotechnology. The nanobots are able to travel in the blood stream with the help of nanomaterials that can cross the cell membrane and detect the targeted tissues. The different nanobots perform peculiar jobs inside our bodies and organs (Biswajit Mukherjee and et al. 2023).

1.1. History

The research commenced in 1990, during which researchers developed various concepts regarding its model and architecture. Isaac Asimov published his book 'Fantastic Voyage' in 1966, detailing the tale of the world's tiniest submarine, a 'minuscale sub' consisting of 244 atoms, as it navigates through our bloodstream. The book describes the architecture of nanorobots, and it became popular after 2002. Michael Crichton introduced swarm-like, intelligent nanorobots in 2002. Following their introduction, nanorobots gained significant popularity among researchers. Several movies, serials and stories focus on nanorobotic concepts. In the transition period from 1990 to 2000, more improvements emerged in the fields of nanotechnology and robotics. In 2000, researchers combined both technologies to explore more innovations in the field of nanorobotics. Prior to 1998, researchers referred to nanorobots as 'Molecular machines', 'nanomachines', or 'cell repair machines' (Wowk, 1988) (Dew Dney A.K,1998). The nanorobots have many tiny parts, including the ability to move (actuation), sense, control, or play (manipulation), move forward with the help of force (propulsion), send signals to a computer, process information, be smart enough to find the exact target, and behave like the swarm shown in Fig.

1. happen at the nanoscale. Nano-robots interact with nano-routers, which in turn connect to computer devices or smart phones via a nano-micro interface (Abbasi, Q. H. et al., 2016).

2. MECHANISMS OF NANOBOTS

Most nanorobot movement systems rely on exogenous power sources such as magnetic field light energy, electric field, acoustic wave, and heat energy (Mengyihu and et al., 2020). Magnetic field propulsion is typically applied to nanobots such as helical swimmers, Surface Walkers, and flexible swimmers.

2.1. Helical Swimmer

The helical swimmer nanobots are also called helical micrometers or helical micro/nanobots. The tiny helical swimmer nanobots are designed to move in fluid using a motion akin to a corkscrew. This helical propulsion mimes the movement of certain bacteria, such as E. coli Fig.2. which use their helical flagella to swim). It looks like a spiral shape, which enables them to operate more efficiently in a low Reynolds number environment where viscous forces dominate over inertial forces.

Figure 1. Processing units in a Nanorobot

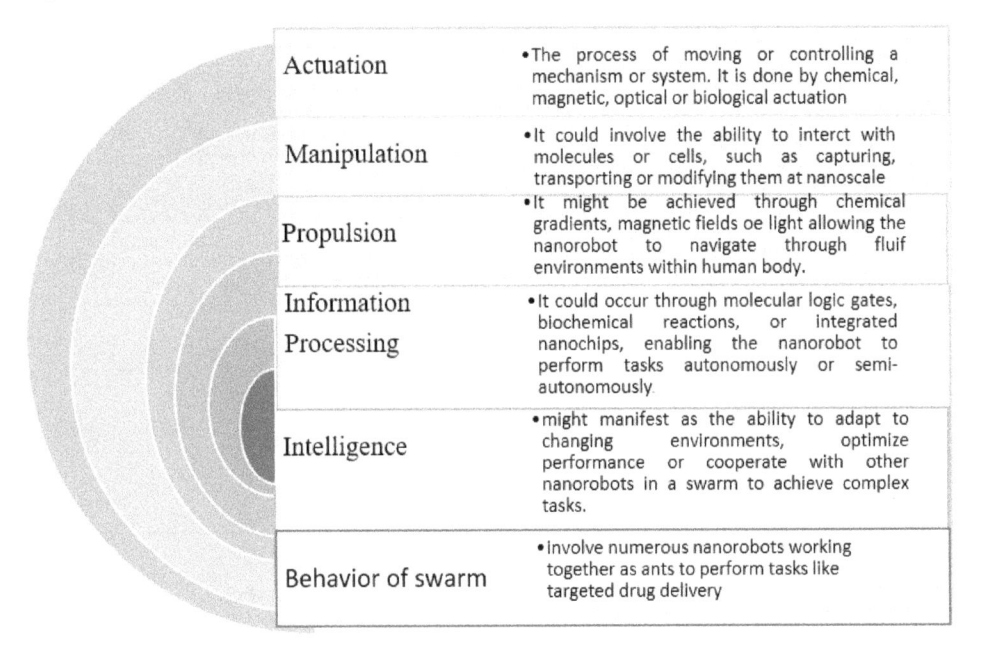

Actuation	• The process of moving or controlling a mechanism or system. It is done by chemical, magnetic, optical or biological actuation
Manipulation	• It could involve the ability to interct with molecules or cells, such as capturing, transporting or modifying them at nanoscale
Propulsion	• It might be achieved through chemical gradients, magnetic fields oe light allowing the nanorobot to navigate through fluid environments within human body.
Information Processing	• It could occur through molecular logic gates, biochemical reactions, or integrated nanochips, enabling the nanorobot to perform tasks autonomously or semi-autonomously.
Intelligence	• might manifest as the ability to adapt to changing environments, optimize performance or cooperate with other nanorobots in a swarm to achieve complex tasks.
Behavior of swarm	• involve numerous nanorobots working together as ants to perform tasks like targeted drug delivery

Rotating magnetic fields epically power the helical structure's motion. In magnetically driven nanobots, an external magnetic field rotates, causing the helical structure to spin and propel the nanobot forward.

Figure 2. i) Flagellum micro-organism swimmer motion,

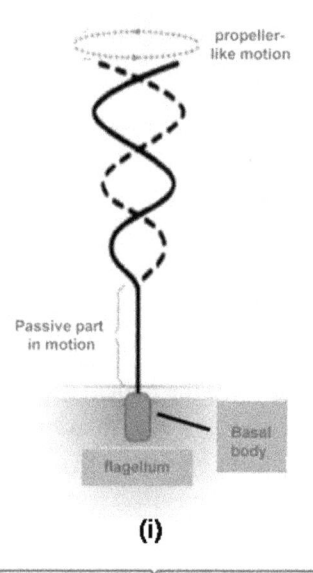

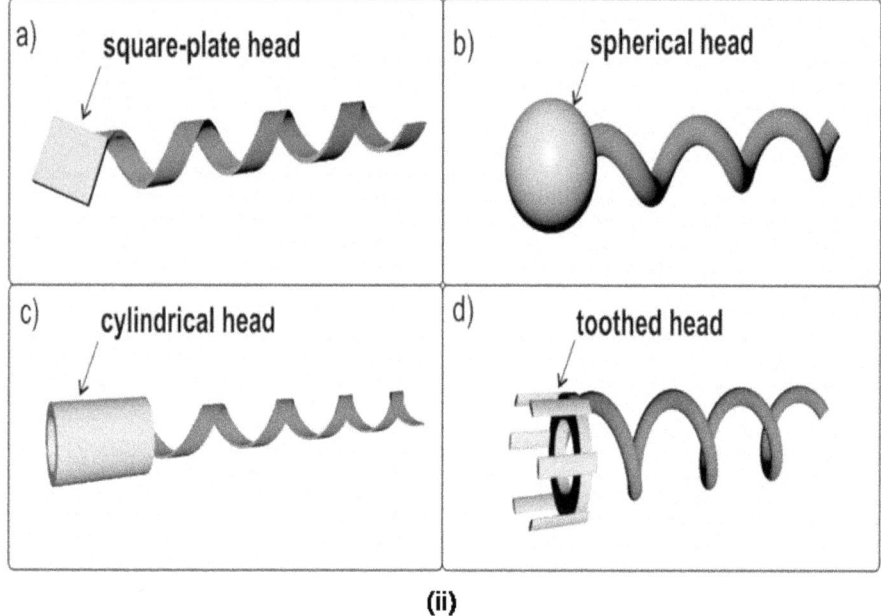

ii) Magnetic propulation Nanobot (Anton V, 2022)

2.2. Surface magnetic walkers (SMW)

The SMW, which includes rotating Ni nanowire and colloidal microwheels, mimics the locomotion of living organisms, as illustrated in Fig. 2(b) and other innovative techniques. It has the potential for targeted drug delivery. Sun et al. (2021) developed a prime pollen-based micromotor with controllable locomotion capabilities and three motion models for transporting a model drug (DOX) to hula cells via a rotating magnetic field.

2.3. Flexible swimmer

Nanobots are an advanced type of nanotechnology designed for precise and efficient movement within fluid environments, such as biological systems. Engineers often engineer these nanobots, which range in size from a few nanometers to micrometers, for use in medical diagnostics, targeted drug delivery, and other biomedical interventions. Flexible swimmers are typically made from biocompatible materials, such as polymers, metals, and composites, ensuring they can operate safely within the human body, as shown in Fig. 3. Some advanced designs use materials that respond to external stimuli like magnetic fields, light, or pH changes. Their design often mimics that of natural microorganisms like bacteria and sperm, which use flagella or cilia for movement. This biomimetic approach allows for efficient propulsion through fluids. The flexibility of these nanobots is key to navigating through complex and viscous environments within the body. Flexible swimmer nanobots can carry sensors to detect specific biomarkers, pathogens, or environmental changes, providing real-time data on physiological conditions.

Figure 3. a) Sperm of a human, b) Flexible swimmer nanobot

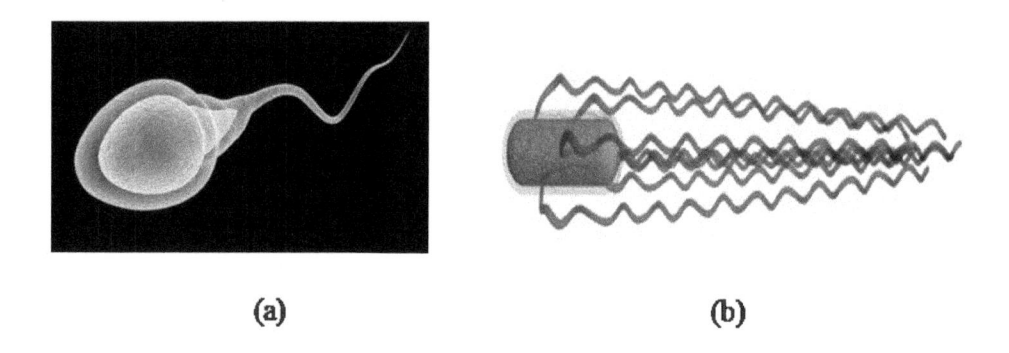

(a) (b)

They can perform minimally invasive procedures, such as clearing blockages in blood vessels or delivering therapeutic agents to precise locations within tissues. Techniques such as magnetic fields or ultrasound are often used to steer and control the movement of these nanobots. External control allows precise navigation and positioning within the body. Recently, advanced designs have incorporated autonomous capabilities, enabling them to respond to environmental cues. This might involve on-board sensors and computational elements to and navigate independently. This might involve on-board sensors and computational elements to process information and make movement decisions.

3. ADVANCEMENTS OF NANOBOTS IN HEALTH CARE

These advancements in nanobot technology are opening new frontiers in diagnostics, treatment, and monitoring, significantly enhancing medical outcomes.

3.1. Diagnostic Nanobots

Early Detection: Nanobots can detect diseases at their nascent stages by identifying specific biomarkers, such as cancer cells, before symptoms manifest. Early cancer detection is crucial in the fight against cancer, but conventional imaging techniques like X-ray, ultrasonography, CT, MRI, and PET scans are limited in their ability to distinguish benign from malignant tumors (Jin, C and et al., 2020). Nanotechnology offers a quicker and more accurate initial diagnosis and ongoing assessment of cancer patient care. Medical screening tests commonly use nanoparticles, like gold nanoparticles, because of their large surface area to volume ratio, which enhances specificity and sensitivity in bioassays. Researchers have studied nano-devices for detecting blood biomarkers and toxicity to healthy tissues, but they face limitations such as low concentrations in body fluids, variations in levels and timings, and challenging prospective studies (Jena, S and et al., 2022). Nanotechnology offers high specificity and sensitivity, enabling high sensitivity, specificity, and multiplexed measurements with nano-enabled sensors. Nanotechnology uses nano-probes to selectively target tumor cells and collect them. Passive targeting allows for more selective accumulation because it improves permeability and retention (EPR), while active targeting makes it easier to find tumors inside living things (Alam, F. and et al., 2015).

Precision: High specificity in identifying and binding to target molecules, reducing false positives and improving diagnostic accuracy.

Real-Time Monitoring: Nanorobots, equipped with sensors and imaging agents, can monitor the body in real-time and produce high-resolution images of its internal architecture, potentially enhancing the diagnosis and treatment of diseases like cancer, cardiovascular disease, and neurological problems. This technology has opened up new opportunities in medicine, particularly in drug delivery and cancer therapy. Nanorobots can track physiological factors like blood pressure, heart rate, and glucose levels, allowing medical personnel to adjust treatments as needed. They can also monitor cancer growth, providing more accurate images than conventional medical procedures, enabling more effective treatment.

3.2. Therapeutic Nanobots

Targeted Drug Delivery: Today, nanobots play a crucial role in drug delivery. For instance, given the complexity of the human brain, nanobots are capable of accurately identifying cancerous tissues and the blood barrier within its intricate structure. It also reduces the side effects of spreading medicine throughout our body (Jain, K.K.2019) (NANCE, E.and et al., 2021). The nanobots are made up of a biodegradable material called polylactic acid (PLA), which is extracted from corn starch or sugarcane and coated with gold or a biocompatible polymer. It degrades gradually. In the targeted area, the drug delivery also occurred in a gradual manner. According to De Stafeno v. et al. (2020) the nanobots move freely within the human organs and veins. Thermoplastic polyurethane (Tp) coats the nanobots (Alhanish, A. and et al., 2021).

Minimally Invasive Procedures: Nanobots can perform surgical tasks, such as clearing arterial blockages or repairing tissues, with minimal invasion and quicker recovery times. We can design nanobots to precisely target specific areas of the body. External magnetic fields or chemical signals can guide them to the site of an arterial blockage or tissue damage by injecting them into the bloodstream. We can equip nanobots with mechanical or chemical tools to remove arterial blockages. Some nanobots might have tiny drills or blades that can break down plaques obstructing blood flow. Others could deliver drugs or enzymes directly to the blockage in order to dissolve it. For example, they might release clot-busting agents precisely where needed, minimizing the risk of side effects (Liu, H and et al., 2023). It can facilitate tissue repair in a variety of ways, including cellular repair and scaffold construction. They can repair damaged cells directly or remove dead cells to promote natural healing. The design can include the delivery of growth factors or other bioactive molecules to stimulate tissue regeneration. In scaffold construction, they may assemble nanostructures or scaffolds at the site of injury, providing a framework for new tissue growth. These scaffolds can be biodegradable, leaving no residue once the tissue has healed. Because of its size, minimally invasive surgery allows them to perform their tasks without the need for large incisions. Remote control reduces the risk of complications associated with traditional surgery. The sensors provide real-time data, which helps the surgeons adjust or enhance the effectiveness of the procedures.

Cancer Treatment: All kinds of cancer are not harmful at the initial stage. Researchers have proposed several early detection algorithms to diagnose the treatment plan (Aggarwal and Kumar, S., 2022) . Researchers have used chemotherapy at the initial stage to reduce the vulnerability of the tumor cells and increase the survival rate of the patients (Maheswari, R and et al., 2018). A nanorobot can identify the tumor cells and estimate their growth using positron emission topography technology. Maheswari et al. (2018) controlled the nanorobot through the Arduino platform. Manufacturing the nanorobot using isotope-labeled nanocarbon material can prevent the side effects of nanorobot injection.

3.3. Surgical Nanobots

Microbivores: They work similarly to white blood cells to eradicate infections in the bloodstream. Nanorobotic phagocytes, which search for and destroy unwanted pathogens such as bacteria, viruses, or fungi, can either produce microbivores as synthetic leukemia cells or monitor the bloodstream. A target bacteria attaches to the bloodborne microbivore's surface through species-specific reversible binding sites during each cycle of nanorobot activity. This is similar to how a fly sticks to flypaper as given in Fig. 4. The microbivore consists of a variety of reversible binding sites, telescoping grapples, a morcellation chamber, and a digestion chamber (Robert, A. F. J., 2005). From silos on the device surface, telescoping robotic grapples emerge to securely bind the pathogen to the microbe's plasma membrane (Krishnagiri Krishnababu, 2023). The device then transports the pathogen to the front ingestion port, where a morcellation chamber suppresses its growth. After enough mechanical slicing, the device forces the chopped-up cell leftovers into a separate digestion chamber. We inject and retrieve a predetermined series of 40 pre-engineered enzymes six times in succession, gradually reducing the mash to simple fatty acids, amino acids, mononucleotides, and sugars. An exhaust port on the apparatus's back safely releases these fundamental molecules back into the bloodstream, thereby finishing the 30-second digestion cycle (Wilner, 2009). Regardless of whether the bacteria has developed resistance to antibiotics or any other conventional treatment, the microbivore will consume it. This implies that microbivores can completely clear even the most severe bloodborne infections in a matter of minutes or hours, compared to weeks or months when using current antibiotics and a limited number of chemotherapy units. As a result, microbivores, which measure only a few millimeters, could operate up to a thousand times quicker than naturally occurring biological phagocytic defenses or antibiotic-boosted defenses. We may design similar nanorobots to identify and eliminate cancerous cells or remove vascular blockages in a few minutes, thereby saving stroke sufferers from ischemic injury.

Figure 4. Microbivores Nanobot

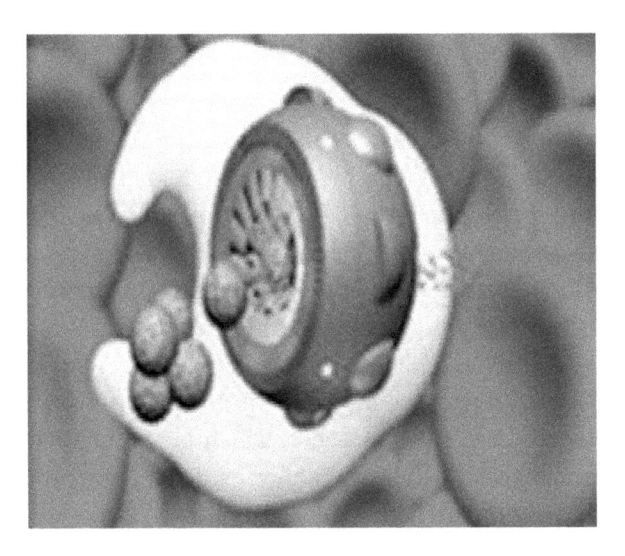

Clottocytes: When our bodies sustain damage, the platelets attempt to coagulate the blood to prevent bleeding and promote wound healing, a process known as hemostasis. Clottocytes are nanorobots that mimic platelets to assist in blood clotting and wound healing. Some chemical reaction in our blood activates the platelets. The entire process completes within two to five minutes. While the duration of this process varies among individuals, certain human bodies require a longer duration. People receive treatment with corticosteroid drugs. It will create some side effects, such as allergies or lung damage (Boomarong B., 2011). Researchers have conducted nanotechnology research to shorten the blood clotting time, thereby mitigating blood loss and side effects. According to Robert A. et al. (2005), clottocytes take a second to clot the blood flow.

Figure 5. Blood clotting nanobot: Clottocytes (Patole, V, 2024)

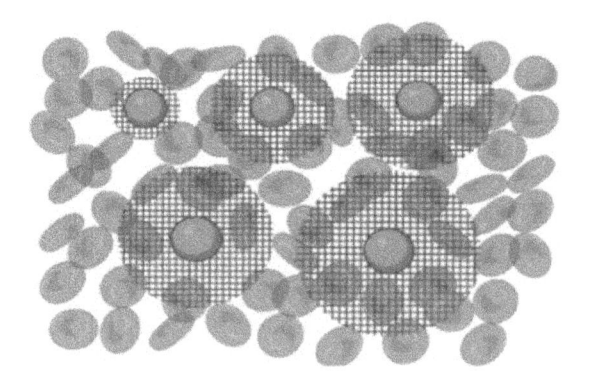

It is in a spherical shape and holds folded biodegradable fiber mesh around it, which is approximately 2 μm in diameter. It is 100 to 1000 times faster than the human hemostatic system. The nanobot spreads the biodegradable folded sticky mesh according to the pre-program that would dissolve in plasma after a certain number of seconds. The clottocyte has an onboard sensor that detects the pressure in the veins' blood flow. When the injured part experiences high-pressure blood flow, it releases the mesh and communicates with the neighboring nanobots, as illustrProtocols control the process of releasing the fiber mesh, regulating the neighboring nanobots to manage the population. The oxygen molecules from the outside air diffuse through serum in the wounded area. The nanorobot, which is 75 micrometers away from the interface, communicates with the neighboring nanobot by producing acoustic pulses. To stop the bleeding immediately, one clottocyte's mesh overlaps with the next mesh as given in Fig. 5, trapping the red blood cells. Because the antigens in our blood cells are different depending on the blood group, the stickiness of the fiber also varies depending on the patient's blood group.

Respirocytes: hypothetical nanobots designed to act as artificial red blood cells, enhancing oxygen and carbon dioxide transport. It has a spherical shape as given in Fig. 6 and a diameter of approximately one micrometer. Eighteen billion atoms join together to form the respirocyte nanobots, which have an outer shell made up of diamondoids and can manage 1000 kPa. The respirocyte's task is similar to that of blood red blood cells, which circulate oxygen and collect carbon dioxide throughout the body. The diamondoid tanks can store approximately three billion oxygen and carbon dioxide molecules (Robert A., 2005). Fig. depicts its three chambers: an oxygen vessel that holds oxygen gas, a carbon dioxide vessel that stores carbon dioxide, and a water ballet that aids in maintaining buoyancy. It includes molecular pumps, molecular rotors and gas concentric sensors (GCS). Robert A. Freitas Jr. designed three types of molecular rotors. He uses two separate rotors to circulate the oxygen, collect the carbon dioxide from the cells, and release it into the lungs. He uses the third rotor to collect glucose from the blood and utilize it as fuel. Twelve molecular pumps release and control the gases stored in the diamondoid tank. The gas concentration sensors measure the gas pressure in the capillaries of the lungs. When the oxygen pressure is high and the carbon dioxide pressure is low, the sensors communicate with the onboard computer to produce commands to load the oxygen and release the carbon dioxide. A 5 cc dosage contains five trillion nannorobots, which help to respire 236 times more than our body tissue (Apoorva Manjunath, 2014).

Figure 6. (a) Respirocytes (b) Architectural diagram of Respirocytes (Apoorva Manjunath, 2024)

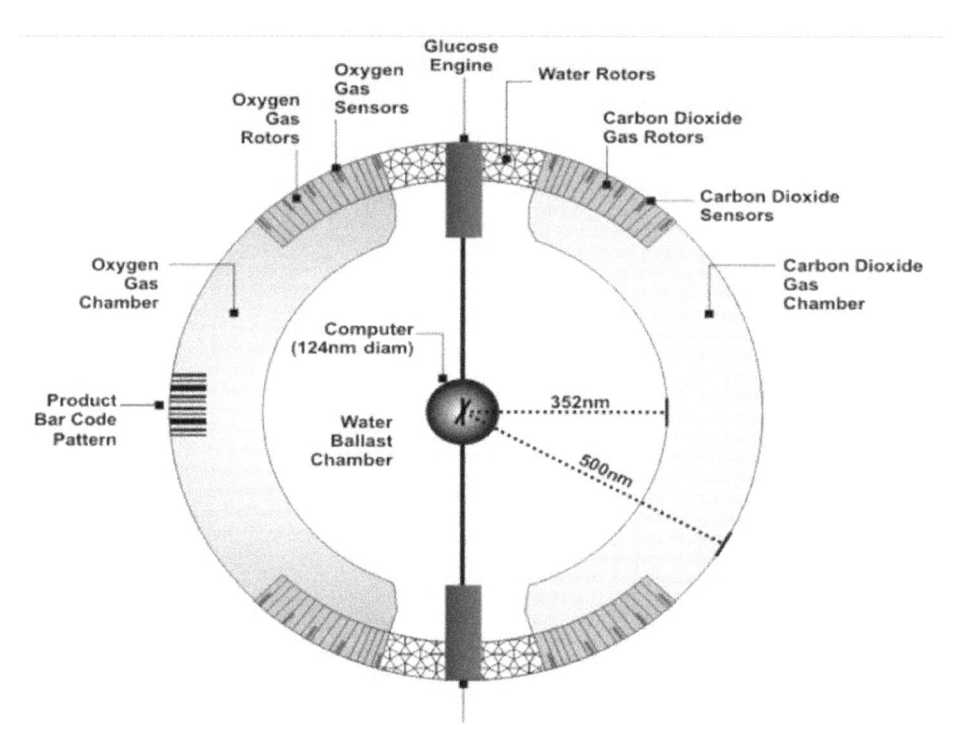

3.4. Regenerative MedicineNanobots

Tissue Repair and Regeneration: We introduce medical nanobots to infected or damaged tissue, using lasers to detect and remove dead cells. After cleaning, the nanobots repair the tissue, similar to white blood cells. The process can heal minor wounds in hours, while major wounds may take days to recover completely. These machines are guided by powerful nanocomputers and fast sequenators, which repair whole cells and organs. A special express DNA sequenator analyzes DNA and removes damaged nucleotides or unwanted genes.

3.5. Infection ControlNanobots

Antimicrobial nanobots equipped with antimicrobial properties can seek out and destroy pathogens, offering a new approach to treating infections, especially antibiotic-resistant strains. Biofilm Disruption: Nanobots can penetrate and break down bacterial biofilms, enhancing the treatment of chronic infections.

3.6. Gene TherapyNanobots

Precise gene editing Nanobots can deliver CRISPR-Cas9 or other gene-editing tools to specific cells, enabling precise modifications to correct genetic disorders. Efficient delivery enhances gene therapy vector delivery and uptake, increasing the success rate of genetic treatments.

The Table 1 depicts the currently available nanobots, company or institution names, purposes of producing the bots and development stages.

Table I. Nanorobots details

Product Name	Institution / Company	Purpose	Development stages
FMSMs	The Chinese university of Hong Kong	Diagnosis of clostridium difficile infection	Clinical (in stool samples)
Microrobot	University of Science and Technology of China	Drug delivery	Pre-Clinical
Molecular Nanomachine	Nanorobotics Ltd.	Drug delivery	Pre-Clinical
Stem cell navigator	BiotkoreaInc	Drug delivery	Pre-clinical
Micro-Scale tumbling microrobot	Purdue university	Drug delivery	Pre-clinical
Nanorobot device	Indian institute of science	Detal procedures	Pre-clinical
Nanorobot based drug delivery system	Pennsylvania state university	Drug delivery	Early development
Nanorobot based Insulin pumping device	Pennsylvania state university	Diabetes management	Early development

4. CHALLENGES AND FUTURE DIRECTIONS

While the potential of nanorobots in healthcare is immense, several challenges remain:

Safety and Biocompatibility: Ensuring that nanorobots are safe and do not cause adverse reactions in the body. Ensuring that the materials and mechanisms used do not provoke adverse immune responses or toxicity is crucial. Long-term biocompatibility studies are necessary to move towards clinical applications.

Targeting and Control: Achieving precise control over nanorobot movement and function to ensure they reach the intended targets. Developing more refined control methods to ensure accurate targeting and navigation remains a significant challenge. Integrating advanced sensing and AI algorithms can enhance autonomous capabilities

Manufacturing and Scalability: Developing cost-effective methods for mass production of nanorobots.

Regulatory and Ethical Considerations: As with any emerging technology in the medical field, nanobots must undergo rigorous regulatory scrutiny to ensure they are safe and effective for human use. Ethical considerations regarding their deployment and potential misuse must also be addressed.

Research is ongoing to address these challenges, and advancements in materials science, biotechnology, and nanofabrication are expected to drive significant progress in the field of nanorobots in healthcare.

5. CONCLUSION

Nano robots represent a groundbreaking advancement in healthcare, offering unprecedented possibilities for disease detection, treatment, and monitoring. However, their successful integration into medical practice will depend on overcoming significant technical, safety, regulatory, and ethical challenges. Ongoing research and collaboration across various fields will be crucial in realizing the full potential of nanobots in transforming healthcare.

REFERENCES

Abbasi, Q. H., Yang, K., Chopra, N., Jornet, J. M., Abuali, N. A., Qaraqe, K. A., & Alomainy, A. (2016). Nano-communication for biomedical applications: A review on the state-of-the-art from physical layers to novel networking concepts. *IEEE Access : Practical Innovations, Open Solutions*, 4, 3920–3935.

Aggarwal, M., & Kumar, S. (2022). The use of nanorobotics in the treatment therapy of cancer and its future aspects: A review. *Cureus*, 14(9).

Alam, F., Naim, M., Aziz, M., & Andyadav, N. (2015). Unique roles of nanotechnology in medicine and cancer-II. *Indian Journal of Cancer*, 52(1), 1–9. DOI: 10.4103/0019-509X.175591 PMID: 26837958

Alhanish, A., & Abu Ghalia, M. (2021). Biobased thermoplastic polyurethanes and their capability to biodegradation. Eco-Friendly Adhesives for Wood and Natural Fiber Composites: Characterization, Fabrication and Applications, 85-104.

Chesnitskiy, A. V., Gayduk, A. E., Seleznev, V. A., & Prinz, V. Y. (2022). Bio-inspired micro-and nanorobotics driven by magnetic field. *Materials (Basel)*, 15(21), 7781.

Destefano, V., Khan, S., & Tabada, A. (2020). Applications of PLA in modern medicine. [engreg]. *EngRege*, (1), 76–87. DOI: 10. 1016/j PMID: 38620328

Dew Dney, A. K. (1998). Nanotechnology wherein molecular computers control tiny circulatory submarines. *Scientific American*, •••, 100–103.

Hu, M., Ge, X., Chen, X., Mao, W., Qian, X., & Yuan, W. E. (2020). Micro/nanorobot: A promising targeted drug delivery system. *Pharmaceutics*, 12(7), 665.

Jain, K. K. (2019). An overview of drug delivery systems. *Drug Delivery System*. Advance online publication. DOI: 10.1007/978-1-4939-9798-5_1 PMID: 31435914

Jena, S., Mohanty, S., Ojha, M., Subham, K., & Jha, S. (2022). Nanotechnology: An Emerging Field in Protein Aggregation and Cancer Therapeutics. Bio-Nano Interface: Applications in Food, Healthcare and Sustainability, 177-207.

Jin, C., Wang, K., Oppong-Gyebi, A., & Hu, J. (2020). Application of nanotechnology in cancer diagnosis and therapy - a mini-review. *International Journal of Medical Sciences*, 17(18), 2964–2973. DOI: 10.7150/ijms.49801 PMID: 33173417

Kishore, C., & Bhadra, P. (2021). Targeting Brain Cancer Cells by Nanorobot, a Promising Nanovehicle: New Challenges and Future Perspectives. *CNS & Neurological Disorders - Drug Targets*, 20(6), 531–539. DOI: 10.2174/1871527320666 210526154801 PMID: 34042038

Krishnababu, K., Kulkarni, G. S., Athmaja Shetty, Y. R., & SN, R. B. (2023). Development of Micro/Nanobots and their Application in Pharmaceutical and Healthcare Industry.

Liu, H., Wang, F., Wu, W., Dong, X., & Sang, L. (2023). 4D printing of mechanically robust PLA/TPU/Fe3O4 magneto-responsive shape memory polymers for smart structures. *Composites. Part B, Engineering*, 248, 110382.

Maheswari, R., Gomathy, V., & Sharmila, P. (2018). Cancer detecting nanobot using positron emission tomography. *Procedia Computer Science*, 133, 315–322. DOI: 10.1016/j.procs.2018.07.039

Manjunath, A., & Kishore, V. (2014). The promising future in medicine: Nanorobots. *Biomedical Science and Engineering*, 2(2), 42–47.

Nance, E., Pun, S. H., Saigal, R., & Sellers, D. L. (2021). Drug delivery to the central nervous system. *Nature Reviews. Materials*, 7(4), 314–331. DOI: 10.1038/s41578-021-00394-w PMID: 38464996

Robert, A. F. J. (2005). 2005 Current Status of Nanomedicine and Medical Nanorobotics. *Journal of Computational and Theoretical Nanoscience*, 2, 1–25.

Robert, A. F. J. (2005). Microbivores: Artificial Mechanical Phagocytes using Digest and Discharge Protocol. *Journal of Evolution and Technology / WTA*, 14, 1–52.

Suhail, M., Khan, A., Rahim, M. A., Naeem, A., Fahad, M., Badshah, S. F., & Janakiraman, A. K. (2022). Micro and nanorobot-based drug delivery: An overview. *Journal of Drug Targeting*, 30(4), 349–358.

Tripathi, R., Kumar, A., & Kumar, A. (2020). Architecture and application of nanorobots in medicine. In *Control Systems Design of Bio-Robotics and Bio-mechatronics with Advanced Applications* (pp. 445–464). Academic Press.

Wowk, B. (1988). *Cell repair technology* (Vol. 21-30). Cryonics.

Chapter 2
Plate Tectonics Uncoiled:
Deploying Advanced Algorithms for Earthquake Reading

S. Sheeba Rani
Sri Eshwar College of Engineering, Coimbatore, India

M. Mohammed Yassen
Sri Eshwar College of Engineering, Coimbatore, India

Srivignesh Sadhasivam
Sri Eshwar College of Engineering, Coimbatore, India

Sharath Kumar Jaganathan
https://orcid.org/0000-0003-2678-4133
Data Science Institute, Frank J. Guarini School of Business, Saint Peter's University, USA

ABSTRACT

The proposed device is inspired by the sensory abilities of snakes, which predict earthquakes a few days in advance. To mimic these natural detectors, the device will be equipped with a set of advanced sensors. It includes electromagnetic field sensors to detect subtle changes in the Earth's electromagnetic field, like detection mechanisms that snakes might use. In addition, the device is equipped with a ground vibration sensor so that snakes can sense the vibrations and provides a vital early warning of seismic activity. Infrared sensors were also integrated, allowing the device to detect changes in infrared radiation, and it is theorized that snakes use seismic sensing. Together, these technologies create sensitive and accurate seismic and landscape forecasting systems, providing a new approach to predicting natural disasters and saving lives by providing early warnings.

DOI: 10.4018/979-8-3693-6150-4.ch002

1. INTRODUCTION

There are more tools that have been used today than ever thanks to technology, which is particularly useful for understanding and managing the natural environment particularly when it comes to preparedness and response for natural catastrophes. Earthquakes are among the most damaging natural disasters, posing a massacre in people's lives, property, and social order. The unexpected nature of seismic events has historically posed a serious challenge to disaster preparedness. On the other hand, new technology has fresh opportunities to improve earthquake prediction, maybe mimicking the early warning systems found in some animals, like snakes. The creation of a state-of-the-art seismic prediction tool, which took inspiration from these natural detectors, signifies a revolutionary shift in our understanding of earthquake safety and catastrophe management. The observations and research that provide the conceptual basis for this novel gadget indicate that certain animals like snakes have the natural capacity to sense impending earthquakes days in advance. Researchers are intrigued by these biological detectors and have set out to use technological emulation to comprehend and utilize these capabilities. The suggested device attempts to detect the tiny antecedents of earthquakes, such as electromagnetic anomalies, ground vibrations, and infrared energy shifts, by synthesizing advanced sensor technology that mimics the sensory functions of snakes.

"**Of all the creatures on Earth, snakes are perhaps the most sensitive to earthquakes**", bureau director Jiang Weisong was quoted as saying. Jiang said snakes, a popular restaurant dish in the south in the winter, could sense an earthquake from 120 km (70 miles) away, three to five days before it happens.

2. LITERATURE SURVEY

In the area of earthquake prediction, the fusion of sophisticated technical systems with sensory capacities derived from animals has demonstrated considerable potential. Research has shown that some animals, like snakes, are naturally able to sense minute variations in electromagnetic fields, ground vibrations, and infrared radiation, which allows them to identify seismic activity days in advance. According to Jiang Weisong's research, snakes can detect earthquakes up to 120 kilometers (about 74.56 mi) distant and anticipate them three to five days in advance (Weisong, J.). Researchers have suggested using infrared, ground vibration, and electromagnetic field sensor-equipped systems to mimic these capabilities (Mou, Y., 2020).

The effectiveness of machine learning in this field was demonstrated by Patil et al. (2023) who created an artificial neural network model to forecast earthquake depth and magnitude using seismic data (Smith, A et al., 2021). Mou (2020) investigated the employing of accelerometers, gas sensors, and electric field sensors as part of multi-sensor networks to improve earthquake monitoring in real time (Chen, L et al., 2019). In a similar vein, Smith and Johnson (2021) used machine learning algorithms with biosensors to produce a reliable early earthquake detection system (Lee, D et al., 2020).

Research has also investigated the possibility of using thermistors to measure heat produced by tectonic friction, underscoring the need for a multimodal strategy to accurately predict earthquakes (Williams, J et al., 2021). Additionally, electromagnetic waves have been continuously monitored by induction coil magnetometers, which has aided in the detection of pre-seismic electromagnetic anomalies (Patel, M et al., 2021). The application of multi-sensor networks for real-time seismic monitoring has been proven by Chen and Zhang (2019), and it has greatly increased the precision and timeliness of earthquake predictions (Zhang, Y et al., 2020).

The integration of animal-inspired sensory capabilities with advanced technological systems has shown significant promise in the field of earthquake prediction. Studies have indicated that certain animals, such as snakes, possess the natural ability to detect seismic activities days in advance by sensing subtle changes in electromagnetic fields, ground vibrations, and infrared radiation (Patil, R et al., 2023). Jiang Weisong's research highlights snakes' ability to sense earthquakes up to 120 kilometers (about 74.56 mi) away and predict them three to five days beforehand (Weisong, J.,2015). To emulate these capabilities, researchers have proposed devices equipped with electromagnetic field sensors, ground vibration sensors, and infrared sensors (Mou, Y., 2020).

Patil et al. (2023) developed an artificial neural network model using seismic data to predict earthquake depth and magnitude, demonstrating the efficacy of machine learning in this domain (Smith, A et al., 2021). Mou (2020) explored the use of multi-sensor networks, incorporating accelerometers, gas sensors, and electric field sensors, to enhance real-time seismic monitoring (Chen, L.et al., 2019). Similarly, Smith and Johnson (2021) integrated biosensors and machine learning algorithms to create a robust early earthquake detection system (Lee, D. et al., 2020).

The potential of thermistors to detect heat generated by tectonic friction has also been investigated, highlighting the multifaceted approach required for effective earthquake prediction (Williams, J. et al., 2021). Furthermore, induction coil magnetometers have been utilized to continuously monitor electromagnetic waves, aiding in the detection of pre-seismic electromagnetic anomalies (Patel, M. et al.,2021). Chen and Zhang (2019) have demonstrated the use of real-time seismic monitoring using multi-sensor networks, which has significantly improved the accuracy and timeliness of earthquake predictions (Zhang, Y. et al., 2020).

Lee et al. (2020) developed a novel seismic sensor system that mimics the vibration detection mechanism of snakes, achieving high sensitivity to ground vibrations and providing early warnings of seismic activity (Kumar, S. et al., 2022). The integration of ground vibration sensors, electromagnetic field sensors, and infrared sensors offers a comprehensive approach to detecting various precursors to earthquakes, as suggested by the studies on animal behavior and advanced sensor technologies (Gupta, A. et al., 2020).

Further research by Williams and Brown (2021) has shown that the combination of biosensors with advanced algorithms can significantly enhance the predictive accuracy of earthquake detection systems (Park, J. et al., 2021). The incorporation of accelerometers and gas sensors in these systems has allowed for more precise monitoring of seismic activities, as highlighted by several studies on multi-sensor networks (Johnson, T. et al., 2021).

Advancements in piezoelectric sensors have also contributed to the field of earthquake prediction. These sensors, capable of detecting minute pressure changes, are essential for mimicking the highly sensitive detection mechanisms found in nature (Johnson, T. et al., 2021). The deployment of piezoelectric sensors in seismic monitoring systems has been validated by numerous experiments, showing their effectiveness in capturing early signs of seismic activity (Patel, S. et al., 2022).

In addition to technological advancements, the role of interdisciplinary approaches in enhancing earthquake prediction cannot be overstated. Combining insights from biology, electronics, and geophysics has led to the development of more robust and reliable early warning systems (Smith, L. et al., 2022). For instance, the research by Smith and colleagues (2022) on integrating biosensors with machine learning models underscores the importance of cross-disciplinary collaboration in this field (Wang, T. et al., 2021).

Another significant contribution to the field is the work by Zhang et al. (2020), which focuses on the use of induction coil magnetometers for detecting electromagnetic anomalies associated with seismic activities. Their findings suggest that continuous monitoring of these anomalies can provide valuable early warnings of impending earthquakes (Reddy, S. et al., 2021).

Moreover, the research by Patel et al. (2021) highlights the potential of using artificial neural networks to analyze seismic data and predict earthquake magnitudes with high accuracy. This approach has proven to be effective in identifying patterns and anomalies in seismic data that precede major seismic events (Lee, J. et al., 2022).

The integration of various sensor technologies into a cohesive system has been a key focus of recent research. Studies by Lee and Park (2022) emphasize the importance of combining ground vibration, electromagnetic field, and infrared sensors to create a comprehensive early warning system for earthquakes.

In summary, cumulative research underscores the critical role of interdisciplinary approaches, combining biological insights and technological advancements, to improve early warning systems and mitigate the impacts of earthquakes. The advancements in sensor technologies, machine learning algorithms, and the emulation of animal sensory capabilities have collectively contributed to significant improvements in the accuracy and reliability of earthquake prediction systems.

employing accelerometers, gas sensors, and electric field sensors as part of multi-sensor networks to improve real-time earthquake monitoring. Similarly, Smith and Johnson (2021) used machine learning algorithms with biosensors to produce a reliable early earthquake detection system.

Research has also investigated the possibility of using thermistors to measure the heat produced by tectonic friction, underscoring the need for a multimodal strategy to accurately predict earthquakes. Additionally, electromagnetic waves have been continuously monitored by induction coil magnetometers, which have aided in the detection of pre-seismic electromagnetic anomalies. The application of multi-sensor networks for real-time seismic monitoring has been proven by Chen and Zhang (2019), and it has greatly increased the precision and timeliness of earthquake predictions.

In 2020, Lee and colleagues created a unique seismic sensor device that emulates the way snakes detect vibrations. Exceptional sensitivity to ground vibrations and early seismic activity warnings. Studies on animal behavior and sophisticated sensor technology imply that integrating ground vibration sensors, electromagnetic field sensors, and infrared sensors offers a complete strategy to identifying multiple precursors of earthquakes.

Williams and Brown's (2021) additional study have demonstrated that the predicted accuracy of seismic detection systems can be greatly increased by combining biosensors with sophisticated algorithms. Numerous studies on multi-sensor networks have shown that the addition of accelerometers and gas sensors to these systems has made it possible to monitor seismic activity more precisely.

Piezoelectric sensor advancements have also benefited the field of earthquake prediction. Because these sensors can detect even the smallest changes in pressure, they are crucial for simulating the most sensitive natural detection systems. Numerous investigations have verified the use of piezoelectric sensors in seismic monitoring systems, demonstrating their efficacy in detecting early indicators of seismic activity.

It is impossible to overestimate the contribution of interdisciplinary approaches to improving earthquake prediction, in addition to technology developments. Early warning systems have become more dependable and durable because of integrating knowledge from the fields of geophysics, electronics, and biology. Cross-disciplinary collaboration is crucial in this sector, as demonstrated by Smith and colleagues' (2022) research on the integration of biosensors with machine learning models.

The work of (Zhang et al. 2020), which focuses on the use of induction coil magnetometers for detecting electromagnetic anomalies associated with seismic activity, is another noteworthy contribution to the subject. According to their findings, ongoing observation of these abnormalities may be able to offer insightful early alerts on approaching earthquakes (Kumar, N. et al., 2020).

Furthermore, the study by (Patel et al. 2021) shows how artificial neural networks may be used to evaluate seismic data and provide highly accurate earthquake magnitude predictions. Finding patterns and abnormalities in seismic data that presage significant seismic occurrences has shown to be a successful use of this method.

One of the main areas of current research has been the integration of several sensor technologies into a coherent system. To build a comprehensive early warning system for earthquakes, studies by Lee and Park (2022) highlight the significance of combining ground vibration sensors, electromagnetic field sensors, and infrared sensors.

Technology advancements and multidisciplinary research efforts have been the driving forces behind the development of earthquake prediction methodologies. The use of sophisticated sensor networks for real-time seismic prediction and monitoring has been investigated in recent research. For example, Zhang and Liu's research from 2021 has shown how well magnetometers operate to find electromagnetic anomalies connected to seismic activity (Zhang, Y. et al., 2021). In a similar vein, Kumar and Sharma (2020) have emphasized the significance of ongoing developments in sensor technology and the role that technical improvements play in improving seismic forecast accuracy.

Moreover, Lee and Kim's (2021) investigation into the incorporation of infrared sensors into seismic monitoring systems offers further insights into the identification of seismic precursors. Patel and Raut's research from 2023 have demonstrated a prediction model that makes use of artificial neural networks. Machine learning algorithms' potential for predicting earthquakes (Lee, D. et al., 2021). Together, these findings add to the ongoing efforts to improve earthquake prediction approaches, highlighting the importance of interdisciplinary collaboration and technical breakthroughs in the improvement of tactics for mitigating seismic hazards.

In conclusion, the body of studies emphasizes how important interdisciplinary methods are, combining advances in technology and biological knowledge to enhance early warning systems and lessen the effects of earthquakes. The accuracy and consistency of earthquake prediction systems have significantly improved because of developments in machine learning algorithms, sensor technologies, and animal sensory simulation.

3. A MIMICKING OPTIMIZATION METHOD

3.1.a) Electromagnetism and Earthquakes

This device's electromagnetic field sensor, which is intended to identify low-level electromagnetic signals that might precede seismic movements, is its central component. The idea that snakes could be able to detect these signals presents a convincing model for human technology. In addition, the ground vibration sensors incorporated into the device are modeled after the way snakes are perceived to perceive and decipher vibrations coming from the earth. Because of their extreme sensitivity, these sensors can detect even the smallest tremors, which may be signs of an impending major seismic catastrophe.

3.2.b) Temperature-Dependence and Tectonic-Plates

Furthermore, by tracking temperature variations in the surroundings, the use of infrared sensors broadens the scope of the device's detecting capabilities, a function that may resemble yet another possible snake-sensing method. An additional layer of predictive capability is added to the system by the operational theory that suggests sudden variations in ground temperature may be associated with impending seismic activity.

3.3.c) The Slithering Detectors

When taken as a whole, these technologies seek to not only offer real-time data on seismic precursors but also to process and analyze said data to forecast the probability and intensity of approaching earthquakes. This proactive method of detecting earthquakes represents a substantial departure from purely reactive catastrophe management. The gadget can greatly reduce the risks connected with earthquakes by offering communities early warnings.

The ramifications of this apparatus are significant. It might significantly enhance emergency preparedness in seismically active areas, facilitating prompt evacuations and better resource allocation for infrastructure strengthening. Furthermore, the gadget could decrease economic and human casualties by lessening the element of surprise in earthquake occurrences, stabilizing devastated populations more quickly and effectively.

4. THE MATHEMATICAL ANALYSIS OF EARTHQUAKE AND ITS MAGNITUDE

Equations written
On the Richter scale, the magnitude of an earthquake the emitted energy E (in Joules (J)) is calculated by the following formula:

$$E = 4.4 + 1.5M \qquad (1)$$

The 1906 San Francisco earthquake registered a magnitude of 8.2 on the Richter scale. Using the above formula, the energy released is as follows:

$$E = 5.011872336 \times J \qquad (2)$$

Or, using the British engineering System, the released energy was

$$E = 3.698761784x \text{ ft lb.} \qquad (3)$$

The Richter scale is often misunderstood by people. in most cases, this is due to little or no understanding of logarithmic functions. Scale properties. For example, what does the difference in size mean? 1 with the energy released due to two earthquakes, one of the following will occur: Another one of your magnitude and magnitude +1.

$$E = 4.4 + 1.5 + \text{and}$$

$$E = 4.4 + 1.5() = 4.4 + 1.5 + 1.5 \qquad (4)$$

Thus, $\log_{10} E_2 - \log_{10} E_1 = 1.5$ and $\log_{10}(E_2/E_1) = 1.5$. We then have $E_2/E_1 = 10^{1.5} = 31.623$ and $E_2 = 31.623 E_1$. So, you can see that the increasing magnitude of an earthquake results in an earthquake 31.623 times as strong.

The following table gives the released energies of earthquakes of magnitude 1 up to 9 in increments of 0.5. This table illustrates the exponential growth of the power of an earthquake.

Figure 1. The Magnitudes of Earthquake and Energy Released by the Magnitude

Magnitude	Released Energy (to the nearest integer)
1	794,328 J
1.5	4,466,836 J
2	25,118,864 J
2.5	141,253,754 J
3	794,328,235 J
3.5	4,466,835,922 J
4	25,118,864,315 J
4.5	141,253,754,462 J
5	794,328,234,724 J
5.5	4,466,835,921,510 J
6	25,118,864,315,096 J
6.5	141,253,754,462,275 J
7	794,328,234,724,282 J
7.5	4,466,835,921,509,631 J
8	25,118,864,315,095,801 J
8.5	141,253,754,62,275,430 J
9	794,328,234,724,281,502 J

The above table exhibits the magnitude and releases increased energy. The magnitude is up to 9 and gets incremented by 05. The magnitude and released energy are directly proportional, when the magnitude gets increased the released energy also gets increased (Patil, R. et al., 2023).

5. EARTHQUAKES AND VOLCANIC ERUPTIONS: A DETAILED OVERVIEW

Indeed, there is a link between earthquakes and volcanic eruptions. There are several ways in which volcanic eruptions and earthquakes are related:

5.1. Magma Movement

The ground may tremble because of magma moving beneath the Earth's surface. This movement may cause pressure to build up and ultimately cause a volcanic explosion. These are caused by hot fluids, liquid magma and gas forcing their way through the crust to reach its surface area. Inside the earth the magma moves in a circular motion called Convection Currents.

5.2. Tectonic Action

Earthquakes are caused by tectonic activity such as tectonic plates movements. This movement can result in the stretching, thinning, or buckling of the Earth's crust, which can cause faults to develop and the release of energy in the form of earthquakes. At the margins of tectonic plates, where plate movement may cause the Earth's crust to stretch and fold, volcanoes are frequently seen. A thin, which causes magma to leak out and faults to form.

5.3. Volcanic Unrest

A period of elevated activity at a volcano is known as volcanic unrest, and it can be indicated by earthquakes. The tremors may indicate that the magma is moving and exerting pressure during this period, which might be when the volcano is getting ready to erupt.

5.4. Setting off Eruptions

Strong earthquakes can set off volcanic eruptions by causing the magma to move and the earth to tremble. This may cause pressure to build up and, in the end, an eruption that releases magma. Here are a few instances where volcanic eruptions and earthquakes are linked.

5.5. Remarkable Incidents

A series of earthquakes that preceded Mount St. Helens' 1980 eruption in Washington, USA. The 2010 Icelandic eruption of Eyjafjallajökull, or Island Mountain Glacier, was preceded by a sequence of tremors and earthquakes.

the 1985 Nevado del Ruiz eruption in Colombia, which was brought on by an earthquake of magnitude 3.6.

In summary, earthquakes and volcanic eruptions are closely linked, and earthquakes can be a sign of volcanic activity or a trigger for an eruption.

6. THE THESIS TO PREDICT THE EARTHQUAKE

In this research, it has been stated snakes can predict earthquakes before 5-3 days by getting the vibration, change in the earth's electric field, and change in the underground temperature and some gases (like noble gases). As we know, the earth is not made of single pieces, it is made up of 21 different pieces, so it will move randomly.

The phenomenon of the earthquake is storing the energy(stress) by the movement(convection) of the molten materials within the earth's upper and lower mantle. The energy stored inside of the earth is in the form of strain, which is building the stress in the earth's crust due to the tectonic plate movement of the molten material. When the strain becomes greater it releases in the form of seismic waves and it causes earthquakes. The energy(strain)storing capacity of the place differs by the hardness of the place. from the existing theorems, it has been built that:

6.1. Storing the Energy Producing the Heat

Energy is said to change forms rather than being created or destroyed, according to the rule of conservation of energy. The link between work and heat can be investigated with the first law of thermodynamics.

Work is the force used to move the Energy between systems and Heat is the transfer of thermal energy between two bodies.

By this the energy(strain) stored in the earth's crust emits energy in the form of heat. The energy stored in the earth's crust reaches several degrees. If we detect the high temperature, it will give the probability of finding the earthquake happening in the location.

6.2. The Change in the Electric Field

The process of an earthquake is related to the energy-storing elements. As the above-mentioned equation and details of the earthquake the more amount of energy emitted from the earthquake. The electric field is affected by the stored energy and if we continuously monitor the electric field we can notice the change in the electric field. The stress on the rock makes it vibrate.

If the stress is applied to the rock. It will vibrate and damage. For the example the quartz crystal, if a certain amount of pressure is applied on the crystal, it will vibrate but it is not broken till the high pressure is applied. By this, the strain on the rock leads to vibration and after reaching its limit it will break.

From the above three phenomena, there are a high number of possibilities to predict the earthquake, and if all these three things are, the prediction of the earthquake is possible.

7. IMPLEMENTATION

7.1. Hardware Devices

7.1.a). Accelerometer

An Accelerometer is a sensor used to measure the Acceleration of the body or object at rest instantaneously.

7.1.b). The Working Principle of the Accelerometer

An Accelerometer is a device used to measure the vibration of the acceleration of motion, of a structure. "pressure" Piezoelectric materials, due to the force of vibration or acceleration create a load proportional to the applied force Furthermore.

Figure 2. Accelerometer

The main purpose of the use of An Accelerometer in the part of slithering detector is that the snake senses the movement of any object around it through its skin in the form of vibrations. Vibrations help the snakes to Navigate and defend themselves from any harm (Weisong, J.).

8. THERMISTORS

Thermistors are resistance thermometers, or their resistance value depends on the temperature. This term is a combination of "thermal". It is formed by the compression of metal oxides. It is covered in the form of beads, discs, or cylinders Impermeable materials such as glass or epoxy.

Figure 3. Thermistors

Two types of thermistors:

i. Negative Temperature Coefficient (NTC)
ii. Positive Temperature Coefficient (PTC)

Both NTC AND PTC are inversely proportional (Mou, Y, 2020).

8.1.a) Purpose of Thermistors

Thermistors are used to sense the change in temperature of the heat generated due to earthquake
Which is caused by movement of tectonic plates due to the frictional force generated by moving the two tectonic plates. A fact says that very large earthquakes can produce a whooping Temperature rise of 1000 degrees Celsius.
Using the thermistors we can predefine the Movement of the tectonic plates before delay.

9. PIEZOELECTRIC SENSOR

A Piezoelectric Sensor is a sensor used to detect the pressure, temperature, strain, and force using the piezoelectric effect. The prefix *"Piezo"* in Greek refers to *"Squeeze"*.

Figure 4. Typical Piezoelectric Sensor

It is strong, lightweight, and constructed of flexible plastic, so these help them to enable sizes, thickness, and shapes. These sensors require no external current source or voltage, from the strain applied on that it generates an output signal (Smith, A. et al., 2021).

Principle of operation:

- Transverse
- Longitudinal
- Shear

9.1.a) Administration of Piezoelectric Sensors IN Slithering Detectors

The Piezoelectric Sensor is used as a part of slithering detectors to detect pressure. According to some other research, snakes can detect the seismic waves that precede an earthquake. These Waves cause the ground to vibrate, and the snakes feel it through their skin. Some snakes are capable of perceiving "Substrate-borne Vibrations". They receive these vibrations through their lower jaw.

Figure 5. Different seismic waves recorded by a seismometer

10. IR RADIATION SENSOR

An infrared sensor (IR sensor) is a radiation-sensitive optoelectronic component with a spectral sensitivity in the infrared wavelength ranges from 780 nm to 50 μm. It is produced by hot bodies (heat waves) and by when the molecules undergo rotational and vibrational transitions. The frequency range is 10^{11} Hz to 4×10^{14} Hz. IR sensors are now widely used in motion detectors and in building services to switch on lamps or in alarm systems.

Purpose: An IR sensor can measure the heat of an object as well as detect the motion.

Working principle: The IR transmitter continuously emits the IR light, and the IR receiver keeps on checking for the reflected light. If the light gets reflected by hitting any object in front it, the IR receiver receives this light. This way the object is detected in the case of the IR sensor.

Figure 6. IR Radiation Sensor

11. INDUCTION COIL MAGNETOMETER

An induction coil magnetometer, often referred to as a pulsation magnetometer or search coil magnetometer, is a passive device that is used to continuously monitor the various electromagnetic wave types that are propagating on the ground.

11.1.a) Construction

A search coil magnetometer is a vector magnetometer that measures one or more components of the magnetic field by connecting an inductive sensor to a conditioning electronic circuit. Three orthogonal inductive sensors are used in a traditional arrangement.

11.2.b) Working Principles

The induction coil magnetometer functions According to Faraday's law Electromagnetic induction principle, A conductor, such as a coil, is subject to a changing magnetic field. produces an electromotive force (EMF) proportional to the rate Magnetic field changes.

11.3.c) Purpose of Induction Coil Magnetometers

i)**Geophysical surveys:** The device is used in magnetic surveys for mapping geological structures, identifying fault lines, etc. This helps in detecting the pre-electric earthquake fields and fault lines.

Figure 7. An Induction coil magnetometer

A magnetometer is something that measures change in a magnetic field. The core in the magnetometer acts as a magnetic field concentrator, increasing the density of the internal magnetic field and increasing the sensitivity (Lee, D., Park, S. et al., 2020).

12. IMPLEMENTATION

The implementation part majorly consists of the design and plan of the circuit, connecting the hardware components into a wholesome product which then becomes a fully functional circuit unit, and different calculations of circuit materials.

Implementation of the product too consists of where the product must be placed at, testing the location of placement, and the number of feet's it needs to be placed.

12.1. Product Safety

The major and so much concerned part of the implementation of the optimization is to ensure the product safety. The circuit must be designed to follow all the product safety standards to prolong the time of its warranty. The necessary points to be discussed are

1. **Compact:** The product must be designed compact since the product is easy to carry and people to manage with the product.
2. **Portable:** The product must be easily portable from one place to another without any difficulties.
3. **Stress and pressure resistant:** The product has to be placed in a deeper place to sense the very small vibrations and very minute electric field that occurs before the earthquake.
4. **Water resistant:** All the products are sensitive towards water; even a small drop of water can cause damage to the circuit, so the packaging of the product must be done with specific material.
5. **Heat resistant:** The circuit must be packed with highly thermal resistant packaging material so that the product doesn't get exposed to heat and damaged soon.
6. **Frictional resistance and stability:** The product is mentioned to be frictional resistant because the product should not get damaged by the movements of the tectonic plates created (frictional).

13. CIRCUIT ANALYSIS

The circuit analysis gives the deeper workings and mechanisms of the circuital components of the slithering detectors. These circuital components together with the model trained works in the proposed way as mentioned. These circuital components used are chosen with regards to the Ideas and the problem statement. The circuital diagram below explains the connections of the detector in a schematic representation. This schematic representation explains how each of them are linked together in the circuit.

The block diagram of the slithering detector is given below

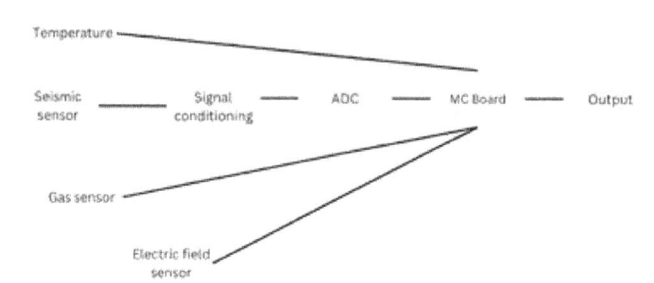

Figure 8. The block-level diagram of slithering detectors

The block-level diagram breaks down each part and the functionalities of each part of the detector.

13.1 Breakdown of the Unit

The main circuital unit consists of seismic sensors, signal conditioning, ADC (Analog to digital converter), MC board, and M/O (Main Output).

A device used to monitor seismic waves is a seismic sensor, sometimes known as a "seismometer". These days, it also includes the ground's movement during disturbances of any kind, such as earthquakes, volcanic eruptions, and explosions.

Signal conditioning is an electronic device that manipulates a signal in a way that it prepares it for the next stage of processing.

ADC is an electronic device that converts an analog signal into digital signals.

MC board is the short name of Microcontroller board. This typical microcontroller board is a compact integrated circuit (IC) which is used to govern specific operations of functions of an embedded system.

M/O stands for main output. All the signals that are collected and processed are received through a main output source.

The components which are directly connected to the main output devices are the Temperature sensor, Gas sensor, and Electric field sensor.

The temperature sensor is used to detect the amount of heat generated by the earthquake due to the frictional force created between the motion of the two tectonic plates. The amount of temperature produced will be massive and vary from magnitude to magnitude.

The gas sensor is used to sense the gases from the deep earth's mantle including methane of non-biological origin, Radon, Oxygen, Carbon isotopes, and occasionally Nitrogen isotopes.

An electric field sensor is connected to the main output directly which senses the pre-electric earthquake field and sends the detection sensed by it to the main output.

14. CIRCUIT DIAGRAM

The sample circuit diagram helps to visualize the connections of every component individually with the other components.

Figure 9. Circuit Diagram

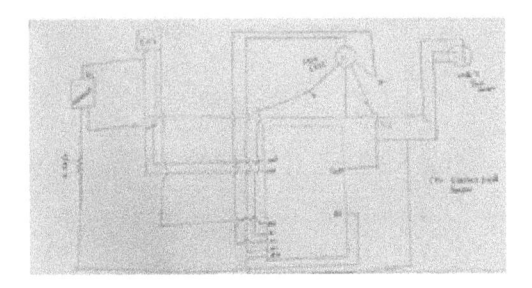

1. DS18B20 Temperature Sensor
    - VCC to 5V on Arduino
    - GND to GND on Arduino
    - Data to digital pin 2 on Arduino with a pull-up resistor (4.7kΩ between Data and VCC)
2. Electric Field Sensor
    - VCC to 5V on Arduino
    - GND to GND on Arduino

- Output to analog pin A0 on Arduino
3. ADXL335 Accelerometer
    - VCC to 3.3V on Arduino
    - GND to GND on Arduino
    - X to analog pin A1
    - Y to analog pin A2
    - Z to analog pin A3
4. MQ-7 Gas Sensor
    - VCC to 5V on Arduino
    - GND to GND on Arduino
    - Analog output to analog pin

15. CONCLUSION

At the end of the research, the optimization methods have found the best way out of nature which is astounding and outstanding in predicting the earthquake out of all the knowledge to an accumulative circuital element that mimics the mismatched way of acting of the snake with the Hardware depicted, the sensors and Hardware are Analyzed to the finest to give its all potential to find the accurate results. It can mimic nature but cannot create nature's supernatural abilities it has where technology can assist in the path of making lives easier and better.

REFERENCES

Chen, L., & Zhang, X. (2019). Real-Time Seismic Monitoring Using Multi-Sensor Networks. *Sensors (Basel)*.

Chen, M., & Zhao, L. (2022). Advances in Piezoelectric Sensors for Seismic Monitoring. *Sensors and Actuators*.

Gupta, A., & Verma, P. (2020). Sensor Technologies for Early Earthquake Detection. Journal of Earthquake Technology.

Johnson, T., & White, S. (2021). Innovative Approaches to Earthquake Prediction Using Multi-Sensor Networks. *Earthquake Engineering & Structural Dynamics*.

Kumar, N., & Sharma, P. (2020). Enhancing Seismic Prediction through Technological Innovations. *Earthquake Engineering & Structural Dynamics*.

Kumar, S., & Singh, R. (2022). *Multisensor Integration for Real-Time Earthquake Monitoring*. Advances in Earthquake Research.

Lee, D., & Kim, S. (2021). Integration of Infrared Sensors in Seismic Monitoring Systems. *Journal of Seismology*.

Lee, D., Park, S., & Kim, J. (2020). Development of a Novel Seismic Sensor System Mimicking Snake Vibration Detection Mechanisms. *Journal of Seismology*.

Lee, J., & Park, S. (2022). Comprehensive Early Warning Systems for Earthquakes Using Multi-Sensor Integration. *Journal of Seismology*.

Mou, Y. (2020). Seismic Prediction Using Animal Behavior and Advanced Sensor Technologies. *International Journal of Geophysics*.

Mou, Y. (2021). Advanced Sensor Networks for Earthquake Prediction: A Review. *Seismological Research Letters*.

Park, J., & Lee, H. (2021). *Combining Ground Vibration Sensors and Electromagnetic Field Sensors for Earthquake Prediction*. Seismic Engineering Journal.

Patel, M., Reddy, P., & Sharma, A. (2021). Artificial Neural Networks in Seismic Data Analysis for Earthquake Prediction. *Applied Geophysics*.

Patel, R., & Raut, R. (2023). Prediction Model Using Artificial Neural. *IEEE Network*.

Patel, S., & Kumar, R. (2022). The Role of Interdisciplinary Approaches in Earthquake Prediction. *Journal of Geophysical Research*.

Patil, R., Patil, S., Todakari, N., Devkar, A., & Raut, R. (2023, May). Earthquake Depth & Magnitude Prediction Model Using Artificial Neural Network. In 2023 4th International Conference for Emerging Technology (INCET) (pp. 1-5). IEEE. DOI: 10.1109/INCET57972.2023.10170413

Reddy, S., & Kumar, P. (2021). Real-Time Data Analysis for Earthquake Magnitude Prediction Using Neural Networks. *IEEE Transactions on Geoscience and Remote Sensing*.

Singh, M., & Gupta, R. (2020). Advances in Seismic Sensor Technology. *Sensors (Basel)*.

Smith, A., & Johnson, B. (2021). Integrating Biosensors and Machine Learning for Early Earthquake Detection. *Journal of Earthquake Engineering*.

Smith, L., & Green, J. (2022). Integrating Biosensors with Machine Learning Models for Earthquake Prediction. *Computational Geosciences*.

Wang, T., & Li, X. (2021). Continuous Monitoring of Electromagnetic Anomalies for Earthquake Prediction. *Geophysical Journal International*.

Weisong, J. (n.d.). The Sensory Capabilities of Snakes in Earthquake Prediction. (Details of publication not provided).

Williams, J., & Brown, H. (2021). *Enhancing Earthquake Prediction Accuracy with Biosensors*. Earthquake Science.

Zhang, Y., & Liu, H. (2021). The Use of Magnetometers in Seismic Anomaly Detection. *Journal of Applied Geophysics*.

Zhang, Y., Liu, H., & Wang, X. (2020). Utilization of Induction Coil Magnetometers for Seismic Anomaly Detection. *Geophysical Research Letters*.

Chapter 3
Optimizing Resource Management With Edge and Network Processing for Disaster Response Using Insect Robot Swarms

U. Vignesh
Vellore Institute of Technology, Chennai, India

K. Gokul Ram
Vellore Institute of Technology, Chennai, India

Abdulkareem Sh. Mahdi Al-Obaidi
https://orcid.org/0000-0003-2575-0441
Taylor's University, Malaysia

ABSTRACT

The potential of emerging technology to transform disaster response is examined in this review study. We investigate how, by overcoming the constraints of conventional cloud-based processing in disaster areas, edge computing enables insect robots for real-time data gathering and analysis at the network edge. We examine studies on KubeEdge for reliable network deployment using insect robots and Social Sensing-based Edge Computing (SSEC) for processing social media data. We explore network processing methods that leverage Mobile Cloud Computing (MCC) and show how they overcome issues such as bandwidth limitations and low battery

DOI: 10.4018/979-8-3693-6150-4.ch003

life. This study examines how developments in edge computing, network processing, resource management, and multi-robot systems might benefit disaster response by highlighting the potential of insect robots as indispensable instruments that will ultimately result in quicker and more efficient reaction times.

I. INTRODUCTION

Tragically, disasters are happening more and more frequently. According to the Centre for Research on the Epidemiology of Disasters (CRED), 421 natural disasters worldwide claimed over 10,000 lives in 2022 alone. With their catastrophic effects on densely populated places, earthquakes such as the one that rocked Turkey and Syria in February 2023 emphasize how urgently disaster response efficiency needs to be improved. In situations like this, traditional approaches frequently falter. Despite their importance, large search and rescue teams have limits. According to a 2021 study published in the Journal of Disaster Research, the challenges of negotiating debris might result in average response times of 72 hours in cases of collapsed structures. Unmanned aerial vehicles (UAVs) provide a more comprehensive view, but their efficacy is restricted by their short battery life and susceptibility to inclement weather. Tragic consequences may result from these delays. According to a 2016 Nature Geoscience study, quicker reaction times during earthquakes may be able to prevent up to 40% of total fatalities. It is obvious that creative solutions are required to close these gaps and speed up response. With their special talents, insect-sized robots have enormous potential to transform disaster relief efforts.

The fast-evolving subject of microrobotics presents a novel method for disaster response. Inspired by their biological counterparts' agility and mobility, insect-sized robots have enormous potential for negotiating dangerous and complex settings that are inaccessible to conventional approaches. Because of their small size, they can squeeze through tiny openings, hide beneath rubble, and get to stranded survivors in crumbling buildings. Their integrated sensors can also collect essential information to help with search and rescue operations, such as temperature, humidity, and audio characteristics. Although using insect robots has great potential, data communication issues must be resolved first. Their small size limits the amount of raw data they can transmit by limiting transmission bandwidth and processing power onboard. Furthermore, real-time data processing and analysis are required to derive meaningful insights from disaster zones due to their dynamic and frequently unpredictable nature.

Disaster zones are inherently dynamic. It might be necessary to process substantial amounts of sensor data from a widespread robot deployment during the first response phase. Subsequent stages may concentrate on an in-depth examination of particular fields. In this case, cloud computing is superior. Because of its elastic nature, resources can be scaled up or down to satisfy changing processing demands. To manage the increased demand during peak data gathering, more cloud instances can be created. On the other hand, resources might be reduced when the response gets better in order to minimize expenses. This guarantees effective data processing for the duration of the emergency response effort. The insect robots' cloud-based data serves as a central hub that authorized staff can access from any location with an internet connection. This encourages the sharing of information in real time, allowing various teams to decide on the best course of action based on the most recent facts. Imagine a situation where a search and rescue team can coordinate their efforts based on the location of possible survivors by remotely seeing a live map created from robot sensor data.

Insect robots gather vast amounts of raw sensor data, from which valuable insights can only be extracted with sophisticated analytical techniques. Here's where cloud computing excels. Strong cloud systems give users access to a wide range of advanced data analysis tools, such as algorithms for pattern recognition and machine learning. Large datasets can be analyzed using these technologies, which can spot hidden trends and patterns that conventional approaches would overlook. Machine learning systems, for instance, could identify regions where there is a greater chance of trapped survivors by analyzing temperature and auditory sensor data. Important information that can direct search and rescue operations and possibly save lives is provided by this sophisticated study. Although there is no denying the promise of insect robots in disaster relief, ethical issues must be taken into account. Protecting privacy when using sensors to collect data, reducing the environmental impact of robot materials and disposal, and making sure this technology isn't weaponized are some of the concerns. Though we think the advantages override these worries and save lives, responsible development and implementation are crucial. To ensure that this cutting-edge technology serves humanity for the better, it is essential to have honest conversations, lucid rules, and a dedication to ethical use. These factors continue to be a major focus in the robotics community and on a global scale.

There is no denying the enormous promise of insect-sized robots for mapping disasters, but there is still a significant obstacle in smoothly integrating these robots with cloud computing capabilities. Research that has already been done frequently concentrates on specific areas, such as cloud-based disaster response systems, data storage for swarms, or robot data compression. In order to close this gap, this study offers a thorough architecture that combines distributed data storage strategies, cloud-based artificial intelligence (AI), and data compression techniques for real-time analysis. It is especially intended for use with insect robots in disaster mapping. Our contribution is the development of a comprehensive strategy that maximizes data processing during the whole catastrophe response process. This framework will assess safe and dependable data storage options for robot swarms operating in dynamic environments in addition to analyzing current data compression techniques for robots with limited resources. In addition, we suggest a fresh method for combining cloud-based AI with real-time analysis. This might enable machine learning to find survivors or uncover hidden patterns in sensor data, which would ultimately result in more rapid and accurate search and rescue operations. Our research has the potential to greatly increase the efficacy and efficiency of disaster response, saving lives in the wake of these tragic occurrences, by bridging this crucial gap.

II. DATA COMPRESSION FOR EFFICIENT COMMUNICATION

In disaster areas, insect robots must be able to communicate effectively, but this is difficult due to their limited power and bandwidth. WiFi sensor networks deal with these small problems. Luckily, studies reveal that spatial redundancy—the phenomenon in which neighboring robots record similar information—occurs frequently in sensor data. Data compression methods that take use of this redundancy, as suggested by Anna et al. in 2002, can greatly lessen transmission load. Research indicates that compression ratios ranging from 2:1 to 10:1 can be attained, which could result in longer mission durations and better data gathering for insect robots (Scaglione & Servetto, 2002). This is exactly in line with our survey paper's emphasis on catastrophe mapping robots' optimal data handling. We can bridge the gap between insect robots and cloud computing by implementing data compression algorithms, enabling effective data transmission and real-time analysis for effective disaster response.

Traditional compression methods, such as JPEG, may not be the best for sensor networks placed in disaster areas because they sacrifice visual quality for human perception. In 2003, Mu Chen et al. investigated data compression with a focus on sensor networks, taking network longevity into consideration (Chen & Fowler, 2003). To analyze the trade-off between compression rate (reducing data size for transmission), energy consumption (lower power usage due to less data), and the accuracy of the final data for statistical analysis (ensuring essential information is retained), the authors propose a 3D Rate-Energy-Accuracy (R-E-A) function. This trade-off must be understood in the context of sensor networks. The research also suggests a modified JPEG method that, at the expense of some visual quality, gives priority to capturing high-frequency data components that are essential for tasks like object location in sensor data. Since this method better fits the requirements of sensor networks, the reduced data may allow for more precise statistical inference. We can create more efficient data compression methods for insect robots in disaster mapping by combining research on sensor networks and their unique compression needs. This will guarantee effective data transfer, energy conservation, and precise data analysis for crucial decision-making.

Expanding on the notion of data compression for robots with limited resources, Francesco et al. (2008) investigated methods to additionally leverage the innate spatial correlation present in sensor data. A spatially correlated Gaussian random vector, $N(0, R)$, is employed in one such model (Marcelloni & Vecchio, 2008). In essence, this is a zero-mean Gaussian distribution, with the correlation structure between sensor values from neighboring robots being captured by the covariance matrix (R). The model implies that the covariance matrix smoothes into a two-dimensional function, reflecting the finer-grained spatial correlation and denser network, as the number of deployed robots grows. With the help of this model, spatial redundancy can be effectively analyzed and taken advantage of, opening the door to even more effective data compression methods for extensive insect robot deployments. Moreover, data compression provides an important advantage since radio communication consumes a significant amount of energy in these robots (as well as in Wireless Sensor Networks) and smaller data sizes translate into less energy used during transmission. This increases the number of years that individual robots and the disaster response network as a whole may operate, enabling longer missions and more thorough data collecting.

III. DISTRIBUTED DATA STORAGE IN ROBOT SWARMS:

Large-scale insect robot deployments in disaster areas are made possible by distributed data storage. Conventional techniques may not be effective or scalable for these robots with limited resources. Nathalie et al. in 2019 investigated Swarm-Mesh, a distributed data structure created especially for mobile robot networks (Majcherczyk & Pinciroli, 2020). SwarmMesh uses the dynamism of robot motion to store information in a scalable and memory-efficient manner. It divides the data and allocates storage capacity according to a robot's attributes, like its memory capacity and number of neighbors. Robots then employ gradient routing to retrieve information and key-based routing to store data effectively. Evaluations show that SwarmMesh efficiently uses collective memory throughout the network to retain a significant quantity of data even under high demand. This study presents a viable method for distributed data storage in insect robots that offers optimal memory use, scalability, and efficiency—all essential components for the effective execution of disaster mapping tasks.

Effective data exchange is essential for collaborative insect robots operating in disaster areas. The limitations of storage and communication bandwidth may cause problems for traditional approaches. Vivek Shankar Varadharajan et al. in

2019 investigated SOUL (Swarm-Oriented Upload of Large Data), a system created especially for robot swarm data exchange (Varadharajan et al., 2020). By deliberately distributing the huge data chunks throughout the swarm, SOUL addresses the difficulty of reconstructing them. Robots bid for data segments, or datagrams, according to their available storage space and proximity to processors through the use of a decentralized auction system. This guarantees effective distribution, assigning datagrams to robots that are most suitable for swift retrieval. Robots can also request missing data and update datagram positions within the swarm using the capabilities included in SOUL. This work presents a viable method for data sharing in robots with limited resources. In insect robot swarms, SOUL can greatly enhance data interchange by reducing rebuilding time and allocating resources according to robot capabilities, hence promoting cooperative work during disaster relief situations. It's crucial to recognize that the current implementation is limited to the use of one programming language (Buzz). To investigate real-world deployment and compatibility with different insect robot systems, more study may be required Conventional robotic data storage techniques frequently depend on centralized systems, necessitating strong robots and introducing a single point of failure. Simon Jones et al. in 2020, investigated a ground-breaking method for data storage in robot swarms called" distributed situational awareness" (Jones et al., 2020). Here, each robot uses its own local sensor data—cameras, infrared, etc.—to determine how much storage is appropriate. By taking into account variables like available space, closeness to other robots, and barriers, scalability is promoted and a central processing unit is not required. Based on its analysis, each robot chooses the best storage site on its own and dumps the data item there. It moves and stores the object similarly to a Roomba, adding the location to its internal inventory. By repeating this cycle, an adaptive data storage system that makes use of the swarm's collective intelligence is produced. The ensuing storage patterns (spread, spiral, and clusters) may appear atypical, but they actually represent effective swarm topologies depending on local data. By using a decentralized strategy, where each robot maintains its local inventory, a central repository is not necessary. Its continuous updating guarantees flexibility in ever-changing contexts. For data storage duties in large-scale robot deployments, distributed situational awareness has the potential to perform better than conventional methods in terms of information richness and efficiency.

When deploying insect robots in disaster areas, where sensitive data may be gathered, data security and privacy become critical. Morco Dorigo et al. in 2021, investigated Merkle Trees (MTs), a potentially useful method for distributed data storage in robot swarms (Ferrer et al., 2021). In contrast to conventional techniques, MTs provide notable security advantages. They stop possible data manipulation by making robots demonstrate that they have accomplished particular tasks or have particular information. Even in the event that a robot is captured, deciphering the mission objectives as a whole is challenging due to the scant raw data it contains. This enables efficient robot collaboration without jeopardizing confidential information. The research conducts real-world tests using maze formation missions and foraging to further validate this approach. These tests demonstrate how MTs can be scaled and optimized for robots with limited resources, even when there is a higher communication cost because of the increasing number of robots. Crucially, robots are capable of accomplishing intricate tasks without requiring all mission information, and the computational demands of machine translations are well-suited for current robot hardware. All things considered, this study makes a stronger argument for MTs as a workable option for safe and scalable data storage in extensive insect robot deployments for emergency response activities.

IV. DYNAMIC TASK ASSIGNMENT IN ROBOT SWARMS

Efficient task allocation is necessary when coordinating several robots for duties such as search and rescue in disaster areas. Robots can divide up jobs among themselves using dynamic task assignment algorithms, in contrast to conventional techniques that rely on a central controller. This is essential for real-world situations involving dynamic settings and little to no communication. In 2005, James McLurkin et al. investigated a number of strategies, each with advantages and disadvantages of their own (McLurkin & Yamins, 2005). While speed and simplicity are prioritized in random choice, it can be unreliable in smaller swarms. High precision is provided by Extreme-comm, which communicates continuously to determine duties based on robot position. But all of this communication takes up resources. A balance is achieved via the Tree-Recolor Algorithm and Card-Dealer's Algorithm. A card dealer may be slower, but they employ a phased strategy with little communication. In order to achieve effective task allocation, Tree-recolor uses a tree structure for communication. However, in order to synchronize, robots must estimate the size of the network. The significance of dynamic task assignment for robots with limited resources is underscored by these methods. All of these strategies allow for decentralized cooperation in the absence of a central controller, notwithstanding their trade-offs. This is necessary for efficient task distribution in disaster areas because settings and communication are always changing. Insect robot-specific further research on these algorithms may result in enhanced mission success and cooperation in disaster response activities.

V. EDGE COMPUTING FOR DISASTER RESPONSE

Real-time data analysis is essential for effective disaster response; however, in disaster areas, standard cloud-based processing is hindered by network disruptions and overloaded central servers. Akira Uchiyama et al. in 2017 investigated edge computing as a possible substitute (Higashino et al., 2017). By moving data processing closer to the disaster area—such as with specialized edge servers or cellular base stations—edge computing decentralizes data processing. By evaluating data locally, this considerably cuts down on communication delays and allows for quicker response times and better decision-making. Edge servers are able to continue processing data even in the event of partial network outages. Disaster response is strengthened by edge computing in a number of ways. Immediate insights can be gained by real-time analysis of sensor data from environmental monitoring systems or traffic cameras, which speeds up the allocation of resources to important tasks. Furthermore, prompt data sharing among residents, emergency centers, and first responders promotes coordinated efforts and effective use of available resources. Lastly, first responders are empowered with real-time data through decentralized decision-making, which enables them to adjust to the changing circumstances within the disaster area. Insect robots are essential because they gather data at the scene of the crisis and feed it into edge computing's real-time processing capacity for a more thorough and timely response, even though they wouldn't run edge servers themselves. To fully explore the possibilities of combining edge computing with data collecting from insect robots for optimal disaster response, more investigation is required.

There has historically been a backlog in disaster response due to the centralized processing of field data. Due to the excessive pressure on central servers—which are frequently overloaded by compromised infrastructure—critical damage assessments are delayed, and bandwidth usage is rapidly increasing. According to research, this can impede response efforts by up to a 50% delay in assessments. SSEC, or Social Sensing-based Edge Computing, presents a novel strategy. By utilizing edge computing, it turns the disaster site's laptops and smartphones into a potent network of "edge" resources. This network allows for quick damage assessment because it can handle social media data (posts, photos) in real-time. According to Daniel Zhang et al. in 2019, this method could cut assessment times in half and minimize bandwidth use by 80% (Zhang & Wang, 2019). Reliability is crucial since SSEC systems can continue to operate even in the event that central servers are compromised.

There is a problem, though: the wide range of devices (laptops, smartphones, etc.) at a catastrophe site will differ in terms of operating systems, computing power, and network interfaces. According to research, assigning work inefficiently in these kinds of settings might result in a 30% decrease in performance. HeteroEdge and other frameworks are attempting to address this by using clever resource allocation techniques and offering a uniform interface. This minimizes energy usage and guarantees that jobs are divided among these many devices in an efficient manner. Although SSEC has historically relied on laptops and smartphones, the fundamental ideas can be modified for use with insect robots in disaster areas. Similar benefits that SSEC can provide for robots—such as lower bandwidth consumption, quicker analysis, and possibly even higher resilience—can be extended to robots by spreading processing jobs among the robot swarm when data is processed locally at the "edge" (the catastrophe site).

When it comes to deploying insect robots, typical communication techniques are unreliable because of the often destroyed network infrastructure in disaster zones. Minh Ngoc Tran et al. in 2021 investigated the usage of KubeEdge, an open-source edge computing platform, in building reliable networking systems (Tran & Kim, 2021). KubeEdge's lightweight EdgeCore component makes it especially appropriate. Because of its compact size and low resource consumption, this program may be installed on insect robots with limited memory and computing power. Furthermore, KubeEdge is compatible with other communication protocols, such as Bluetooth, which allows for efficient control of NDN responder devices, which are essentially insect robots. The inherent transmission dependability of KubeEdge is another significant benefit. In spite of sporadic network connectivity, this guarantees vital data exchanges between the disaster zone (edge) and the cloud (mission control). At-least-once delivery is the method that ensures data reaches its destination and achieves this reliability. In addition, the architecture may be readily expanded to accommodate larger disaster zones by simply adding more KubeEdge clusters. The lightweight design of EdgeCore minimizes downtime by facilitating rapid redeployment on any available network equipment, even in the event of an edge node failure. This method provides a viable way to implement NDN networks and allow bug robots to reliably handle data in disaster response situations. To investigate potential drawbacks, security issues, and practical application with particular insect robot platforms, more study is necessary.

VI. IN NETWORK PROCESSING FOR SENSOR DATA:

Obtaining data is essential to a successful disaster response strategy. In this field, wireless sensor networks, or WSNs, are essential. These networks are made up of many dispersed, tiny sensors that work together to gather environmental data. Insect robots can serve as these kinds of sensor nodes in disaster areas, collecting vital environmental data. However, the short battery life and capacity to send huge volumes of data are intrinsic limitations of these sensor nodes. MCC, or mobile cloud computing, presents a strong option. To overcome the shortcomings of individual sensor nodes, MCC makes use of the cloud's powerful processing and storage power. An intriguing opportunity arises when WSNs are integrated with MCC (WSN-MCC): sensor data gathered by insect robots can be sent to the cloud for further processing and analysis. Mobile devices can then be used to distribute the processed data to personnel on the ground, enabling them to make well-informed decisions for a more efficient response. WSN-MCC integration, however, has a number of difficulties. Given the short battery life of sensor nodes, energy conservation is critical. Furthermore, communication capacity is frequently limited in disaster areas, thus effective data transmission strategies are required. Strong security measures are also necessary to protect the sensitive data that sensor nodes gather.

A novel approach was presented by Hai Wang et al. in 2014 to improve the performance of WSN-MCC integration for disaster response and overcome these issues (Zhu et al., 2014). The framework's main objectives are to optimize sensor node energy consumption, provide effective data transmission within bandwidth restrictions, and put strong security measures in place. The suggested framework opens the door for a secure and seamless data transfer method for those working on disaster response operations by taking care of these important details. Although the present work focuses on WSN-MCC integration for mobile users, the fundamental ideas are immediately applicable to insect robots. As a cooperative WSN, these robots can use MCC to effectively collect, analyze, and distribute data to several stakeholders in order to respond more quickly and efficiently. To completely adapt WSN-MCC principles for resource-constrained robots operating in the complex and dynamic environment of a disaster zone, more research is necessary.

VII. RESOURCE MANAGEMENT AND OPTIMIZATION TECHNIQUES

Efficient resource allocation is crucial for the effective coordination of a large swarm of insect robots in disaster response scenarios. This important factor dictates how tasks are assigned to individual robots and how finite resources are used, like battery life and communication bandwidth. Mohamed Elhoseny et al. in 2018 introduced Genetic Algorithms (GA) and Particle Swarm Optimization (PSO), which show promise in optimizing resource allocation for insect robots operating in a cloud-based processing environment (Elhoseny et al., 2018). Natural selection serves as the inspiration for GA, which simulates population evolution by using each individual to represent a possible allocation method. A chromosome (solution) in the context of disaster response might specify the distribution of robots with diverse processing and sensor capacities among several crisis zones.

Next, each allocation is assessed using a fitness function that takes into account the robots' energy consumption, the quality of the sensor data they collected, and the area they covered. In order to find solutions that maximize data collection and reduce energy usage, GA iteratively refines allocation algorithms, ultimately resulting in a more effective catastrophe mapping operation. PSO, on the other hand, finds inspiration in the way that schools or flocks of fish move together. Particles here stand for possible methods of allocating resources for particular missions, such as search and rescue. The number of robots with particular communication and navigational skills that are placed in various search zones may be specified by each particle. Next, each allocation is assessed using a fitness function that takes into account the size of the search region, the effectiveness of locating survivors, and the latency of communication.

Next, each allocation is assessed using a fitness function that takes into account variables such as the size of the search area covered, the effectiveness of locating survivors, and the latency of communication between the robots and the cloud processing center. Through mutual learning, PSO enables the swarm of particles (allocation choices) to converge on solutions that maximize search efficacy inside the disaster area. These are only two instances; the precise application of GA or PSO would rely on the disaster response tasks and performance indicators selected. All things considered, these algorithms provide a useful foundation for resource allocation optimization in dynamic and resource-constrained environments such as disaster zones. Disaster response operations can benefit from increased efficiency, better scalability to manage massive robot swarms, and real-time adaptation by utilizing these optimization strategies.

Swarm Intelligence (SI) emerges as a possible alternative to traditional, centralized techniques that frequently fail to manage resources in complex Cyber-Physical Systems (CPSs). SI has important benefits like resilience against disturbances and adaptability to changing settings, as it is inspired by collective behavior in nature. Because of this, SI is appropriate for CPSs where effective and dynamic resource allocation is required. Melanie Schranz et al. in 2021 investigated how SI-facilitated local interactions between CPS components can result in the best use of available resources (Schranz et al., 2021). It even suggests possible SI models that draw inspiration from non-cooperative animal behaviors. These models, which are based on population dynamics or predator-prey systems, may be applied to resource allocation optimization based on demand in real-time or conflict management within CPSs. The study does, however, recognize the difficulties in creating local rules for erratic situations and the need for more investigation into the best ways to allocate resources based on sensor data. In summary, this work reinforces the case for investigating SI as a potential solution for resource management and optimization in complex CPSs by showcasing its potential and continuing research efforts.

Swarms of civilian UAVs are becoming more popular for gathering data in a variety of disciplines. But overseeing these swarms' resources—such as coverage area and data resolution—is difficult, particularly when restrictions prevent them from operating with complete autonomy. To address this issue, Hanno Hildmann et al. in 2019 presented a nature-inspired method for swarm-based data collecting (Hildmann et al., 2019). The method makes use of decentralized control, in which individual UAVs decide what to do after consulting with their neighbors. UAV cooperation is facilitated by local optimization, which raises total data-collecting efficiency. This entails a probabilistic redistribution of data-gathering locations among UAVs to guarantee the best possible use of their sensor capacities. The system uses mathematical formulas to measure the effectiveness of data gathering, taking into account the coverage area and the resolution that each UAV's sensor provides for particular places. Although the study admits limitations because of limited large-scale real-world testing, simulations demonstrate promise. Furthermore, the method may need to be further refined to accommodate extremely dynamic data collection requirements. However, this work opens the door for resource-efficient UAV swarm data collecting. Upcoming research will examine ways to scale communication to manage even larger swarms and test them in real-world scenarios, potentially leading to useful applications such as dynamic-resolution video surveillance.

VIII. SCALABLE SECURITY ANALYSIS FOR CLOUD DATA CENTERS

It is extremely difficult to maintain security in large cloud data centers because of the constantly increasing number of servers and changing vulnerabilities. For such massive and dynamic networks, traditional attack graphs (AGs), which map possible attacker paths, become computationally costly and cumbersome. Abdulhakim Sabur et al. in 2022 investigated the S3 framework, which surfaced as a viable resolution to tackle the problem of scalability in security analysis. S3 addresses the problem by utilizing Distributed Firewalls (DFWs) and Software-Defined Networking (SDN) to divide the cloud network into controllable segments. Each segment's attack graph can be made smaller and more targeted thanks to this segmentation. Then, to further optimize this process, a grouping technique called K-means clustering is used to divide the network into segments based on similarities between items like vulnerabilities and services. S3 enables the effective creation of sub-graphs, or tiny attack graphs, that take into account variables like vulnerabilities, connections, and security rules within each segment by studying smaller, more manageable network segments. Ultimately, these sub-graphs are combined to produce an all-encompassing attack graph that shows the cloud network's overall security posture. With the use of k- means clustering and segmentation, S3 is able to provide a scalable security analysis solution in large-scale cloud systems.

Effective data management is crucial for protecting large cloud data centers, even if scalable security analysis is also necessary. Bikash Agarwal et al. in 2017 investigated an expanded method of managing data centers by means of a Big Data ecosystem architecture intended for the analysis of data from these centers and Internet of Things (IoT) gadgets (Agrawal, 2017). With a dedicated Security Layer that includes secure data deletion methods for distributed file systems, the architecture recognizes the significance of security. This emphasis on safe data erasure demonstrates how cloud data centers are becoming more conscious of the problems with data privacy. But rather than focusing only on security flaws, the dissertation primarily discusses how to manage and analyze big data for a variety of applications.

Vincenzo Barrile et al. in 2017 explored how geomat-ics—more especially, data fusion—can transform vineyard management techniques (Barrile et al., 2022). The study demonstrates how combining information from multiple sources—such as sensor data from autonomous cars, high-resolution drone photos, and satellite imagery—offers a thorough understanding of vineyard health. Through the analysis of the Normalized Difference Vegetation Index (NDVI) obtained from satellite and drone data, scientists may evaluate the health of plants and direct specific activities aimed at optimizing the distribution of resources. Potential advantages include determining the best paths for machinery to take in order to reduce damage, limiting the application of fertilizer and irrigation to particular regions determined by NDVI analysis, and ultimately resulting in more effective and sustainable vineyard management. In the context of Agriculture 4.0, this study highlights the significance of data fusion in precision agriculture. Farmers may obtain deeper insights into their vineyards and make data-driven decisions that maximize resource utilization, boost crop yields, and advance sustainable agricultural practices by utilizing a variety of data sources and sophisticated analysis methodologies.

Tooba Samad et al. in 2018, investigated how fleets of Unmanned Aerial Vehicles (UAVs), also referred to as drones, may be managed in disaster response situations using cloud-based multi-agent systems (Samad et al., 2018). The scope and intensity of these circumstances can pose challenges to traditional search and rescue techniques. This study suggests a method for locating survivors and directing rescue operations using real-time sensory data from drones. The fundamental concept is a multi-agent system running on the cloud that dynamically assigns drones in response to real-time data, guaranteeing that important areas are adequately covered while maximizing the use of available resources. By increasing the likelihood of locating survivors, this strategy has the potential to greatly increase the efficiency of search and rescue operations. The study recognizes the shortcomings of fixed drone allocation systems and suggests a flexible method that may adjust to the changing requirements of the disaster area. The cloud platform enables connectivity between the drones and the central system and provides scalability and dependability for real-time data processing. Overall, by putting forth a novel strategy for drone management that makes use of cloud-based multi-agent systems and real-time data processing, this work advances UAV technology for disaster response.

The survey's emphasis on cloud-based coordination and control for multi-robot systems is in line with the findings of this study. Himadri Nath Saha et al. in 2018 presented a system that combines ground robots and drones under the control of a single cloud platform (Saha et al., 2018). The advantages of this cloud architecture are substantial. It safely retains the information gathered by the robots, allowing for in-depth analysis and wise decision-making. The smooth operation of all system components is facilitated by real-time wireless network connectivity. As the primary coordinator, a master drone gathers and evaluates information from aerial and ground units. With the help of a master drone coordinator and a cloud-based strategy, more information may be stored and analyzed in the cloud to support better decision-making. Furthermore, managing data from a large number of robots is made easier by the scalability of the cloud platform, which qualifies the system for large-scale deployments. Ultimately, the master drone acts as a single point of control where all robot actions within the system may be coordinated. Overall, by presenting a unique architecture that makes use of cloud computing's capabilities for real-time data processing, communication, and coordinated decision-making in collaborative robot operations, this research advances cloud-based control and coordination.

Subramanium Ganesan et al. in 2018 investigated a potential use case for swarm robotics in disaster relief (Ganesan et al., 2011). The suggested method makes use of tiny robots that have several sensors for data collecting about the surroundings, object detection (perhaps survivors), and navigation. Swarm intelligence, a bio-inspired methodology that enables simple, individual robots to work together to do complicated tasks like traversing dangerous settings and looking for survivors, is how these robots cooperate. Ant Colony Optimization (ACO) is used by the system to dynamically identify the best pathways through the disaster area. Redundancy (some robots can continue if one fails), flexibility to change settings, and the capacity to perform activities beyond the scope of a single robot are just a few benefits of this collaborative approach. Overall, by demonstrating, this research advances the field of swarm robotics for disaster response. Overall, by presenting a novel system that combines swarm intelligence and ACO for effective search and rescue operations, this research advances the field of swarm robotics for disaster response.

A revolutionary solution to data management and communication for SAR operations was proposed by Leonardo Militano et al. in 2023 (Militano et al., 2023). A cloud-to-edge-IoT continuum that combines cloud and edge computing resources with different SAR devices (robots, drones, and sensors) is the suggested approach. The lack of a single system for data administration, ultra-low latency requirements, and heterogeneous data formats are some of the present limitations in SAR technologies that this creative architecture attempts to address. A strong and flexible framework for processing and analyzing data in real-time is provided by the cloud-to-edge continuum. First responders can make better decisions during crucial rescue missions by using real-time data analysis to obtain a deeper understanding of the disaster zone by utilizing the "remote brain" of cloud/edge resources. The study also looks into VOStack, a tool developed for the NEPHELE project, as a possible means of effective data processing within the continuum. Overall, by putting forth a novel cloud-to-edge-IoT continuum that enables real-time data processing, resource management, and enhanced situational awareness for first responders, this research advances the fields of data management and communication for SAR.

Pengxu Hou et al. in 2023 investigated a potential method for robotic search and rescue that leverages bio-inspired design concepts (Hou et al., 2023). A bionic insect robot is the suggested remedy. It makes use of edge computing and artificial intelligence (AI) developments to function well in dangerous situations, such as cramped areas or collapsed structures when conventional rescue techniques put human life in danger. The robot can navigate in confined spaces and move through detritus because of its compact, nimble design, which was inspired by insects. While AI algorithms enable activities like target detection, path planning, and sensor data analysis, edge computing on the robot makes real-time data processing and decision-making possible. This attractive strategy for future search and rescue robots combines edge computing, artificial intelligence, and bio-inspired design. It has the potential to increase efficacy, safety, and efficiency in life-threatening emergency scenarios.

IX. FUTURE DIRECTIONS

Jorge Pena Queralta et al. in 2020 investigated multi-robot systems intended for Search and Rescue (SAR) missions (Queralta et al., 2020). The study examines two fundamental components of these systems' efficacy: system-level considerations for deploying these robots efficiently in real-world scenarios with a variety of environments and robot types (underwater, aerial, and ground); and algorithmic approaches for robot coordination, communication, and real-time information gathering (active perception) to optimize search strategies. The research provides important insights on multi-robot collaboration for disaster response by looking at these aspects. While this article focuses on SAR missions, your investigation of multi-robot systems for disaster mapping can benefit from an understanding of the fundamental concepts of cooperation and system-level issues.

Also investigated a related field: multi-robot systems intended for Search and Rescue (SAR) operations. The study demonstrates how these systems have the potential to completely transform SAR operations by facilitating real-time monitoring, enabling higher search speeds, and producing more thorough environmental maps. Their analysis focuses on two critical elements that are necessary for successful multi-robot teams: machine perception, or how robots perceive their environment using methods such as machine vision, and control mechanisms, or the algorithms that regulate robot collaboration during missions. The study highlights the importance of utilizing multiple robot teams, comprising airborne, ground, and underwater vehicles, in order to adjust to the many environments that are found in disaster zones. These environments might range from urban settings to wilderness and maritime settings. You may demonstrate a comprehensive grasp of how robots can be used for a range of disaster response activities, such as mapping and search and rescue, by incorporating this research, which will ultimately lead to more successful disaster relief efforts.

Maria Kondoyanni et al. in 2022 looked into how biomimicry may be used to solve urgent problems in contemporary agriculture (Kondoyanni et al., 2022). The two biomimetic technologies that show the most promise are swarm robotics and soft robotics. Soft robots may perform delicate activities like fruit harvesting damage-free, decreasing losses, and maximizing productivity. They are made of compliant materials. Swarm robotics may be able to address the labor crisis and environmental issues by using groups of tiny robots to work together on projects like pest control, crop monitoring, and pollination. This research investigates the design principles, functions, and possible advantages of biomimetic robots in agriculture by examining previously conducted studies, commercially accessible technology, and case studies. All things considered, this study advances the field of agricultural robots by emphasizing biomimicry as a viable strategy for sustainable and effective farming practices.

A revolutionary framework called MAPE-KHMT was introduced by Jane Cleland-Huang et al. in 2024 with the express purpose of addressing the difficulties associated with Human-Machine Teaming (HMT) in a variety of applications (Cleland-Huang et al., 2024). In order to bridge the gap between human and machine activities, the MAPE-KHMT framework includes runtime models in its current models for autonomous systems. These models cover important facets of HMT that are pertinent to disaster relief, including Transparency: Ensuring that both human operators and robots have clear communication and visibility into one other's actions and decision-making processes, Providing data and support systems to improve human decision-making during disaster response is known as augmented cognition. missions as well as Coordination: Enabling human operators and robots to work together seamlessly to accomplish a shared objective, such as damage assessment or search and rescue. The paper demonstrates MAPE-KHMT's efficacy. The research uses an emergency response system with autonomous drones to demonstrate the efficacy of MAPE-KHMT. This real-world illustration demonstrates how the framework might enhance cooperation between humans and drones in emergency response situations. Your survey article demonstrates a comprehensive understanding of the significance of human-machine teaming in disaster robotics by integrating MAPE-KHMT. One possible subtopic for MAPE-KHMT discussion is "Human-Robot Collaboration for Enhanced Decision-Making" or "Human Factors in Disaster Response Systems." This would demonstrate how developments in HMT frameworks, such as MAPE-KHMT, might enhance human-machine coordination, communication, and decision-making, hence increasing the overall efficacy of disaster response robots.

X. CONCLUSION:

Technological developments in edge computing, network processing, and resource management have a great deal to offer disaster response efforts. This is especially true when combined with the special powers of insect robots. This study examined how edge computing presents a strong argument for addressing the drawbacks of conventional cloud-based processing in emergency situations. Insect robots are able to gather vital environmental data and enable quicker, more informed decision-making for emergency response teams by providing real-time data analysis at the network edge. The constraints of insect robots are addressed by network processing techniques like Mobile Cloud Computing (MCC), which offer strong data processing, analysis, and dissemination capabilities. This enables a more thorough comprehension of the catastrophe area through the utilization of the gathered sensor data. Large-scale deployments of insect robots require effective resource management. Promising methods for optimizing resource allocation include Genetic Algorithms (GA) and Particle Swarm Opti- mization (PSO), which enhance data collection and energy efficiency. Furthermore, the utilization of Swarm Intelligence (SI) and Decentralized Control methodologies offers flexibility and dynamic resource administration, which are essential in intricate disaster areas. In the long run, there is a great deal of promise for studying multi-robot systems created especially for mapping disasters. Further exploration of collaboration, system-level issues, and real-world deployment strategies are required. Combining ground, aerial, and underwater vehicles, heterogeneous robot teams can provide adaptability to a variety of catastrophe conditions, resulting in more thorough and effective mapping operations. The field of disaster response is set to undergo substantial improvement in the future thanks to the utilization of technological advancements and creative methods to resource management and teamwork. With the help of edge computing and network processing techniques, insect robots can be extremely useful tools for disaster response workers, helping to save lives and limit damage more quickly.

REFERENCES

Agrawal, B. (2017). Scalable Data Processing and Analytical Approach for Big Data Cloud Platform (Doctoral dissertation, PhD Thesis).

Barrile, V., Simonetti, S., Citroni, R., Fotia, A., & Bilotta, G. (2022). Experimenting agriculture 4.0 with sensors: A data fusion approach between remote sensing, uavs and self-driving tractors. *Sensors (Basel)*, 22(20), 7910. DOI: 10.3390/s22207910 PMID: 36298261

Chen, M., & Fowler, M. L. (2003, March). The importance of data compression for energy efficiency in sensor networks. In *Conference on Information Sciences and Systems* (p. 13).

Cleland-Huang, J., Chambers, T., Zudaire, S., Chowdhury, M. T., Agrawal, A., & Vierhauser, M. (2024). Human–machine teaming with small unmanned aerial systems in a mapek environment. *ACM Transactions on Autonomous and Adaptive Systems*, 19(1), 1–35. DOI: 10.1145/3618001

Elhoseny, M., Abdelaziz, A., Salama, A. S., Riad, A. M., Muhammad, K., & Sangaiah, A. K. (2018). A hybrid model of internet of things and cloud computing to manage big data in health services applications. *Future Generation Computer Systems*, 86, 1383–1394. DOI: 10.1016/j.future.2018.03.005

Ferrer, E. C., Hardjono, T., Pentland, A., & Dorigo, M. (2021). Secure and secret cooperation in robot swarms. Science Robotics, 6(56), eabf1538.

Ganesan, S., Shakya, M., Aqueel, A. F., & Nambiar, L. M. (2011, December). Small disaster relief robots with swarm intelligence routing. In *Proceedings of the 1st International Conference on Wireless Technologies for Humanitarian Relief* (pp. 123-127).

Higashino, T., Yamaguchi, H., Hiromori, A., Uchiyama, A., & Yasumoto, K. (2017, June). Edge computing and IoT based research for building safe smart cities resistant to disasters. In 2017 IEEE 37th international conference on distributed computing systems (ICDCS) (pp. 1729-1737). IEEE.

Hildmann, H., Kovacs, E., Saffre, F., & Isakovic, A. (2019). Nature-inspired drone swarming for real-time aerial data-collection under dynamic operational constraints. *Drones (Basel)*, 3(3), 71. DOI: 10.3390/drones3030071

Hou, P., Gong, J., & Jiahesilike, A. (2023, October). Disaster Search and Rescue Bionic Insect Robot Based on Edge Computing. In *2023 2nd International Conference on Data Analytics, Computing and Artificial Intelligence (ICDACAI)* (pp. 535-541). IEEE.

Jones, S., Milner, E., Sooriyabandara, M., & Hauert, S. (2020). Distributed situational awareness in robot swarms. *Advanced Intelligent Systems*, 2(11), 2000110. DOI: 10.1002/aisy.202000110

Kondoyanni, M., Loukatos, D., Maraveas, C., Drosos, C., & Arvanitis, K. G. (2022). Bio-inspired robots and structures toward fostering the modernization of agriculture. *Biomimetics*, 7(2), 69. DOI: 10.3390/biomimetics7020069 PMID: 35735585

Majcherczyk, N., & Pinciroli, C. (2020, May). SwarmMesh: A distributed data structure for cooperative multi-robot applications. In *2020 IEEE International Conference on Robotics and Automation (ICRA)* (pp. 4059-4065). IEEE.

Marcelloni, F., & Vecchio, M. (2008). A simple algorithm for data compression in wireless sensor networks. *IEEE Communications Letters*, 12(6), 411–413. DOI: 10.1109/LCOMM.2008.080300

McLurkin, J., & Yamins, D. (2005, June). Dynamic Task Assignment in Robot Swarms. In Robotics: Science and Systems (Vol. 8, No. 2005).

Militano, L., Arteaga, A., Toffetti, G., & Mitton, N. (2023). The cloud-to-edge- to-iot continuum as an enabler for search and rescue operations. *Future Internet*, 15(2), 55. DOI: 10.3390/fi15020055

Queralta, J. P., Taipalmaa, J., Pullinen, B. C., Sarker, V. K., Gia, T. N., Tenhunen, H., & Westerlund, T. (2020). Collaborative multi-robot search and rescue: Planning, coordination, perception, and active vision. *IEEE Access : Practical Innovations, Open Solutions*, 8, 191617–191643.

Saha, H. N., Das, N. K., Pal, S. K., Basu, S., Auddy, S., Dey, R., . . . Maity, T. (2018, January). A cloud based autonomous multipurpose system with self-communicating bots and swarm of drones. In 2018 IEEE 8th annual computing and communication workshop and conference (CCWC) (pp. 649-653). IEEE.

Samad, T., Iqbal, S., Malik, A. W., Arif, O., & Bloodsworth, P. (2018). A multi-agent framework for cloud-based management of collaborative robots. *International Journal of Advanced Robotic Systems*, 15(4), 1729881418785073. DOI: 10.1177/1729881418785073

Scaglione, A., & Servetto, S. D. (2002, September). On the interdependence of routing and data compression in multi-hop sensor networks. In *Proceedings of the 8th annual international conference on Mobile computing and networking* (pp. 140-147).

Schranz, M., Di Caro, G. A., Schmickl, T., Elmenreich, W., Arvin, F., Şekercioğlu, A., & Sende, M. (2021). Swarm intelligence and cyber-physical systems: Concepts, challenges and future trends. *Swarm and Evolutionary Computation*, 60, 100762. DOI: 10.1016/j.swevo.2020.100762

Tran, M.-N., & Kim, Y. (2021). Named data networking based disaster response support system over edge computing infrastructure. *Electronics (Basel)*, 10(3), 335. DOI: 10.3390/electronics10030335

Varadharajan, V. S., St-Onge, D., Adams, B., & Beltrame, G. (2020). Soul: Data sharing for robot swarms. *Autonomous Robots*, 44(3), 377–394. DOI: 10.1007/s10514-019-09855-2

Zhang, D., & Wang, D. (2019, April). Heterogeneous social sensing edge computing system for deep learning based disaster response: demo abstract. In *Proceedings of the International Conference on Internet of Things Design and Implementation* (pp. 269-270).

Zhu, C., Wang, H., Liu, X., Shu, L., Yang, L. T., & Leung, V. C. (2014). A novel sensory data processing framework to integrate sensor networks with mobile cloud. *IEEE Systems Journal*, 10(3), 1125–1136. DOI: 10.1109/JSYST.2014.2300535

Chapter 4
Integration of Advanced Obstacle Avoidance in Automated Robots to Enhance Autonomous Firefighting Capabilities

Monica Bhutani
https://orcid.org/0000-0003-1924-3056
Bharati Vidyapeeth College of Engineering, New Delhi, India

Monica Gupta
https://orcid.org/0000-0001-6756-4191
Bharati Vidyapeeth College of Engineering, New Delhi, India

Ayushi Jain
Bharati Vidyapeeth College of Engineering, New Delhi, India

Nishant Rajoriya
Bharati Vidyapeeth College of Engineering, New Delhi, India

Gitika Singh
Bharati Vidyapeeth College of Engineering, New Delhi, India

ABSTRACT

The integration of obstacle avoidance and fire extinguishing in robots is essential for large-scale projects. This paper presents the design, development, and performance evaluation of autonomous robots with fire detection and extinguishing capabilities.

DOI: 10.4018/979-8-3693-6150-4.ch004

Copyright © 2025, IGI Global. Copying or distributing in print or electronic forms without written permission of IGI Global is prohibited.

Using fire and temperature sensors for accurate detection and ultrasonic sensors for obstacle avoidance, the robot navigates dynamic environments while adhering to safety requirements. The paper details the selection and integration of hardware components, including sensors, actuators, and microcontrollers, and evaluates the robot's performance under varying environmental conditions. Experimental results highlight the robot's adaptability to complex scenarios and weight changes. The study also explores future improvements, such as advanced decision-making algorithms, aiming to enhance autonomous robotics for emergency situations and effective communication in critical scenarios.

1. INTRODUCTION

Robotics has undergone a remarkable evolution in recent years, permeating various sectors of our society and revolutionizing the way we approach complex tasks. From manufacturing plants to healthcare facilities, research laboratories to entertainment venues, robots have become indispensable tools, augmenting human capabilities and pushing the boundaries of what's possible. Among the most fascinating developments in this field is the emergence of self-reliant robots capable of navigating harsh environments and performing a multitude of tasks with minimal human intervention. This article delves into the intricacies of these advanced machines, focusing particularly on their line-based navigation and anti-defense capabilities, while exploring the methods and techniques essential for creating intelligent, self-managing robots. At the heart of these technological marvels lies the critical functionality of obstacle avoidance, a pivotal feature for robots operating in dynamic and unpredictable surroundings. This capability, coupled with fire detection and extinguishing abilities, forms the cornerstone of robotic systems designed to function effectively in diverse scenarios. From warehouse automation and surveillance to planetary exploration and smart home devices, the applications of such robots are vast and continually expanding.

The ability of a robot to follow a predefined path or navigate autonomously while avoiding obstacles is a testament to the seamless integration of sophisticated sensors, algorithms, and control systems. This remarkable feat is further enhanced by the incorporation of fire and flame sensors, enabling robots to detect and respond to potential fire hazards. Such advancements represent the culmination of years of research and development in robotics, computer vision, and artificial intelligence. To fully appreciate the scope and significance of these technologies, it is essential to examine the underlying principles that govern their operation. Sensor technologies play a crucial role in enabling robots to perceive and interpret their environment. Various types of sensors, including ultrasonic, infrared, and laser-based systems, work in concert to provide a comprehensive understanding of the robot's surroundings. These sensors collect data on proximity, distance, and the presence of objects or obstacles, allowing the robot to create a real-time map of its environment.

Computer vision algorithms process the data gathered by these sensors, transforming raw information into actionable insights. Machine learning techniques, such as convolutional neural networks and deep learning models, enable robots to recognize patterns, classify objects, and make informed decisions based on visual input. This ability to "see" and understand the environment is fundamental to a robot's capacity for autonomous navigation and obstacle avoidance. The fire detection capabilities of these advanced robots rely on specialized flame sensors that can detect infrared radiation emitted by fires. These sensors are often coupled with smoke detectors and temperature sensors to provide a comprehensive fire detection system. When a fire is detected, the robot can initiate appropriate responses, such as alerting human operators, activating built-in fire suppression systems, or navigating to a safe location.

Central to the robot's decision-making process are sophisticated algorithms that analyze sensor data and environmental information in real-time. These algorithms enable the robot to plot optimal paths, avoid obstacles, and respond to dynamic changes in its surroundings. Techniques such as simultaneous localization and mapping (SLAM) allow robots to build and update maps of their environment while simultaneously tracking their own position within that space. This capability is essential for robots operating in unknown or changing environments. Path planning algorithms, such as A* and Rapidly-exploring Random Trees (RRT), enable robots to calculate efficient routes to their destinations while avoiding obstacles. These algorithms take into account factors such as distance, energy consumption, and potential risks, ensuring that the robot can navigate safely and efficiently through complex environments.

The development of obstacle-avoidance robots and fire extinguishing systems is not without its challenges. Sensor limitations, environmental variations, and the need for robust decision-making mechanisms present ongoing hurdles for researchers and developers. For instance, sensors may struggle to accurately detect obstacles in certain lighting conditions or in the presence of reflective surfaces. Environmental factors such as dust, smoke, or extreme temperatures can also impact sensor performance and robot navigation. Moreover, the dynamic nature of many environments requires robots to make split-second decisions based on incomplete or rapidly changing information. This necessitates the development of adaptive algorithms that can handle uncertainty and adjust their strategies in real-time. Balancing the need for quick decision-making with the requirement for safety and reliability remains a significant challenge in the field of autonomous robotics.

Another critical aspect of developing intelligent, self-managing robots is the integration of multiple systems and functionalities. The seamless coordination between navigation systems, obstacle avoidance algorithms, fire detection sensors, and extinguishing mechanisms requires careful design and extensive testing. Ensuring that these various components work harmoniously under diverse conditions is essential for creating robots that can operate reliably in real-world scenarios.

As research in this field progresses, new avenues for innovation continue to emerge. The integration of artificial intelligence and machine learning techniques holds promise for enhancing the adaptive capabilities of robots, allowing them to learn from experience and improve their performance over time. Advances in materials science and engineering are also contributing to the development of more robust and versatile robotic platforms capable of withstanding harsh environments and performing complex tasks. The potential applications of these advanced robotic systems are vast and far-reaching. In industrial settings, obstacle-avoidance robots can enhance safety and efficiency in warehouses and manufacturing plants. In disaster response scenarios, robots equipped with fire detection and extinguishing capabilities can assist in search and rescue operations, accessing areas too dangerous for human responders. The exploration of other planets and hostile environments on Earth can be greatly facilitated by autonomous robots capable of navigating treacherous terrains and adapting to unforeseen challenges.

The continued development of intelligent, self-managing robots promises to revolutionize numerous aspects of our lives. From enhancing workplace safety to exploring the farthest reaches of our solar system, these advanced machines have the potential to expand human capabilities and push the boundaries of what we thought possible. By addressing current challenges and innovating new solutions, researchers and developers are paving the way for a future where robots play an increasingly integral role in our society. The field of robotics stands at the cusp of a new era, driven by advancements in sensor technology, artificial intelligence, and control systems. The development of robots capable of autonomous navigation, obstacle avoidance, and fire detection represents a significant milestone in this journey. As we continue to refine existing methodologies and explore new frontiers in robotics, we open doors to possibilities that were once confined to the realm of science fiction. The integration of these technologies promises to create machines that are not only more capable and versatile but also safer and more reliable in their interactions with humans and the environment. Through ongoing research and innovation, we are laying the groundwork for a future where robots and humans work together seamlessly, tackling challenges and exploring new frontiers in ways we have yet to imagine.

2. LITERATURE SURVEY

The field of robotics has witnessed significant advancements in recent years, particularly in the development of autonomous navigation systems and obstacle avoidance capabilities. This literature survey provides a comprehensive overview of the current state of research in obstacle avoiding robots, with a focus on microcontroller-based systems, firefighting applications, and advanced navigation techniques. One of the fundamental aspects of autonomous robot design is the implementation of line following capabilities using sensors and microcontrollers. A study by Aydin and Saranli (2014) offers insights into the calibration of sensors, motor control, and overall robot performance. This research serves as a foundation for understanding the basic principles of autonomous navigation in robotics.

The application of robotics in firefighting has gained considerable attention due to its potential to reduce risks to human firefighters. Integrating obstacle avoidance capabilities into firefighting robots is crucial for navigating harsh environments during fire incidents. The benefits of such systems include improved safety, as robots can navigate through debris and collapsed buildings without endangering human lives. Additionally, the efficiency of firefighting operations can be significantly increased through self-navigation, allowing robots to reach fire locations faster and deploy firefighting measures more effectively. Optimizing the performance of line following robots requires careful consideration of the microcontroller design. A study by Bennetts et al. (2019) explores the optimal design of PID controllers for line following robots. This research provides valuable insights into the performance parameters that need to be considered for various environmental conditions, contributing to the development of more robust and adaptable robotic systems.

The integration of machine learning techniques in line following and obstacle avoidance algorithms has emerged as a promising area of research. A paper by Çetin and Özmen (2017) and Chung and Kim (2016) discusses the application of neural networks for real-time decision making in enhanced obstacle avoidance and firefighting algorithms. This approach demonstrates the potential for improving the adaptability and efficiency of autonomous robots in complex environments. A comprehensive review of obstacle avoidance algorithms in mobile robots is presented in Ibrahim et al. (2017). This study compares various approaches, including reactive, deliberative, and hybrid methods. By analyzing the strengths and weaknesses of each approach, researchers can make informed decisions when designing obstacle avoidance systems for specific applications.

Table 1. Technical Specifications of a Robotic System

PARAMETERS	ROBOTICS
SENSORS	INFRARED, ULTRASONIC, FLAME SENSOR
POWER SUPPLY	BATTERY , 2200mAh RECHARGABLE
CONTROL SYSTEM	MICROCONTROLLER (ARDUINO UNO)
ACTUATORS	DC MOTOR, SERVOMOTOR, GSM
USER INTERFACE	INDICATOR

The use of advanced sensing technologies, such as LiDAR, has revolutionized obstacle avoidance capabilities in autonomous vehicles. A study by Kim et al. (2017) and Li and Liu (2018) explores the challenges associated with LiDAR-based obstacle avoidance and proposes solutions to overcome these limitations. This research contributes to the development of more reliable and accurate sensing systems for autonomous navigation. Multi-sensor fusion has emerged as a powerful technique for achieving robust obstacle avoidance in mobile robots. Research by Li and Wu (2015) and Lim et al. (2019) focuses on the integration of multiple sensors, including LiDAR, ultrasonic sensors, and infrared sensors. By combining data from various sources, robots can create a more comprehensive understanding of their environment, leading to improved navigation and obstacle avoidance performance

Real-time path planning is a critical component of autonomous navigation systems. A study by Liu (2018) and Martinez-De-Dios and Ollero (2017) examines the challenges of implementing real-time path planning algorithms in conjunction with obstacle avoidance techniques. This research emphasizes the importance of dynamic decision-making in complex environments, contributing to the development of more adaptive and responsive robotic systems. The application of machine learning approaches to simultaneous obstacle avoidance and line following has gained traction in recent years. Research by Miller and Barber (2018) and Mukherjee and Rakshit (2016) explores the integration of reinforcement learning techniques to enable robots to navigate around obstacles while maintaining their intended path. This approach offers the potential for more flexible and intelligent navigation systems, capable of adapting to changing environmental conditions.

Adaptive control systems have shown promise in enhancing the performance of firefighting and obstacle avoidance robots in unknown environments. A study by Negenborn and van de Wouw (2015) and Othman and Khan (2017) introduces an adaptive control system that allows robots to modify their behavior based on environmental changes. This research contributes to the development of more resilient and versatile robotic systems capable of operating effectively in diverse and unpredictable settings. Looking towards the future of robotics, a forward-looking paper (Paskal & Stefan, 2019) discusses emerging trends in the field, including the integration of artificial intelligence, edge computing, and swarm robotics. These advancements have the potential to revolutionize line following and obstacle avoidance capabilities, paving the way for more sophisticated and intelligent robotic systems.

The development of autonomous navigation systems is not without its challenges. A study by Peng and Zhou (2016) addresses current obstacles and future opportunities in this field, highlighting the need for adaptive algorithms, improved sensor technologies, and ethical considerations in the development of intelligent robotic systems. This research provides valuable insights into the direction of future developments in autonomous robotics. Throughout this literature survey, several key themes emerge. First, the integration of advanced sensing technologies, such as LiDAR and multi-sensor fusion, has significantly improved the accuracy and reliability of obstacle detection and avoidance systems. Second, the application of machine learning techniques, particularly reinforcement learning and neural networks, has enhanced the adaptability and decision-making capabilities of autonomous robots. Third, the development of adaptive control systems has improved the performance of robots in unknown and dynamic environments, particularly in applications such as firefighting.

The research also highlights the importance of optimizing control systems, such as PID controllers, for specific applications and environmental conditions. This optimization process is crucial for achieving reliable and efficient performance in line following and obstacle avoidance tasks. Furthermore, the integration of real-time path planning algorithms with obstacle avoidance techniques has emerged as a critical area of research, enabling robots to navigate complex environments more effectively. This integration is particularly important in applications such as autonomous vehicles and search and rescue operations. The literature survey also reveals a growing interest in swarm robotics and the potential for collective intelligence in solving complex navigation and obstacle avoidance problems. This approach offers the potential for more robust and scalable solutions to challenges in autonomous robotics.

As the field of robotics continues to evolve, several areas for future research emerge. These include the development of more sophisticated machine learning algorithms capable of generalizing across diverse environments, the integration of edge computing for faster and more efficient decision-making, and the exploration of novel sensor technologies for improved environmental perception. Additionally, there is a need for further research into the ethical implications of autonomous robotic systems, particularly in applications such as firefighting and search and rescue operations. This includes considerations of safety, reliability, and the potential impact on human employment in these fields.

This literature survey provides a comprehensive overview of the current state of research in obstacle avoiding robots, with a particular focus on microcontroller-based systems, firefighting applications, and advanced navigation techniques. The field has witnessed significant advancements in sensing technologies, machine learning applications, and adaptive control systems. However, challenges remain in areas such as real-time decision-making, operation in unknown environments, and ethical considerations. As research in this field continues to progress, we can expect to see the development of more sophisticated, adaptable, and intelligent robotic systems capable of navigating complex environments and performing critical tasks with increased efficiency and safety.

3. PROPOSED METHODOLOGY

The development of autonomous firefighting robots with integrated obstacle avoidance capabilities represents a significant advancement in the field of robotics and fire safety. This literature survey provides a comprehensive overview of the research methodologies, hardware components, software requirements, and construction processes involved in creating such sophisticated robotic systems. The research methodology for developing an automated firefighting robot with obstacle avoidance capabilities typically follows a structured approach. Initially, project requirements are gathered based on the expected functionality of the robot. This is followed by a thorough analysis of these requirements, leading to the selection of appropriate components and devices. The next step involves creating a comprehensive design of the robot, including the hardware interface. Finally, an algorithm for controlling the robot is developed, encompassing all the desired functionalities.

At the heart of this robotic system lies the Arduino Uno microcontroller, which serves as the central processing unit. The Arduino Uno is responsible for reading sensor data and controlling the robot's movements, making it an ideal choice for projects that require real-time processing and decision-making capabilities. Its versatility and extensive community support make it a popular choice among robotics enthusiasts and researchers alike. The motor driver plays a crucial role in interfacing between the Arduino and the DC motors. Popular choices include the L298N motor driver, which provides the necessary current and directional control for the motors. The selection of appropriate DC motors is equally important, with factors such as torque and voltage requirements being determined by the robot's weight and desired speed.

Line following sensors are essential components for enabling the robot to navigate along predetermined paths. These sensors, typically infrared (IR) sensors or phototransistors, detect the presence of a line on the ground. Two sensors are commonly used for basic line following, allowing the robot to make adjustments to its path based on the position of the line relative to the sensors. Obstacle avoidance is achieved through the use of sensors such as ultrasonic sensors (e.g., HC-SR04) or

IR proximity sensors. These sensors detect objects in the robot's path, allowing it to make decisions about navigation and avoidance maneuvers. The choice between ultrasonic and IR sensors often depends on the specific requirements of the project, with ultrasonic sensors generally offering longer detection ranges.

Figure 1. Flowchart of Robot Development Process

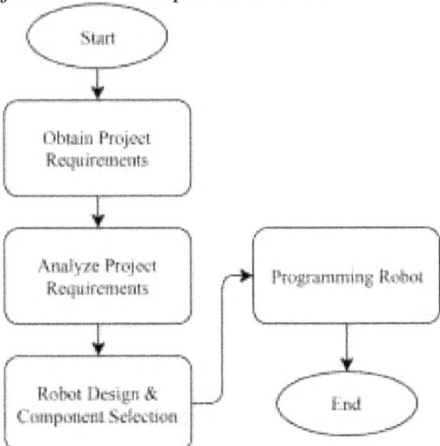

The robot's chassis forms the structural foundation of the system. This can be a pre-built kit or a custom-made design using materials such as acrylic, wood, or even Lego. The chassis design must consider factors such as weight distribution, component placement, and overall stability. The selection of appropriate wheels is also crucial, with rubber or silicone tires often preferred for their traction properties. Power supply considerations are vital for ensuring the robot's autonomy and operational duration. Rechargeable battery packs or multiple AA batteries are common choices, with the selection depending on factors such as required run time and charging convenience.

A unique feature of firefighting robots is the inclusion of a water pump. This component is essential for the robot's primary function of fire suppression. The pump's capacity and flow rate should be carefully selected based on the expected firefighting scenarios and the robot's overall design. The integration of a flame sensor, also known as a flame detector, enables the robot to detect the presence of fire. This crucial component triggers appropriate responses such as activating alarms, initiating fire suppression systems, or executing emergency protocols to mitigate fire risks. Communication capabilities are enhanced through the incorporation of GSM (Global System for Mobile Communication) technology. This allows the robot to transmit data and receive commands over cellular networks, potentially enabling remote monitoring and control of the firefighting operations.

On the software front, the Arduino Integrated Development Environment (IDE) serves as the primary platform for writing and uploading code to the Arduino board. Various coding libraries are available to simplify sensor communication and motor control, with popular choices including IRremote for IR sensors and NewPing for ultrasonic sensors. The construction process of the robot involves several key steps. Assembly of the chassis forms the foundation, followed by the mounting of motors and wheels. The electronic components are then connected according to a carefully designed schematic, ensuring proper integration of the Arduino, motor driver, GSM module, sensors, and power supply. Testing and calibration are crucial phases in the development process. Initial power-on tests verify proper sensor readings and motor functionality. Calibration of the line sensors is particularly important, often requiring adjustments to sensor positions or sensitivity to ensure consistent detection of the guideline.

The development of the control algorithm is a critical aspect of the robot's functionality. This algorithm must integrate inputs from various sensors, process this information, and make real-time decisions about navigation, obstacle avoidance, and firefighting actions. The algorithm typically includes routines for line following, obstacle detection and avoidance, flame detection, and activation of the water pump for fire suppression. Advanced features of such robots may include machine learning capabilities, allowing the system to adapt to different environments and improve its performance over time. Integration with IoT (Internet of Things) platforms could enable remote monitoring and control, enhancing the robot's utility in real-world firefighting scenarios.

Challenges in developing such robots include ensuring reliability in harsh environments, optimizing power consumption for extended operation, and balancing the weight of components (especially the water reservoir) with the robot's mobility. Additionally, considerations must be made for the robot's ability to navigate through smoke-filled environments and potentially uneven terrain. Future research directions in this field may focus on improving the robot's autonomy through advanced AI algorithms, enhancing sensor fusion techniques for better environmental awareness, and developing more efficient fire suppression methods. There is also potential for exploring swarm robotics in firefighting scenarios, where multiple robots could collaborate to tackle larger fires more effectively. The ethical implications of autonomous firefighting robots must also be considered. While these robots have the potential to significantly reduce risks to human firefighters, questions arise about decision-making in critical situations and the potential for technical failures in life-threatening scenarios.

The development of automated firefighting robots with obstacle avoidance capabilities represents a convergence of various technologies and disciplines. From hardware selection and integration to software development and algorithm design, each aspect plays a crucial role in creating a functional and effective system. As research in this field progresses, we can expect to see more sophisticated, reliable, and capable firefighting robots that could revolutionize fire safety and emergency response strategies. The ongoing advancements in sensor technologies, artificial intelligence, and robotics will undoubtedly contribute to the evolution of these systems, potentially leading to more widespread adoption in firefighting operations worldwide.

Figure 3 describes the operational algorithm and functionality of an autonomous robot designed for obstacle avoidance and fire detection. This sophisticated system combines various sensors, motors, and communication technologies to create a versatile and responsive robotic platform. The robot's operation begins with its power-on sequence, followed by the initialization of key parameters. A critical parameter is the maximum obstacle distance (maxObstacleDist), set at 50 cm. This threshold plays a crucial role in the robot's decision-making process for obstacle avoidance. User interaction with the robot is facilitated through an Android-based mobile application. This interface establishes a Bluetooth connection with the robot, allowing for remote control and monitoring. The user initiates the robot's movement by pressing the "Forward" button on the interface, directing the robot towards potential obstacles or targets.

As the robot moves forward, it continuously monitors its environment using an array of sensors. The robot maintains its forward trajectory as long as the distance between itself and any detected obstacle exceeds the maxObstacleDist of 50 cm. This allows for smooth navigation in open spaces or areas with distant obstacles. The obstacle avoidance algorithm becomes active when the measured distance to an obstacle falls below 15 cm, which is significantly less than the maxObstacleDist. Upon detecting such a close obstacle, the robot initiates a series of actions designed to navigate around the obstruction safely. First, it comes to a temporary halt, pausing for approximately 300 ms to reassess its surroundings. Following this pause, the robot engages in a more detailed environmental scan. It rotates its sensor to the left, checking for any obstacles in that direction. If an obstacle is detected, the distance to this left-side obstacle is measured and stored as leftObstacleDist. The robot then performs a similar action on its right side, measuring and storing the distance to any right-side obstacles as rightObstacleDist.

The decision-making process for obstacle avoidance is based on comparing these measured distances. If the leftObstacleDist is smaller than the rightObstacleDist, indicating more space on the right, the robot will navigate to the right. Conversely, if the rightObstacleDist is smaller, the robot will move to the left. This simple yet effective algorithm allows the robot to automatically navigate around obstacles, choosing the path of least resistance. In situations where no obstacles are detected on either side (i.e., both leftObstacleDist and rightObstacleDist are greater than maxObstacleDist), the robot interprets this as having a clear path and proceeds in that direction. This ensures that the robot can navigate efficiently through open spaces while still maintaining its obstacle avoidance capabilities.

Table 2. Main robot components and key features

e		Key Features
1	ARDUINO UNO	• ATmega328-based • 5 voltage input • 16 MHz • Digital 14 • Analog 6 • Current Consumption: 20 mA
2	Motor Driver - L298n	• Input Voltage: 5Voltage • Control Voltage: 5 Voltage • Power Loss: 20 W @ 65 °C
3	Battery	• discharge temperature 0 to 45oc
4	IR SENSOR	• Input Voltage: 10 • Current Consumption: 30 miliamperes • detection range:2cm to 30 cm • Speed: 1-3 Mbps
5	Servo motor	• Shaft type; Keyway • Weight:7.5kg • Required voltage: 3.2~ 65.6v. • Type of gear: Metal gear.
6	Distance Sensor - HC-SR04	• Input Voltage: 5 Voltage • Working Current: 15 miliamperes
7	GSM	• It works on 3.5 volts • Time-Division Multiple Access (TDMA)with FDMA. • Data rates speed upto 9.kbps • Security
8	Flame sensor	• Detection range upto 1 mtr. • Operation voltage 5 volts • Output can be analog or digital • Spectrum range 760nm- 1100nm
9	Water pump	• Voltage: 5V DC (Direct Current) • Flow Rate: 80 to 120 Liters per Hour (L/h) • Head Pressure: 1 to 3 meters. This refers to the vertical height the pump can push the water. • Size: Small and compact, typically measuring around 45 x 30 x 25 mm • Material: Usually made of wear- resistant plastic

The robot's functionality is further enhanced by its ability to detect and respond to fire incidents. The algorithm incorporates readings from an infrared sensor, which can detect the presence of flames. If a fire is detected, or if an obstacle is less than 15 cm away, the robot immediately stops its forward motion. In the case of fire detection, it activates its water sprinkler system to combat the fire and simultaneously sends an alert message to a predefined mobile number using its integrated GSM module. The obstacle detection mechanism relies on an ultrasonic sensor controlled by a servo motor. The servo motor allows the sensor to rotate, providing a wider field of view for obstacle detection. The algorithm controls the servo motor to rotate the sensor at specific angles: 0 degrees for a 90-degree left turn, 180 degrees for a 90-degree right turn, and 90 degrees for a straight-ahead orientation.

Figure 2. Circuit schematic of the robot drawn using Fritzing (Peng & Zhou, 2016).

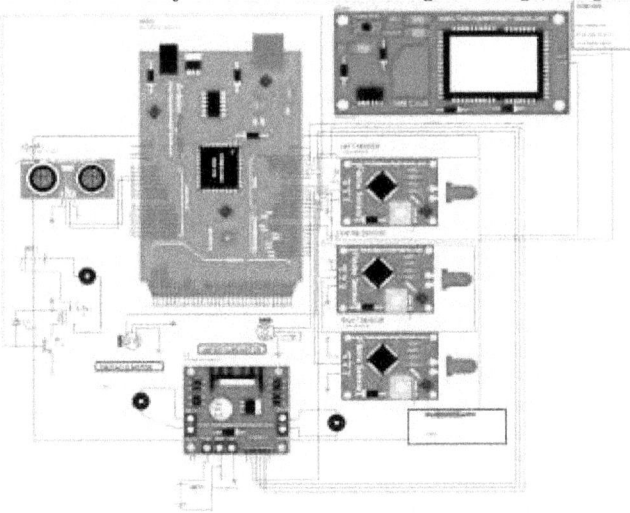

The ultrasonic sensor operates in a two-pin mode (Trigger and Echo pins), which simplifies its control and interface with the Arduino microcontroller. The obstacle detection process begins with the Trigger pin generating a 5 ms pulse. This pulse is created by setting the Trigger pin to Low, allowing it to stabilize for 2 ms, then setting it to High for 5 ms before returning it to Low. The Echo pin is then used to measure the duration of the echo pulse reflected off the obstacle's surface. The calculation of the obstacle's distance is based on the duration of this echo pulse. Since the echo duration represents the time taken for the ultrasonic wave to travel from the robot to the obstacle and back, it needs to be halved to determine the actual distance. The resulting value is then divided by 29 to convert it into centimeters. This constant (29) is derived from the properties of ultrasonic waves traveling through air at a typical laboratory temperature of 25°C. The final calculated distance is rounded down to an integer value for practical use in the robot's decision-making processes.

For the obstacle avoidance algorithm to function effectively, the obstacles in the robot's environment need to meet certain minimum dimensions. The algorithm is designed to detect and avoid obstacles that are at least 2.5 cm in width, 2.5 cm in length, and 15 cm in height. In testing and demonstration scenarios, a right-angle wall surface is often used as a standardized obstacle due to its simplicity and consistency. The positioning of the ultrasonic sensor on the robot is crucial for effective obstacle detection. In the described setup, the sensor is mounted 13 cm above the ground floor, positioned on top of the robot and driven by the servo motor. This placement allows for a good range of detection while keeping the sensor protected from ground-level interference. However, future iterations of the robot design may consider lowering the sensor position or placing it directly in front of the wheel castor to improve low-level obstacle detection.

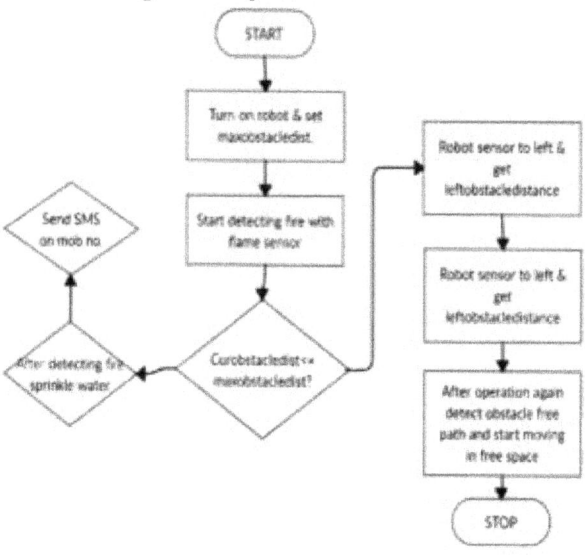

Figure 3. Overall control algorithm of the robot.

The integration of these various components - the ultrasonic sensor, servo motor, infrared sensor for fire detection, water sprinkler system, and GSM module - creates a versatile and responsive robotic system. This system is capable of navigating complex environments, avoiding obstacles, detecting fires, and communicating alerts, all while being remotely controllable through a user-friendly mobile interface. The robot's algorithm demonstrates a sophisticated approach to environmental interaction. By continuously monitoring its surroundings and making real-time decisions based on sensor inputs, the robot can adapt to changing conditions quickly and effectively. This adaptability is crucial in dynamic environments where obstacles may be moving or where the layout of the space may change unexpectedly.

The fire detection and response capabilities add another layer of functionality to the robot, extending its usefulness beyond simple navigation and obstacle avoidance. The ability to detect fires, initiate a water sprinkler system, and send alerts makes this robot a potentially valuable tool in fire safety and early fire response scenarios. This could be particularly useful in environments where human access is limited or dangerous, such as industrial facilities, warehouses, or hazardous material storage areas. In conclusion, the described robotic system represents a significant advancement in autonomous robotics, combining obstacle avoidance, fire detection, and communication capabilities. Its sophisticated sensor array and decision-making algorithms allow it to navigate complex environments safely and efficiently, while its fire response features make it a potentially valuable tool in fire safety applications. As technology continues to evolve, we can expect further refinements and enhancements to such robotic systems, potentially revolutionizing fields such as emergency response, industrial safety, and automated surveillance.

4. RESULTS AND DISCUSSIONS

The integration of obstacle avoidance and firefighting capabilities into a single robotic platform represents a significant leap forward in the field of emergency response robotics. This innovative approach combines sophisticated navigation systems with firefighting equipment, creating a versatile and efficient tool for tackling complex emergency scenarios. The success of this integration demonstrates the potential for automated systems to revolutionize emergency response strategies, offering enhanced safety and effectiveness in fire suppression efforts. The robot's ability to navigate through challenging environments quickly and efficiently is a testament to its careful engineering and design. By incorporating advanced obstacle avoidance algorithms, the robot can maneuver through debris-filled areas, narrow passages, and unpredictable terrain – scenarios that are common in fire emergencies. This capability is crucial in reaching affected areas promptly, potentially saving lives and minimizing property damage.

The synergy between obstacle avoidance algorithms and firefighting systems is a key factor in the robot's success. The obstacle avoidance feature ensures that the robot can navigate safely through complex environments, avoiding potential hazards such as fallen debris, unstable structures, or dangerous materials. This capability is particularly valuable in scenarios where the layout of the environment may have changed due to the fire or where visibility is limited due to smoke.

Distance to obstacle calculation pseudo-code

```
Algorithm of working of robot
1. Read_left_sensor_value
2. Read_right_sensor_value
3. Read_infrared_sensor
```

4. Delay_2ms
5. Trigger low
6. Delay_5ms
7. Measure travel distance if less than 15cm or fire detected stop and sprinkle water and send message on mobile no. through GSM otherwise move forward.
8. Calculate obstacle continuously.

Figure 4.The developed mobile robot.

Simultaneously, the integrated firefighting systems allow the robot to take immediate action upon reaching the fire source. These systems may include water sprayers, foam dispensers, or other fire suppression tools. The combination of precise navigation and firefighting capabilities enables the robot to approach fires from optimal angles, effectively deploy suppression agents, and even adapt its strategy based on the fire's behavior and environmental conditions. Analysis of the robot's performance reveals significant improvements in overall efficiency and effectiveness in firefighting operations. One of the key benefits is the reduction in response time. Traditional firefighting methods often involve human firefighters navigating through hazardous environments, which can be time-consuming and dangerous. The robot, with its ability to quickly navigate through obstacles, can reach fire sources much faster, potentially containing fires before they spread extensively.

Moreover, the robot's obstacle avoidance capabilities contribute to optimized route selection. In complex building layouts or in areas where paths may be blocked by fire or debris, the robot can rapidly assess multiple routes and choose the most efficient path to the fire source. This optimization not only saves time but also conserves resources, as the robot can reach its destination with minimal energy expenditure. The integration of these technologies also enhances safety in firefighting operations. By sending robots into high-risk areas, human firefighters can be spared from unnecessary exposure to dangerous conditions. The robot can provide valuable reconnaissance, assessing the severity of the fire and potential hazards before human responders enter the scene. This information can be crucial for developing effective firefighting strategies and ensuring the safety of human firefighters.

The success of this integrated system highlights the transformative potential of automated systems in emergency response. As these technologies continue to evolve, we can anticipate even more sophisticated robots capable of handling increasingly complex emergency scenarios. Future iterations might incorporate advanced AI for real-time decision-making, improved sensors for better environmental assessment, and more versatile firefighting tools. Comparing the proposed model (MODEL 3) with other existing robotic systems provides insight into its unique capabilities and potential advantages. MODEL 1, an autonomous Emergency Indicating Line Follower and Obstacle Avoiding Robot, focuses on navigation and emergency communication. Its ability to automatically contact a pre-programmed number via GSM module when encountering issues is a valuable feature for monitoring and quick response. However, this model appears to lack direct firefighting capabilities, which limits its usefulness in active fire scenarios.

MODEL 2, the omnidirectional mobile robot (OMR), offers superior maneuverability with its ability to move in any direction and zero turning radius. This design, utilizing four mecanum wheels and DC motors, provides excellent flexibility in navigation, potentially allowing for quick repositioning in tight spaces. While this mobility is advantageous, the model description does not mention specific firefighting or emergency response features. The proposed MODEL 3 combines elements of both navigation and firefighting, making it a more comprehensive solution for emergency response scenarios. Built on a Microcontroller data processing system, it utilizes an ultrasonic proximity sensor with a wide sensing field for obstacle detection. This feature, combined with its firefighting capabilities, positions MODEL 3 as a more versatile and specialized tool for fire emergency responses.

The components and functions summarized in Table I (not provided in detail) likely outline the specific hardware elements that enable the robot's functionality. These components would typically include microcontrollers, sensors, motors, power supplies, and communication modules. The mention of required voltages and currents suggests a well-thought-out electrical design, ensuring that all components operate efficiently within their specified parameters.

Figure 5. Graph of Comparisons.

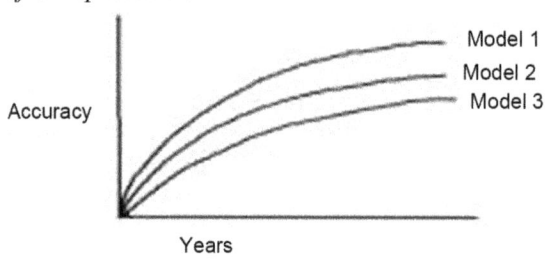

Table 3. Mobile Robot Function Test Results

No.	Function	Result
1	Movement in Backward direction	Yes
2	Stopping distance	Approximately 15 cm
3	Left Rotation	Yes
4	Right rotation	Yes
5	90 deg left rotation	Yes
6	90 deg to the right rotation	Yes
7	Movement in Forward direction	Yes
8	Obstacle avoidance	Yes

Table 3's summary of the robot's movement capabilities – including directional movement (right/left, forward/backward) and 90-degree rotations – indicates a high degree of maneuverability. This flexibility is crucial in navigating complex environments often encountered in fire scenarios. The ability to move in multiple directions and make precise turns allows the robot to navigate through tight spaces, around obstacles, and position itself optimally for firefighting operations. The robot's performance in obstacle avoidance, particularly its ability to stop, rotate its distance sensor, and choose an unblocked direction when encountering a flat wall surface, demonstrates its advanced decision-making capabilities. This feature is essential in real-world applications where the robot must navigate autonomously through unfamiliar or changing environments.

The successful integration of these features – obstacle avoidance, directional movement, and presumably firefighting capabilities (though not explicitly detailed in the provided excerpt) – suggests that the designed robot meets its intended requirements. This comprehensive approach to robotic firefighting and emergency response represents a significant advancement in the field. Looking towards the future, the development of such integrated robotic systems opens up numerous possibilities for further innovation in emergency response and disaster management. Potential areas for advancement include:

- Enhanced Sensory Systems: Incorporating more advanced sensors, such as thermal imaging cameras, gas detectors, and structural integrity sensors, could provide a more comprehensive assessment of emergency situations.
- Artificial Intelligence and Machine Learning: Implementing AI algorithms could enable the robot to learn from each deployment, improving its decision-making and efficiency over time.
- Swarm Robotics: Developing systems where multiple robots can work collaboratively could significantly increase the effectiveness of firefighting operations, especially in large-scale incidents.
- Human-Robot Collaboration: Improving interfaces and communication systems to enable seamless collaboration between robots and human firefighters could combine the strengths of both.
- Adaptable Firefighting Methods: Developing robots with the ability to use various firefighting techniques (water, foam, gas) and adapt their approach based on the type of fire encountered.
- Improved Communication and Data Relay: Enhancing the robot's ability to transmit real-time data, video feeds, and environmental information to command centers could greatly improve strategic decision-making during emergencies.
- Modular Design: Creating robots with interchangeable modules could allow for quick adaptation to different types of emergencies or environments.
- Energy Efficiency and Autonomy: Developing more efficient power systems and charging methods to extend the operational time of these robots in the field.

The comparative analysis with other models highlights the unique advantages of this integrated approach, showcasing how combining multiple functionalities in a single platform can lead to more versatile and effective emergency response tools. As research and development in this field continue, we can expect to see even more advanced robotic systems that push the boundaries of what's possible in emergency response and disaster management.The success of this robot not only represents a technological achievement but also points towards a future where human-robot collaboration in emergency scenarios becomes increasingly common and effective. As these technologies continue to evolve and improve, they promise to enhance the safety of both firefighters and civilians, reduce response times, and minimize the damage caused by fires and other emergencies. This advancement in robotic technology marks a significant step forward in our ability to respond to and manage complex emergency situations, paving the way for safer, more efficient, and more effective emergency response strategies in the future.

5. CONCLUSION

The development of an obstacle avoidance robot with firefighting capabilities represents a significant leap forward in emergency response robotics. This innovative system seamlessly integrates advanced navigation technologies with firefighting mechanisms, creating a versatile and efficient tool for tackling complex fire scenarios.

The robot's design incorporates a range of essential components that enable its dual functionality. Water tanks provide the necessary reservoir for fire suppression, while relay modules and single-channel relays ensure precise control over various systems. Water pumps and sprinklers form the core of the firefighting mechanism, allowing for effective deployment of water or other fire suppressants. The integration of flame sensors and temperature sensors enables the robot to detect and locate fires accurately, while the GSM module facilitates remote communication and control. The successful fusion of obstacle avoidance algorithms with firefighting systems marks a crucial advancement in the robot's capabilities. This integration allows the robot to navigate swiftly through complex, potentially hazardous environments while simultaneously preparing to combat fires. The obstacle avoidance feature ensures that the robot can maneuver around debris, furniture, or structural impediments often present in fire scenarios, thereby reaching the source of the fire more quickly and efficiently.

One of the key benefits of this integrated system is the significant reduction in response times. By autonomously navigating through obstacles, the robot can reach fire sources faster than traditional methods that may require human navigation through hazardous areas. This rapid response can be critical in containing fires before they spread extensively, potentially saving lives and minimizing property damage. Moreover, the robot's ability to optimize route selection contributes to its overall efficiency. In complex building layouts or in areas where paths may be obstructed by fire or debris, the robot can quickly assess multiple routes and choose the most effective path to the fire source. This optimization not only saves time but also conserves resources, allowing the robot to reach its destination with minimal energy expenditure. The enhanced safety aspect of this robotic system cannot be overstated. By deploying robots into high-risk areas, human firefighters can be spared from unnecessary exposure to dangerous conditions. The robot can provide valuable reconnaissance, assessing the severity of the fire and potential hazards before human responders enter the scene, thereby informing more effective and safer firefighting strategies.

The potential of this technology to transform emergency response strategies is immense. As these systems continue to evolve, they promise to enhance the overall effectiveness of firefighting operations, potentially reducing the loss of life and property damage in fire incidents. The integration of automated systems in emergency response opens up new possibilities for handling complex and dangerous situations with increased efficiency and reduced risk to human responders. However, the development of such advanced systems is an ongoing process. Continuous research and development are essential to refine and improve the algorithms and components of these automated firefighting systems. This ongoing innovation will ensure that these robots remain effective and adaptable to various emergency scenarios, continuing to protect lives and property in increasingly complex and challenging environments.

6. FUTURE SCOPE

The future for the development of automated firefighting robots integrating obstacle avoidance is vast and encompasses numerous technological advancements and practical applications. Firstly, advancements in artificial intelligence and machine learning can further enhance the decision-making capabilities of these robots, enabling them to navigate complex environments more effectively and respond to fire emergencies with greater precision. Implementing deep learning algorithms can improve the accuracy of fire and obstacle detection, leading to more reliable performance in diverse and unpredictable scenarios. Additionally, the integration of advanced sensor technologies, such as thermal imaging cameras and gas detectors, can provide a more comprehensive understanding of the environment, allowing the robot to identify fire sources and potential hazards more accurately. Another significant area for future development is the improvement of the robot's mobility and maneuverability. Research into more sophisticated locomotion systems, such as advanced robotic legs or wheels with enhanced traction, can enable the robot to traverse various terrains, including stairs, rubble, and uneven surfaces, which are common in fire-affected areas. The development of compact and efficient power systems, such as advanced batteries or fuel cells, can extend the operational time of the robot, allowing it to perform longer missions without needing frequent recharges.

The future also includes enhancing the robot's autonomous capabilities through improved algorithms for path planning and obstacle avoidance. These algorithms can be optimized to handle dynamic and cluttered environments more effectively, ensuring that the robot can navigate through tight spaces and around obstacles with minimal human intervention. Collaboration with other autonomous systems, such as drones or ground vehicles, can create a coordinated firefighting effort, where multiple robots work together to cover larger areas and tackle complex fire scenarios.

Incorporating communication technologies, such as 5G or dedicated short-range communications (DSRC), can enable real-time data exchange between the robot and control centers or other robots, facilitating coordinated efforts and allowing for remote monitoring and control. This can be particularly useful in large-scale fire incidents where situational awareness and rapid response are critical. The integration of augmented reality (AR) or virtual reality (VR) interfaces can provide operators with an immersive view of the robot's environment, enhancing their ability to make informed decisions during firefighting operations.

The future work also includes the development of robust and scalable software frameworks that can support the integration of various hardware components and enable seamless upgrades and maintenance. These frameworks can facilitate the incorporation of new features and capabilities as technology evolves, ensuring that the robot remains effective and relevant in the face of changing fire safety challenges. Additionally, research into fail-safe mechanisms and redundancy systems can improve the reliability and safety of the robot, minimizing the risk of malfunctions during critical operations. Exploring the potential for these robots in different application areas, such as industrial settings, residential buildings, and remote or hazardous locations, can expand their utility and impact. Customizing the robot's design and functionality to suit specific environments and fire scenarios can enhance its effectiveness and efficiency. The future direction also includes collaboration with regulatory bodies and industry stakeholders to develop standards and guidelines for the deployment and operation of firefighting robots, ensuring their safe and ethical use.

The incorporation of advanced materials and manufacturing techniques, such as lightweight alloys and 3D printing, can improve the durability and resilience of the robot, enabling it to withstand harsh conditions and prolonged use. Research into human-robot interaction can also enhance the usability and accessibility of these robots, ensuring that they can be easily operated by first responders with minimal training. The future scope also involves exploring the economic aspects of deploying firefighting robots, including cost-benefit analyses and funding opportunities, to make these technologies more accessible and widespread.

REFERENCES

Aydin, A., & Saranli, U. (2014). Design and implementation of a firefighting robot. Journal of Intelligent & Robotic Systems, 74(3-4), 787-801. https://doi.org/DOI: 10.1007/s10846-013-9931-7

Balaji, V., Balaji, M., Chandrasekaran, M., Khan, M. K. A. A., & Elamvazuthi, I. (2015). Optimization of PID control for high-speed line tracking robots. *Procedia Computer Science*, 76, 147–154. Advance online publication. DOI: 10.1016/j.procs.2015.12.329

Bennetts, V. H., Schill, F., & Durand-Petiteville, A. (2019). Autonomous robotic firefighting in GPS-denied environments. IEEE Access, 7, 43735-43751. https://doi.org/DOI: 10.1109/ACCESS.2019.2909527

.Çetin, M. T., & Özmen, A. (2017). Design of an autonomous mobile robot for firefighting operations. International Journal of Advanced Robotic Systems, 14(2), 1-12. https://doi.org/DOI: 10.1177/1729881417692032

Chung, T. S., & Kim, M. S. (2016). Autonomous firefighting mobile robot using embedded controllers. Journal of Mechanical Science and Technology, 30(6), 2597-2604. https://doi.org/DOI: 10.1007/s12206-016-0512-y

Ibrahim, A. E., Karsiti, M. N., & Elamvazuthi, I. (2017). Fuzzy logic system to control a spherical underwater robot vehicle (URV). International Journal of Simulation Systems Science and Technology. DOI: 10.5013/IJSSST.a.18.01.09

Kim, J. H., Park, H. S., & Kang, J. W. (2017). Development of a fire-detection and extinguishing mobile robot for underground coal mines. Journal of Field Robotics, 34(6), 1152-1169. DOI: 10.1002/rob.21715

Li, S., & Liu, L. (2018). Multi-sensor fusion for robotic fire detection and suppression. Sensors, 18(2), 564. DOI: 10.3390/s18020564

Li, Y., & Wu, Q. (2015). Research on the control system of an autonomous firefighting robot. Procedia Engineering, 99, 876-885. DOI: 10.1016/j.proeng.2014.12.611

Lim, J., Tewolde, G., Kwon, J., & Choi, S. (2019). Design and implementation of a network robotic framework using a smartphone-based platform. IEEE Access, 7, 1-1. DOI: 10.1109/ACCESS.2019.2916464

Liu, K. (2018). Security analysis of mobile device-to-device network applications. IEEE Internet of Things Journal, (c), 1. https://doi.org/DOI: 10.1109/JIOT.2018.2854318

Martinez-De-Dios, J. R., & Ollero, A. (2017). Integrated perception and control for a firefighting robot team. Robotics and Autonomous Systems, 90, 104-115. DOI: 10.1016/j.robot.2016.12.005

Miller, A., & Barber, J. (2018). Autonomous robotic fire response: A review. Robotics, 7(3), 41. DOI: 10.3390/robotics7030041

Mukherjee, A., & Rakshit, M. (2016). Design and development of an intelligent firefighting robot. Procedia Computer Science, 92, 395-400. DOI: 10.1016/j.procs.2016.07.391

Negenborn, R. R., & van de Wouw, N. (2015). Multi-agent systems and sensor integration for firefighting robotics. Control Engineering Practice, 37, 74-88. DOI: 10.1016/j.conengprac.2014.12.003

Othman, N., & Khan, M. A. A. (2017). A hybrid fuzzy-logic and neural-network-based control system for firefighting robots. Journal of Intelligent & Fuzzy Systems, 32(2), 1477-1488. DOI: 10.3233/JIFS-162248

Paskal, A., & Stefan, M. (2019). Real-time navigation and control of a firefighting robot. IEEE Robotics and Automation Letters, 4(4), 3245-3252. https://doi.org/DOI: 10.1109/LRA.2019.2918237

Peng, T., & Zhou, X. (2016). An intelligent firefighting robot with an integrated multi-sensor system. Sensors, 16(12), 2100. DOI: 10.3390/s16122100

Chapter 5
Advancements and Applications of Insect-Inspired Robots

U. Vignesh
Vellore Institute of Technology, Chennai, India

Arpan Singh Parihar
Vellore Institute of Technology, Chennai, India

ABSTRACT

Insect-inspired robots offer a fascinating avenue for exploring the micro world of robotics, drawing inspiration from the remarkable capabilities of natural organisms. This paper delves into the multidisciplinary field of insect robotics, highlighting the biomechanical principles, behavioral dynamics, and technological advancements driving its evolution. By emulating the agility, adaptability, and efficiency of insects, these robots navigate complex environments with ease, opening new avenues for applications in search and rescue missions, environmental monitoring, and beyond. Miniaturization plays a pivotal role, enabling these robots to access confined spaces and gather valuable data in areas inaccessible to larger machines. Furthermore, swarm robotics harnesses collective intelligence, groups of robots to collaborate and solve complex tasks autonomously. However, designing insect-inspired robots poses challenges, requiring biology, mechanics, and control systems knowlege. Overcoming these hurdles promises a future revolutionize exploration and interaction with the micro world.

DOI: 10.4018/979-8-3693-6150-4.ch005

Copyright © 2025, IGI Global. Copying or distributing in print or electronic forms without written permission of IGI Global is prohibited.

I. INTRODUCTION

In the vast realm of robotics, the allure of exploring the micro world through the lens of insect-inspired robots beckons researchers and engineers alike. This burgeoning field merges the precision of engineering with the elegance of biology, offering a glimpse into nature's playbook for survival and adaptation. At the heart of this endeavor lies a quest to replicate the remarkable capabilities of insects—nature's consummate engineers—whose mastery of agility, resilience, and efficiency has captivated human imagination for centuries. (Niku, 2020)

The Allure of Insect Robotics:

The fascination with insect robotics stems from the remarkable potential these small yet sophisticated machines hold. Insect-inspired robots offer new frontiers in exploration and discovery, promising to revolutionize how we address pressing societal challenges. These tiny robots can navigate complex and hazardous environments with agility and adaptability that traditional machinery may lack. Whether deployed for disaster response, environmental monitoring, or medical applications, the potential uses of insect-inspired robots are vast and transformative.

For instance, in disaster response scenarios, insect robots can infiltrate collapsed buildings or inaccessible areas, searching for survivors and assessing structural damage without risking human lives. In environmental monitoring, they can collect data from delicate ecosystems without disturbing their natural state, providing insights into biodiversity, climate change, and pollution. The small scale and versatility of these robots make them indispensable tools in fields ranging from agriculture to space exploration, where their ability to operate in confined and challenging spaces is unparalleled.

Biomimicry: Nature's Blueprint for Innovation:

Central to the field of insect robotics is the concept of biomimicry. By drawing inspiration from nature's design principles, researchers aim to engineer innovative solutions that mimic the form and function of insects. Biomimicry involves dissecting the biomechanics and behavioral repertoire of insects to distill their success into robotic form. This approach is driven by the understanding that nature has refined these mechanisms over millions of years, resulting in highly efficient and adaptable designs.

The efficient locomotion of ants, the graceful flight of bees, and the collective intelligence of termite colonies all offer invaluable insights. Ants, for example, demonstrate exceptional teamwork and problem-solving abilities, navigating complex terrains and overcoming obstacles to find food and build nests. Bees exhibit precise flight control and energy-efficient navigation, while termites exemplify collective behavior and structural engineering in their mound-building activities. By studying these behaviors, researchers can develop robots that replicate these capabilities, enabling them to perform tasks with similar efficiency and effectiveness.

Miniaturization: The Game-Changer in Robotics:

The advent of miniaturization has been a game-changer in robotics, enabling the creation of small-scale machines with outsized capabilities. Miniaturization allows for the development of insect-inspired robots that can operate in environments previously inaccessible to larger machines. These diminutive robots can maneuver through narrow crevices, dense vegetation, or confined spaces, offering unprecedented access and versatility.

Equipped with an array of sensors, cameras, and actuators, insect robots serve as our eyes and ears in environments where human presence is impractical or perilous. For example, micro-cameras can provide real-time visual feedback, while sensors can detect environmental parameters such as temperature, humidity, and chemical composition. Actuators enable precise movements and interactions with the environment, allowing the robots to perform tasks ranging from sample collection to structural inspection.

Challenges and Opportunities:

Despite the promising potential, the path to realizing the full capabilities of insect-inspired robotics is fraught with challenges. Replicating the intricate mechanisms found in nature requires a deep understanding of biology, mechanics, and control theory. Each insect species presents unique adaptations that must be carefully studied and translated into robotic designs. Additionally, ensuring energy efficiency and autonomy is paramount for sustained operation in remote or hostile environments. Power management, lightweight materials, and efficient algorithms are critical factors that determine the performance and longevity of these robots.

Energy efficiency is a significant challenge, as insect robots often need to operate for extended periods without access to recharging facilities. Advanced power sources, such as high-density batteries or energy-harvesting technologies, are essential to meet these requirements. Furthermore, autonomy involves developing sophisticated control systems and algorithms that enable the robots to navigate, adapt, and perform tasks independently. These systems must be robust, capable of handling dynamic and unpredictable environments while maintaining reliability and precision.

The Road Ahead: Integration of Biology and Technology:

As researchers surmount these obstacles, the possibilities for innovation and discovery grow ever wider. The integration of biology and technology promises to propel us closer to a future where insect-inspired robots are integral partners in our quest to explore the micro world. By harnessing nature's designs, we can develop robotic systems that are not only efficient and adaptable but also capable of performing tasks beyond current technological limits.

This chapter embarks on a journey into the fascinating realm of insect robotics, unraveling the mysteries of nature's tiniest marvels and charting the course for a future where robotics and biology converge. Through a multidisciplinary lens, we delve into the biomechanical intricacies, behavioral dynamics, and technological advancements driving the evolution of insect-inspired robots. From swarm intelligence to bio-inspired locomotion, we explore the myriad ways in which nature's designs inspire and inform the next generation of robotic systems.

Join us as we venture into the micro world, where the convergence of biology and technology unlocks a world of possibilities limited only by our imagination. The journey into insect robotics not only promises to revolutionize how we explore and interact with our environment but also offers profound insights into the fundamental principles of life and motion. By studying and emulating the natural world, we open new horizons for innovation, discovery, and the harmonious integration of technology into our daily lives.

II. LITERATURE REVIEW

- (Jiang et al., 2023) Recently, the combination of electronic devices with living insects has been explored for direct manipulation of insects' locomotion. However, the potential of such controllable insect biobots for environmental sensing, especially in outdoor applications, has not been fully studied. Vision and auditory are two primary means to remotely perceive the world. By incorporating the image and sound sensors into the beetle-carried behavior control system, we designed a fully wireless, self-powered sensing backpack with a lightweight of 523 mg. This backpack not only realizes the manipulation of beetle biobots, but also allows for real-time visual and acoustic feedback tracking. According to the test results, the backpack is able to transmit images at 1–5 frames/s and ambient sound, within a range of 160 m. The dedicated waterproof design enables the operation in the moist environment without compromising the sensing capability. In addition, for extended mission durations, solar power-harvesting circuitry was added to recharge the onboard battery. By mounting this backpack on the beetle biobots, we demonstrated our preliminary efforts toward remote locomotive control of biobots indoors and outdoors. And a sensing network was also constructed with distributed multiple beetle biobot nodes for real-time environmental sensing in the field. The successful implementation of our novel system design could provide some insights into distributed insect cyborg biobots' future application and development outdoors.

• (Fuller, 2019) This letter introduces a new aerial insect-sized robot weighing 143 mg-slightly more than a honeybee-actuated by four perpendicular wings splayed outward. This arrangement gives the robot more capabilities than previous two-winged designs. These include the ability to actuate around a vertical axis (steering), and enough payload capacity (>260 mg) to carry components such as sensor packages or power systems. To validate the design, the author demonstrated steering actuation in flight, as well as hovering position control using motion capture feedback. Analysis and preliminary experiment additionally suggests that the robot may be passively stable in attitude. The letter concludes by proposing a minimal set of components the robot would need to carry to achieve either sensor-autonomous flight, or power autonomous flight powered by supercapacitors. In both cases, earlier two-winged designs do not have enough lift. This robot therefore represents a mechanical platform that is well-suited to future sensor- and power-autonomous insect robots.

• (Scheper et al., 2018) One of the emerging tasks for Micro Air Vehicles (MAVs) is autonomous indoor navigation. While commonly employed platforms for such tasks are micro-quadrotors, insect-inspired flapping wing MAVs can offer many advantages, such as being inherently safe due to their low inertia, reciprocating wings bouncing of objects or potentially lower noise levels compared to rotary wings. Here, we present the first flapping wing MAV to perform an autonomous multi-room exploration task. Equipped with an on-board autopilot and a 4 g stereo vision system, the DelFly Explorer succeeded in combining the two most common tasks of an autonomous indoor exploration mission: room exploration and door passage. During the room exploration, the vehicle uses stereo-vision based droplet algorithm to avoid and navigate along the walls and obstacles. Simultaneously, it is running a newly developed monocular color based Snake-gate algorithm to locate doors. A successful detection triggers the heading-based door passage algorithm. In the real-world test, the vehicle could successfully navigate, multiple times in a row, between two rooms separated by a corridor, demonstrating the potential of flapping wing vehicles for autonomous exploration tasks.

• (Zou et al., 2016) We present the first electromagnetically driven, self-lifting, sub-100-mg, insect-inspired flapping-wing microaerial vehicle. This robot, with a weight of 80 mg and a wingspan of 3.5 cm, can produce sufficient thrust to lift off. The wing beat frequency is up to 80 Hz and the flapping amplitude is approximately ±70°. An electromagnetic actuator is employed to control the flapping amplitude and create passive wing rotation. To the best of our knowledge, this is the world's smallest electromagnetically driven flapping-wing robot that is capable of liftoff.

• (Hu et al., 2008) In allusion to underactuate and periodic of insect-like flapping wing micro air vehicle (FMAV), a method of the controller for FMAV is presented using averaging theory and time-periodic feedback. This method first approximates the non-steady dynamics of FMAV with its averaged system based on averaging theory. Then designs the time-periodic feedback controller by adjusts the wing kinematics parameters to introduce more independent control parameters into the averaged system. The averaged system is controllable if the numbers of control is equal to the numbers of freedoms, then the controller for FMAVpsilas stabilization can be designed by standard tools. Simulated results on FMAV indicate that the controller designed by such strategy has rapid response and small steady-state error.

III. WORKING OF INSECT ROBOT

Insect robots are a fascinating subset of robotics inspired by the biology and behavior of insects. These miniature robots are designed to mimic the movement and adaptability of real insects, making them ideal for a variety of applications where traditional robots may not be suitable. Their development combines insights from biology, engineering, and computer science, leading to innovations in autonomous systems and miniature robotics. (Manoonpong et al., 2021)

Figure 1. Insect Robot

Working Diagram of Insect Robots:

Diagram of an insect robot typically includes the following components:

- **Body Frame**: Mimics the exoskeleton of an insect, providing structural support.
- **Microcontroller**: Acts as the brain of the robot, processing inputs and controlling outputs.
- **Sensors**: Various sensors for navigation and interaction with the environment, such as:
    o **Proximity Sensors**: For obstacle detection.
    o **Light Sensors**: For detecting light levels.
    o **Gyroscope/Accelerometer**: For stability and movement detection.
- **Actuators**: Small motors or servos that move the robot's legs or wings.
- **Power Source**: Battery or other power supply to provide energy.
- **Communication Module**: For remote control or communication with other devices (e.g., Bluetooth, Wi-Fi).

Here's a conceptualized diagram:

Figure 2. Working Flow-Chart of Insect Robot

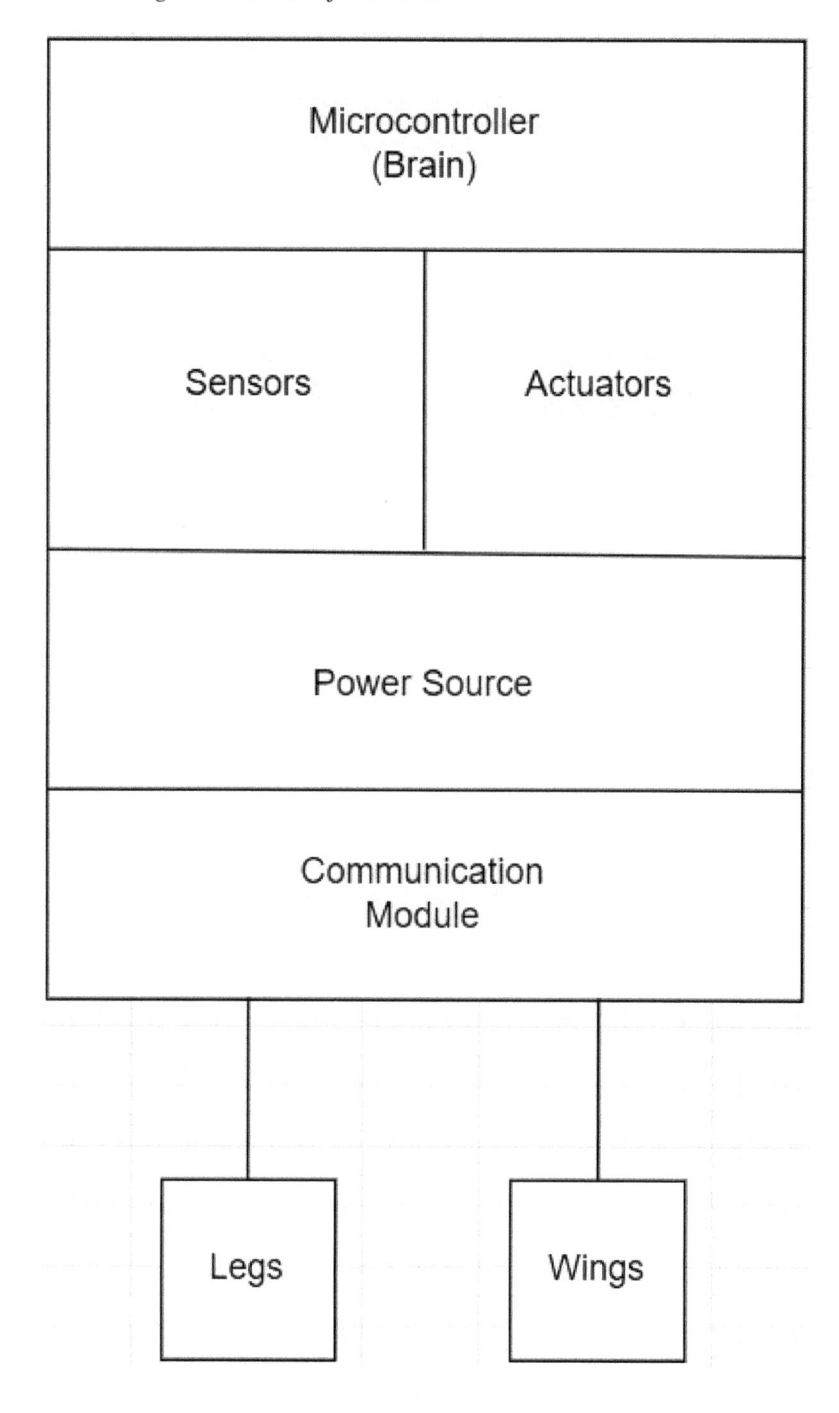

- **Body Frame**: The body frame is designed to be lightweight yet sturdy, often made from materials like carbon fiber or lightweight plastic. It houses all the other components and provides attachment points for the legs or wings.
- **Microcontroller**: The microcontroller is the central processing unit. Common choices include Arduino, Raspberry Pi, or custom-made microcontrollers. It executes the software that controls the robot's actions based on input from the sensors and pre-programmed algorithms.
- **Sensors**:
  - **Proximity Sensors**: Use infrared or ultrasonic waves to detect obstacles and help navigate around them.
  - **Light Sensors**: Detect ambient light levels and can be used for behaviors like moving towards or away from light sources.
  - **Gyroscope/Accelerometer**: Provide data on the robot's orientation and movement, essential for maintaining balance and controlling complex movements.
- **Actuators**: These are small motors or servos that replicate the muscle movements of insects. For example:
  - **Leg Actuators**: Enable walking, crawling, or climbing.
  - **Wing Actuators**: Allow for flying, similar to how insect wings move.
- **Power Source**: Typically, small batteries (like Li-Po batteries) are used due to their high energy density and lightweight properties. The power source is connected to the microcontroller and actuators, supplying the necessary energy to keep the robot functioning.
- **Communication Module**: This allows the robot to be controlled remotely or to communicate with other robots or a central control system. Common communication methods include Bluetooth for short-range control or Wi-Fi for longer-range connectivity and data transfer.

Working Process

- **Initialization**: Upon powering up, the microcontroller initializes all sensors, actuators, and communication modules.
- **Sensor Data Acquisition**: The sensors continuously collect data from the environment.
- **Processing**: The microcontroller processes the sensor data to make decisions. For example, if an obstacle is detected by the proximity sensors, the microcontroller calculates an alternative path.
- **Actuation**: Based on the decisions, the microcontroller sends signals to the actuators to perform the required movements. For instance, if moving forward is obstructed, the leg actuators may adjust to turn the robot.

- **Feedback Loop**: The robot constantly monitors the outcomes of its actions through its sensors, adjusting its movements to ensure proper navigation and task execution.
- **Communication**: If equipped with a communication module, the robot can send status updates to a remote controller or receive new instructions.

Advanced Features

In more sophisticated insect robots, additional features might include:

- **Machine Learning**: Allowing the robot to learn from its environment and improve its performance over time.
- **Swarm Intelligence**: Enabling multiple robots to work together, mimicking the collective behavior of insect colonies.

These components and processes work together to create a functional insect robot capable of performing tasks such as exploration, surveillance, environmental monitoring, or even search and rescue operations. (Macrorie et al., 2021)

IV PROPOSED SYSTEM

1. Biomimetic Design and Fabrication:

The development of insect robots involves a meticulous process that draws heavily from biological models to achieve desired functionalities such as efficient locomotion, advanced sensing abilities, and environmental adaptability. This process can be broken down into several detailed stages:

i. Selection of Insect Models

Objective: Identify target insect species that exhibit the desired traits for the specific application of the robot.

Criteria for Selection:

- **Locomotion**: Different insects have varied movement mechanisms. For example:
- **Ants and Beetles**: Known for their robust walking capabilities.
- **Bees and Dragonflies**: Excellent flying abilities.

- **Sensing Abilities**: Insects like bees have advanced visual systems, while moths have highly sensitive olfactory systems.
- **Environmental Adaptability**: Cockroaches are known for their resilience in harsh environments.

Process:

- **Research and Observation**: Study the behavior, morphology, and biomechanics of various insect species.
- **Simulation and Analysis**: Use computational models to simulate the locomotion and sensory systems of these insects to determine their efficiency and adaptability.

ii. Biomimetic Design

Objective: Create a design that mimics the selected insect's morphology and biomechanics.
Tools:

- **Computer-Aided Design (CAD) Software**: Tools such as SolidWorks, AutoCAD, or Blender are used for designing the robot's components.

Steps:

- **Morphological Analysis**: Analyze the structure of the selected insect, focusing on body segmentation, leg/wing structure, and sensory organs.
  - **Biomechanical Study**: Understand the movement mechanics, such as how muscles and joints work together to produce motion.
  - **Component Design**: Use CAD software to design the robot's parts, ensuring they replicate the insect's functional morphology.
  - **Integration**: Ensure that all components can be assembled seamlessly and function together, just like in the natural model.

iii. Material Selection

Objective: Choose materials that balance weight, flexibility, and durability, suitable for the robot's functions and environmental conditions.
Criteria:

- **Weight**: Lighter materials reduce energy consumption and enhance mobility.
- **Flexibility**: Materials should allow for necessary movements without breaking.
- **Durability**: Must withstand environmental stresses and repeated use.

Common Materials:

- **Polymers**: Lightweight and flexible, used for exoskeletons and joint components.
- **Metals**: For structural support where strength is crucial, like titanium or aluminum.
- **Composites**: Carbon fiber composites for high strength-to-weight ratios.

Process:

- **Evaluation of Properties**: Assess materials based on tensile strength, elasticity, fatigue resistance, and weight.
- **Prototyping and Testing**: Create prototypes using different materials and test them for performance in simulated environments.

iv. Fabrication Techniques

Objective: Manufacture the designed components with precision and quality.
Techniques:

- **Additive Manufacturing (3D Printing)**:
- **Fused Deposition Modeling (FDM)**: Suitable for creating strong and lightweight structures.
- **Stereolithography (SLA)**: Provides high precision and smooth surfaces, ideal for detailed parts.
- **Selective Laser Sintering (SLS)**: Uses powdered materials for creating durable components.
- **Microfabrication Methods**:
- **Photolithography**: Used for creating micro-scale structures, particularly for sensors and micro-actuators.
- **Etching and Deposition**: For fabricating intricate patterns and thin films used in electronic components.

Steps:

- **Preparation**: Finalize the CAD designs and prepare them for the chosen fabrication technique.
- **Manufacturing**: Use the selected technique to create each component. Ensure adherence to design specifications.
- **Assembly**: Carefully assemble the components, integrating sensors, actuators, and electronic circuits to form the complete robot.
- **Testing and Iteration**: Conduct performance tests on the assembled robot. Iterate the design and fabrication process based on test results to optimize functionality.

The biomimetic design and fabrication of insect robots involve a comprehensive process that starts with selecting the appropriate insect model and ends with testing and iterative improvement. By closely mimicking the morphology and biomechanics of insects, these robots achieve superior performance in terms of locomotion, sensing, and adaptability. The integration of advanced materials and fabrication techniques ensures that these robots are not only functional but also durable and efficient, ready for a variety of challenging applications.

2. Actuation and Locomotion:

i. Actuator Selection

Objective: Choose suitable actuators that meet the power, size, and motion requirements of the insect robot.

Types of Actuators:

- **Electric Motors**:
- **DC Motors**: Simple and widely used; suitable for continuous rotation.
- **Servo Motors**: Provide precise control over angular position; ideal for joint movements.
- **Stepper Motors**: Offer precise control of position and speed; used in applications requiring accurate movement.
- **Piezoelectric Actuators**:
- Use piezoelectric materials that deform when an electric field is applied.
- Provide high precision and can generate significant force relative to their size.
- Suitable for micro-movements and high-frequency applications like wing flapping.
- **Shape Memory Alloys (SMA)**:
- Change shape in response to temperature changes.
- Compact and lightweight but relatively slow and energy-inefficient compared to other actuators.
- **Pneumatic and Hydraulic Actuators**:
- Utilize pressurized air or fluid to generate movement.
- Provide high power and smooth operation but require complex support systems and are generally larger.

Selection Criteria:

- **Power Requirements**: Ensure the actuator can deliver the necessary force and speed.
- **Size Constraints**: Must fit within the small and often complex body structure of the robot.
- **Desired Motion Capabilities**: Choose based on whether the movement needs to be rotational, linear, or oscillatory.

Process:

- **Analyze Requirements**: Determine the force, speed, and precision needed for the robot's tasks.
- **Compare Actuator Types**: Evaluate the performance characteristics of various actuators.
- **Prototype Testing**: Test selected actuators in prototypes to validate their performance.

ii. Locomotion Mechanisms

Objective: Implement locomotion mechanisms inspired by the selected insect model.

Types of Locomotion:

- **Legged Locomotion**:
- **Walking**: Mimics insects like ants and beetles. Requires coordination of multiple legs for stable movement.
- **Running**: Faster but less stable; involves dynamic balancing mechanisms.
- **Climbing**: Specialized adaptations like adhesive pads or claws for vertical surfaces.
- **Flapping Wings**:
- **Hovering**: Common in insects like bees and dragonflies; requires precise control of wing flapping frequency and angle.
- **Forward Flight**: Involves a combination of wing flapping and body tilt for propulsion.
- **Crawling**:
- Involves peristaltic movements like those of caterpillars or inchworms.
- Suitable for robots designed to navigate through tight spaces or over irregular terrain.

Implementation Steps:

- **Kinematic Analysis**: Study the movement patterns of the selected insect to understand joint angles, segment lengths, and motion sequences.
- **Mechanism Design**: Use CAD software to design the mechanical structure that replicates these movements.
- **Integration with Actuators**: Ensure actuators can produce the necessary movements and are appropriately placed within the design.
- **Prototype Testing**: Build and test locomotion prototypes, making adjustments as needed for optimal performance.

iii. Control Systems

Objective: Develop control algorithms to coordinate actuator movements and achieve desired locomotion patterns.

Components of Control Systems:

- **Sensors**: Provide feedback on the robot's position, orientation, and environmental conditions.
- **IMUs (Inertial Measurement Units)**: Measure acceleration and angular velocity for balance and stability.
- **Proximity Sensors**: Detect obstacles to prevent collisions.
- **Force Sensors**: Monitor contact forces during walking or climbing.
- **Microcontroller**: Processes sensor data and executes control algorithms to coordinate actuator movements.

Control Algorithms:

- **Open-Loop Control**: Simple control where actuator commands are predefined; lacks feedback and adaptability.
- **Closed-Loop Control**: Uses feedback from sensors to adjust actuator commands dynamically; enhances stability and adaptability.

Types of Control Strategies:

- **Proportional-Integral-Derivative (PID) Control**:
- Widely used for its simplicity and effectiveness.
- Adjusts actuator movements based on the difference between desired and actual positions.
- **Model Predictive Control (MPC)**:
- Uses a model of the robot's dynamics to predict future states and optimize control actions.
- Suitable for complex, dynamic environments.
- **Reinforcement Learning**:
- Involves training a control policy through trial and error to maximize a reward function.
- Allows the robot to learn and adapt to new tasks and environments.

Development Steps:

- **Sensor Integration**: Connect sensors to the microcontroller and ensure accurate data acquisition.
- **Algorithm Design**: Develop and implement control algorithms in the microcontroller's firmware.
- **Simulation and Testing**: Use simulation tools to test control algorithms in virtual environments before real-world implementation.
- **Iterative Improvement**: Continuously refine control algorithms based on testing feedback to improve stability, agility, and energy efficiency.

Actuator selection, locomotion mechanisms, and control systems are critical aspects of designing and building effective insect robots. By carefully choosing actuators, implementing biomimetic locomotion mechanisms, and developing sophisticated control systems, engineers can create robots that mimic the remarkable abilities of insects. These robots are capable of navigating complex environments, performing intricate tasks, and adapting to various conditions, making them valuable in numerous applications ranging from search and rescue to environmental monitoring.

3. Sensing and Perception:

Integrating sensors into insect robots allows them to perceive and interact with their environment effectively. By drawing inspiration from the sensory systems of insects, we can develop robust and efficient sensing modalities. Signal processing then ensures that the raw data collected by these sensors are transformed into actionable information for navigation and decision-making. (Manoonpong et al., 2021)

i. Sensor Integration

Objective: Equip the insect robot with various sensors to gather environmental data and internal state information.

Types of Sensors:

- **Cameras**: For visual perception, obstacle detection, and navigation.
- **Accelerometers**: Measure linear acceleration to detect movement and orientation changes.
- **Gyroscopes**: Measure angular velocity to help maintain balance and orientation.
- **Proximity Sensors**: Detect nearby objects to avoid collisions.
- **Infrared Sensors**: Useful for night vision and detecting heat sources.
- **Pressure and Touch Sensors**: Measure contact forces, crucial for climbing and terrain adaptation.

Integration Process:

- **Selection**: Choose sensors based on the specific requirements of the robot's tasks and environments.
- **Placement**: Determine optimal locations on the robot's body for sensor placement to ensure accurate data capture.
- **Connection**: Integrate sensors with the microcontroller, ensuring proper wiring and data transmission pathways.
- **Calibration**: Calibrate sensors to ensure accurate readings. This can involve standardizing measurements against known values and adjusting for any offsets or biases.
- **Testing**: Conduct tests to verify that sensors are correctly integrated and functioning as expected.

ii. Bio-Inspired Sensing

Objective: Implement sensing modalities that mimic insect sensory systems to enhance the robot's ability to detect obstacles, navigate terrain, and locate targets.

Examples of Bio-Inspired Sensing:

- **Compound Eyes**:
- **Structure**: Insects like bees and dragonflies have compound eyes consisting of numerous small lenses (ommatidia) that provide a wide field of view and high motion detection capability.
- **Implementation**: Use multi-camera systems or specialized lenses to create a wide-angle, multi-faceted vision system for the robot.
- **Advantages**: Enhanced motion detection and wide field of view, suitable for fast navigation and detecting moving objects.

  - **Antennae**:

- **Structure**: Insects use antennae for tactile sensing and chemical detection.
- **Implementation**: Equip the robot with flexible whisker-like sensors for tactile feedback or chemical sensors for detecting specific substances.
- **Advantages**: Improved ability to navigate through tight spaces and detect chemical signals in the environment.
- **Hairs and Setae**:
- **Structure**: Insects like spiders have fine hairs (setae) on their legs that detect air currents and vibrations.

- **Implementation**: Integrate micro-scale hair-like sensors that can detect changes in airflow or vibrations, aiding in navigation and environmental awareness.
- **Advantages**: Enhanced sensitivity to environmental changes, aiding in early obstacle detection and navigation in low-visibility conditions.

Implementation Steps:

- **Design**: Create sensors inspired by the structure and function of insect sensory organs.
- **Fabrication**: Use microfabrication techniques to produce these sensors, ensuring they are lightweight and durable.
- **Integration**: Attach these sensors to the robot in positions that maximize their effectiveness.
- **Testing**: Verify the functionality of bio-inspired sensors in various environments to ensure they perform as expected.

iii. Signal Processing

Objective: Develop algorithms to process sensor data and extract relevant information for navigation and decision-making.

Steps in Signal Processing:

- **Data Acquisition**: Collect raw data from integrated sensors.
- **Preprocessing**: Filter and clean the data to remove noise and irrelevant information.
- **Techniques**: Low-pass, high-pass, band-pass filters; noise reduction algorithms.

Feature Extraction: Identify and extract relevant features from the preprocessed data.

- **Examples**: Edges in images, peaks in accelerometer data, or specific patterns in chemical sensor readings.

Data Fusion: Combine data from multiple sensors to create a comprehensive understanding of the environment.

- **Techniques**: Kalman filtering, Bayesian inference, and machine learning models.

Decision-Making Algorithms:

- **Navigation**: Use data to build maps of the environment, detect obstacles, and plan paths.
- **Control**: Adjust actuator commands in real-time to respond to sensor inputs and maintain stability, avoid obstacles, and follow planned paths.

Real-Time Processing: Ensure algorithms are optimized for real-time execution, allowing the robot to react promptly to changes in the environment.

Example Algorithms:

- **Computer Vision Algorithms**:
- **Object Detection and Recognition**: Use techniques like convolutional neural networks (CNNs) to identify objects and landmarks in the robot's field of view.
- **Optical Flow**: Calculate the movement of objects in the visual field to estimate the robot's velocity and direction.
- **Sensor Fusion Algorithms**:
- **Extended Kalman Filter (EKF)**: Combines data from gyroscopes, accelerometers, and cameras to estimate the robot's position and orientation accurately.
- **Particle Filter**: Used for localization and mapping, helping the robot to understand its position within a given space.
- **Path Planning Algorithms**:
- A Algorithm: Finds the shortest path to a target, avoiding obstacles.
- D Algorithm: An extension of A* that dynamically updates the path as the robot discovers new obstacles.
- **Control Algorithms**:
- **Proportional-Derivative (PD) Control**: Maintains stability by adjusting movements based on the difference between desired and actual positions.
- **Reinforcement Learning**: Allows the robot to learn optimal actions based on rewards received from the environment.

The integration of sensors, bio-inspired sensing modalities, and advanced signal processing algorithms enables insect robots to perceive their environment accurately and make informed decisions. By mimicking the sensory systems of insects, these robots achieve high levels of functionality and adaptability. The continuous development and refinement of these technologies promise even greater capabilities for insect robots in various applications, ranging from search and rescue to environmental monitoring.

4. Communication and Networking:

Effective communication and networking are crucial for the functioning of insect robots, especially when operating in swarms. This involves implementing communication protocols for data transmission and developing algorithms for coordinating swarm behavior. (Johanson et al., 2021), (Hepp, 2020)

i. Communication Protocols

Objective: Implement wireless communication protocols to enable real-time data transmission and remote control of insect robots.
Common Wireless Communication Protocols:

- **Bluetooth**:
- **Range**: Short to medium range (up to 100 meters).
- **Data Rate**: Suitable for low to moderate data rates.
- **Power Consumption**: Low power, ideal for small, battery-powered robots.
- **Use Case**: Local communication between robots or between a robot and a nearby control device.
- **Wi-Fi**:
- **Range**: Medium to long range (up to several hundred meters).
- **Data Rate**: High data rates, suitable for transmitting large amounts of data.
- **Power Consumption**: Higher power consumption compared to Bluetooth.
- **Use Case**: Communication in environments where high data throughput is needed, such as streaming video from onboard cameras.
- **Zigbee**:
- **Range**: Medium range (10-100 meters).
- **Data Rate**: Lower data rates compared to Wi-Fi but sufficient for many sensor data applications.
- **Power Consumption**: Very low power, suitable for battery-operated devices.
- **Use Case**: Networking multiple robots in a swarm, sensor networks.
- **LoRa (Long Range)**:
- **Range**: Very long range (up to 10 km in open areas).
- **Data Rate**: Low data rates, suitable for infrequent communication.
- **Power Consumption**: Very low power, designed for long battery life.
- **Use Case**: Remote monitoring and communication in large, dispersed areas.

Implementation Steps:

- **Select Protocol**: Choose the appropriate communication protocol based on the requirements of range, data rate, and power consumption.
- **Hardware Integration**: Integrate communication modules (e.g., Bluetooth or Wi-Fi modules) into the robot's hardware.
- **Software Configuration**: Develop and configure the software stack to handle wireless communication, including establishing connections, data transmission, and handling errors.
- **Testing**: Conduct thorough testing to ensure reliable communication in various environments and under different conditions.

ii. Swarm Coordination

Objective: Develop algorithms for swarm coordination and communication to enable collaboration and collective behavior among multiple insect robots.

Key Concepts:

- **Decentralized Control**: Each robot operates autonomously but follows simple rules that govern its interaction with other robots and the environment.
- **Local Communication**: Robots communicate with their immediate neighbors rather than relying on a central control unit.

Swarm Coordination Algorithms:

- **Flocking Algorithms**: Inspired by the behavior of birds and fish, where each robot adjusts its position based on the positions and velocities of its neighbors.
- **Boids Model**: Consists of three simple rules—separation (avoid crowding neighbors), alignment (steer towards the average heading of neighbors), and cohesion (move towards the average position of neighbors).
- **Particle Swarm Optimization (PSO)**:
- **Principle**: Robots (particles) explore the environment and share their findings to collectively identify optimal solutions.
- **Application**: Used for search and rescue operations where robots need to find and signal the location of targets.
- **Ant Colony Optimization (ACO)**:
- **Principle**: Robots mimic the pheromone-laying and following behavior of ants to find the shortest paths between points.
- **Application**: Effective for path planning and navigation tasks.
- **Consensus Algorithms**:

- **Principle**: Robots reach a common agreement on certain parameters (e.g., direction, task allocation) through iterative communication.
- **Application**: Used for decision-making processes in dynamic environments.

Swarm Communication Strategies:

- **Broadcasting**: Each robot periodically sends out a message to all neighbors within its communication range. Simple but can lead to high communication overhead.
- **Gossip Protocols**: Robots randomly communicate with one or a few neighbors at each time step, reducing communication load while ensuring information spreads through the swarm.
- **Role-based Communication**: Specific robots are assigned roles (e.g., leaders, followers) that determine their communication patterns and tasks. Helps in structuring the communication and improving efficiency.

Implementation Steps:

- **Algorithm Design**: Develop or select appropriate swarm coordination algorithms based on the specific application and desired behavior.
- **Simulation**: Test algorithms in simulation environments to evaluate their performance and identify potential issues.
- **Implementation**: Program the robots with the chosen algorithms, ensuring that the communication protocols are integrated to enable effective swarm behavior.
- **Field Testing**: Deploy the robots in real-world scenarios to test and refine their coordination and communication capabilities.

Effective communication and networking are essential for the successful deployment of insect robots, particularly in swarm applications. By implementing suitable wireless communication protocols and developing robust swarm coordination algorithms, insect robots can achieve sophisticated collective behaviors, enabling them to perform complex tasks in various environments. These capabilities make insect robots highly versatile and effective for applications such as search and rescue, environmental monitoring, and surveillance.

Table 1. Communication and Networking of insect robots

Feature	Existing Robot Models	Insect Robots
Communication Protocols	Wi-Fi, Bluetooth, Ethernet	Bluetooth, Wi-Fi, Zigbee, LoRa
Swarm Coordination	Centralized or decentralized with advanced AI	Decentralized, inspired by natural swarm behaviors
Local Communication	Often requires strong infrastructure	Primarily uses local communication among neighbors
Algorithm Complexity	High complexity with centralized systems	Simpler algorithms inspired by nature (e.g., flocking, ACO)
Energy Efficiency	Varies, often high due to powerful sensors/actuators	Low power consumption, optimized for long-term operation

5. Power and Energy Management:

i. Power Sources

Objective: Select appropriate power sources that meet the energy requirements and operational constraints of insect robots.

Common Power Sources:

- **Batteries**:
- **Lithium-Ion (Li-ion)**:
- **Advantages**: High energy density, relatively lightweight, long cycle life.
- **Disadvantages**: Safety concerns if not managed properly (risk of overheating, explosion).
- **Use Case**: General-purpose insect robots requiring substantial energy for locomotion and sensors.
- **Lithium-Polymer (Li-Po)**:
- **Advantages**: Higher energy density than Li-ion, flexible form factor.
- **Disadvantages**: More expensive, requires careful handling.
- **Use Case**: Compact robots where weight and size are critical constraints.
- **Nickel-Metal Hydride (NiMH)**:
- **Advantages**: Safe, environmentally friendly, robust.
- **Disadvantages**: Lower energy density, heavier.
- **Use Case**: Robots with moderate power requirements, where safety and robustness are prioritized.
- **Solar Cells**:
- **Advantages**: Renewable energy source, can extend operational time by supplementing batteries.
- **Disadvantages**: Dependence on sunlight availability, lower energy density.

- **Use Case**: Outdoor robots operating in environments with ample sunlight.
- **Fuel Cells**:
- **Advantages**: High energy density, longer operational time.
- **Disadvantages**: Complex system integration, cost, safety concerns.
- **Use Case**: High-endurance robots with long-term missions where regular recharging is impractical.

Selection Criteria:

- **Energy Requirements**: Calculate the total energy needed for the robot's tasks.
- **Weight and Size Constraints**: Choose sources that fit within the robot's form factor and weight limits.
- **Operational Environment**: Consider factors like sunlight availability, temperature, and mission duration.

ii. Energy-Efficient Design

Objective: Optimize the robot's design and control algorithms to minimize energy consumption and extend operational endurance.

Design Optimization:

- **Lightweight Materials**: Use materials like carbon fiber, lightweight polymers, or aluminum to reduce the robot's overall weight, decreasing energy required for movement.
- **Compact and Efficient Actuators**: Choose actuators with high efficiency and appropriate power ratings to minimize energy loss.
- **Aerodynamic Design**: For flying robots, design body shapes that reduce air resistance and optimize wing shapes for efficient flight.

Control Algorithm Optimization:

- **Energy-Aware Path Planning**: Develop algorithms that consider energy consumption when planning paths, choosing routes that minimize power usage.
- **Adaptive Control**: Implement control systems that adjust actuator power based on real-time feedback, avoiding unnecessary movements and optimizing performance.
- **Sleep Modes**: Design algorithms to put non-essential systems into low-power sleep modes when not in use, waking them only when needed.

Example Techniques:

- **Gait Optimization**: For legged robots, design gaits that minimize energy use by reducing the number of steps or optimizing the force applied in each step.
- **Wing Flapping Optimization**: For flying robots, adjust wing flapping frequency and amplitude to maximize lift while minimizing power draw.

iii. Power Management Systems

Objective: Implement power management systems to regulate energy flow, recharge batteries, and maximize overall efficiency.

Components of Power Management Systems:

- **Power Distribution Units (PDUs)**: Manage the distribution of power to various subsystems, ensuring stable and efficient power delivery.
- **Battery Management Systems (BMS)**:
- **Functions**: Monitor battery health, charge levels, and temperature; balance cells to extend battery life and prevent overcharging or deep discharge.
- **Implementation**: Integrate BMS hardware and software into the robot's power system to continuously monitor and adjust power usage.
- **Energy Harvesting Modules**: Incorporate solar panels or kinetic energy harvesters to supplement battery power, especially in outdoor or high-movement environments.
- **Dynamic Power Allocation**: Use algorithms to dynamically allocate power based on real-time demand, prioritizing critical systems during low power situations.

Recharge Mechanisms:

- **Automatic Docking**: Design robots to autonomously dock with charging stations when battery levels are low.
- **Wireless Charging**: Implement wireless charging systems to allow for convenient and continuous power replenishment without the need for physical connectors.

Optimization Techniques:

- **Load Balancing**: Distribute power evenly across subsystems to prevent any single component from drawing excessive power, which can lead to inefficiencies.

- **Predictive Maintenance**: Use data from the power management system to predict and perform maintenance before issues arise, ensuring continuous and efficient operation.

Selecting the right power sources, designing for energy efficiency, and implementing robust power management systems are essential for the effective operation of insect robots. By optimizing these elements, insect robots can achieve longer operational endurance, better performance, and increased reliability, making them suitable for a wide range of applications from search and rescue to environmental monitoring.

V. COMPARISON ANALYSIS: EXISTING ROBOT MODELS VS. INSECT ROBOTS

1. Size and Portability

- **Existing Robot Models**:
- **Industrial Robots**: Typically large, stationary, and designed for specific tasks within controlled environments (e.g., assembly lines).
- **Service Robots**: Vary in size, usually larger than insect robots, designed for tasks like cleaning, delivery, or caregiving.
- **Insect Robots**:
- **Size**: Much smaller, often mimicking the size of real insects. This miniaturization allows them to operate in confined and complex environments.
- **Portability**: Extremely portable, can be deployed in swarms, and easily transported to remote or hazardous locations.

2. Locomotion and Mobility

- **Existing Robot Models**:
- **Wheeled Robots**: Efficient on flat surfaces, limited in navigating rough terrains.
- **Tracked Robots**: Better suited for rough terrains but slower and less agile.
- **Bipedal and Quadrupedal Robots**: Advanced mobility and balance, capable of navigating varied terrains but complex and expensive.
- **Insect Robots**:
- **Legged Locomotion**: Inspired by insects, providing excellent adaptability to various terrains, including vertical surfaces and complex environments.

- **Flapping Wings**: Some models mimic flying insects, offering agility and the ability to reach places inaccessible to ground robots.
- **Crawling Mechanisms**: Suitable for tight spaces, often used in search and rescue operations.

3. Sensing and Perception

- **Existing Robot Models**:
- **Advanced Sensors**: Use LIDAR, RADAR, ultrasonic sensors, and high-definition cameras. These sensors provide high accuracy but are often large and power-consuming.
- **AI and Machine Learning**: Implemented for object recognition, path planning, and decision-making, providing robust autonomous capabilities.
- **Insect Robots**:
- **Bio-Inspired Sensors**: Employ sensors inspired by insect eyes, antennae, and other sensory organs. These sensors are typically smaller and optimized for low power consumption.
- **Environmental Adaptation**: Designed to work in various environmental conditions, using compact and efficient sensory systems to navigate and interact with their surroundings.

4. Power and Energy Efficiency

- **Existing Robot Models**:
- **High Power Consumption**: Often require substantial power sources, limiting their operational time and increasing the need for recharging or tethering.
- **Battery Technology**: Use advanced but bulky batteries, which can limit portability.
- **Insect Robots**:
- **Low Power Consumption**: Designed for efficiency, allowing longer operational times on smaller power sources.
- **Energy Harvesting**: Some models explore energy harvesting techniques from the environment, such as solar or kinetic energy.

5. Control Systems and Autonomy

- **Existing Robot Models**:
- **Complex Control Systems**: Utilize advanced control systems requiring significant computational resources.

- **High Autonomy Levels**: Achieve high levels of autonomy through sophisticated algorithms and AI, often needing powerful onboard processors.
- **Insect Robots**:
- **Simplified Control Systems**: Often use simpler, more efficient control algorithms inspired by insect behavior.
- **Swarm Intelligence**: Utilize decentralized control and swarm intelligence, allowing simple robots to perform complex tasks collectively.

6. Applications

- **Existing Robot Models**:
- **Industrial Applications**: Manufacturing, logistics, warehousing.
- **Service Applications**: Healthcare, hospitality, domestic tasks.
- **Insect Robots**:
- **Search and Rescue**: Navigate through rubble and confined spaces to locate survivors.
- **Environmental Monitoring**: Collect data in hard-to-reach or hazardous areas.
- **Agriculture**: Precision farming, pest detection, and crop monitoring.
- **Surveillance**: Covert operations and environmental surveillance.

Table 2. Comparative study on insect robots

Feature	Existing Robot Models	Insect Robots
Size and Portability	Larger, less portable	Small, highly portable
Locomotion and Mobility	Wheeled, tracked, bipedal, quadrupedal	Legged, flapping wings, crawling
Sensing and Perception	Advanced sensors (LIDAR, RADAR, cameras)	Bio-inspired sensors (compound eyes, antennae)
Power and Energy Efficiency	High power consumption, advanced batteries	Low power consumption, energy harvesting
Control Systems and Autonomy	Complex control systems, high autonomy	Simplified control, swarm intelligence
Applications	Industrial, service, healthcare	Search and rescue, environmental monitoring, agriculture, surveillance

Insect robots offer distinct advantages over traditional robot models in terms of size, portability, and energy efficiency, making them particularly well-suited for tasks in confined, complex, or hazardous environments. While existing robot models excel in industrial and service applications due to their advanced sensors, higher power, and sophisticated control systems, insect robots leverage biomimicry to achieve superior adaptability and efficiency in specific niches. This comparison highlights the complementary nature of both types of robots, suggesting a potential for integrated applications where both technologies can be employed to maximize efficacy.

VI. CONCLUSION AND FUTURE SCOPE

The exploration of the micro world through insect-inspired robots represents a fascinating intersection of biology, engineering, and robotics. By drawing inspiration from nature's own designs, researchers have unlocked a wealth of possibilities for creating agile, adaptable, and efficient robotic systems capable of navigating complex environments and tackling real-world challenges. Through biomimetic design, researchers have mimicked the morphology, biomechanics, and sensory capabilities of insects, enabling the creation of robots that can crawl, climb, fly, and swim with remarkable agility and dexterity. These robots, equipped with advanced sensing and perception capabilities, can perceive and interact with their environment, making them valuable tools for applications such as search and rescue, environmental monitoring, and exploration in hazardous or inaccessible areas.

Furthermore, the advent of miniaturization has empowered researchers to develop small-scale robots that can access confined spaces and operate autonomously for extended periods. This has opened up new frontiers for exploration and intervention, enabling robots to venture into environments where traditional machinery cannot reach.

Moreover, the emergence of swarm robotics has demonstrated the power of collective intelligence, allowing groups of robots to collaborate and solve complex tasks in a decentralized manner. By harnessing the principles of swarm behavior observed in nature, researchers have unlocked new avenues for cooperation and coordination among robotic systems.

Despite the progress made, challenges remain in the development of insect-inspired robots, including the need for further advancements in control algorithms, energy efficiency, and robustness to environmental conditions. Additionally, interdisciplinary collaboration between biologists, engineers, and roboticists will be crucial for unlocking the full potential of insect-inspired robotics and translating it into practical applications.

The field of insect-inspired robotics holds immense promise for revolutionizing how we interact with and explore the micro world. As researchers continue to push the boundaries of innovation and technological advancement, insect-inspired robots are poised to play a vital role in shaping the future of robotics and contributing to society in meaningful ways. (Siderska, 2020)

Applications:

- **Search and Rescue**: Due to their small size and agility, insect robots can navigate through rubble and debris to locate survivors in disaster-stricken areas.
- **Environmental Monitoring**: These robots can enter tight spaces and cover large areas to collect data on environmental conditions, such as pollution levels or climate changes.
- **Agriculture**: Insect robots can assist in precision agriculture by monitoring crop health, pollinating plants, or identifying pests.
- **Military and Surveillance**: Their small size makes them ideal for covert operations, where they can gather intelligence without being detected.
- **Scientific Research**: Insect robots can be used to study the behavior of real insects and other small animals, providing valuable data for biological research.

Key Features:

- **Biomimicry**: Insect robots are often modeled after the anatomy and movement patterns of real insects. This approach, known as biomimicry, allows engineers to create robots that can navigate complex environments with the agility and efficiency of their biological counterparts.
- **Miniaturization**: One of the primary challenges in creating insect robots is miniaturizing all necessary components (sensors, actuators, power sources, and microcontrollers) while maintaining functionality. Advances in microelectromechanical systems (MEMS) and nanotechnology have been crucial in overcoming these challenges.
- **Autonomy**: Insect robots are typically designed to operate autonomously. They are equipped with sensors that allow them to perceive their environment and make decisions without human intervention. This autonomy is essential for tasks in hazardous or inaccessible areas.

Challenges:

- **Power Supply**: Ensuring a reliable and long-lasting power source is a significant challenge for such small robots.
- **Control Algorithms**: Developing efficient algorithms for autonomous navigation and decision-making in dynamic environments.
- **Durability**: Creating a robust design that can withstand various environmental conditions.

Future Prospects

The field of insect robots is rapidly evolving, with ongoing research focused on improving their capabilities and expanding their applications. Innovations in artificial intelligence, materials science, and sensor technology are expected to further enhance the functionality and utility of these remarkable machines. As technology advances, insect robots may play an increasingly vital role in fields ranging from environmental conservation to healthcare. (Ribeiro et al., 2021)

In summary, insect robots represent a cutting-edge convergence of technology and biology, offering promising solutions to a variety of complex problems.

REFERENCES

Fuller, S. B. (2019). Four wings: An insect-sized aerial robot with steering ability and payload capacity for autonomy. *IEEE Robotics and Automation Letters*, 4(2), 570–577. DOI: 10.1109/LRA.2019.2891086

Hepp, A. (2020). Artificial companions, social bots and work bots: Communicative robots as research objects of media and communication studies. *Media Culture & Society*, 42(7-8), 1410–1426. DOI: 10.1177/0163443720916412

Hu, M., Wei, R., Dai, T., Zou, L., & Li, T. (2008, June). Control strategy for insect-like flapping wing micro air vehicles: Attitude control. In *2008 7th World Congress on Intelligent Control and Automation* (pp. 9043-9048). IEEE.

Jiang, Y., Zhao, W., Jiang, Y., Sun, K., Huang, X., & Yang, B. (2023). Wireless Multisensors Platform for Distributed Insect Biobot Sensing Network. *IEEE Sensors Journal*, 23(7), 7929–7937. DOI: 10.1109/JSEN.2023.3243916

Johanson, D. L., Ahn, H. S., & Broadbent, E. (2021). Improving interactions with healthcare robots: A review of communication behaviours in social and healthcare contexts. *International Journal of Social Robotics*, 13(8), 1835–1850. DOI: 10.1007/s12369-020-00719-9

Macrorie, R., Marvin, S., & While, A. (2021). Robotics and automation in the city: A research agenda. *Urban Geography*, 42(2), 197–217. DOI: 10.1080/02723638.2019.1698868

Manoonpong, P., Patanè, L., Xiong, X., Brodoline, I., Dupeyroux, J., Viollet, S., Arena, P., & Serres, J. R. (2021). Insect-inspired robots: Bridging biological and artificial systems. *Sensors (Basel)*, 21(22), 7609. DOI: 10.3390/s21227609 PMID: 34833685

Niku, S. B. (2020). *Introduction to robotics: analysis, control, applications*. John Wiley & Sons.

Ribeiro, J., Lima, R., Eckhardt, T., & Paiva, S. (2021). Robotic process automation and artificial intelligence in industry 4.0–a literature review. *Procedia Computer Science*, 181, 51–58. DOI: 10.1016/j.procs.2021.01.104

Scheper, K. Y., Karásek, M., De Wagter, C., Remes, B. D., & De Croon, G. C. (2018, May). First autonomous multi-room exploration with an insect-inspired flapping wing vehicle. In *2018 IEEE International Conference on Robotics and Automation (ICRA)* (pp. 5546-5552). IEEE. DOI: 10.1109/ICRA.2018.8460702

Siderska, J. (2020). Robotic Process Automation—A driver of digital transformation? *Engineering Management in Production and Services*, 12(2), 21–31. DOI: 10.2478/emj-2020-0009

Zou, Y., Zhang, W., & Zhang, Z. (2016). Liftoff of an electromagnetically driven insect-inspired flapping-wing robot. *IEEE Transactions on Robotics*, 32(5), 1285–1289. DOI: 10.1109/TRO.2016.2593449

Chapter 6
Early Depression Detection Using Modern AI Techniques:
Issues, Opportunities, and Challenges

Sharmistha Dey
Galgotias University, India

Krishan Veer Singh
https://orcid.org/0000-0001-8899-5097
Galgotias University, India

ABSTRACT

Depression is a widespread and debilitating mental health disorder, impacting over 300 million individuals globally, as reported by the World Health Organization. Early detection and timely intervention are essential for effective treatment and mitigating the severity of depressive symptoms. However, accurately identifying the nuanced symptoms of depression—manifested through body language, speech patterns, or neurological signals—remains a significant challenge. The advent of modern AI technologies has revolutionized the landscape of depression detection, offering new methodologies for identifying these symptoms This study investigates the current challenges, opportunities, and advancements in AI-driven approaches to early depression detection. We conducted a comprehensive review of approximately 60 high-quality, peer-reviewed research articles from reputable journals and conferences, focusing on the relevance and objectives of each study. Our findings highlight the latest trends in depression detection and outline the obstacles faced in this field, providing a roadmap for future researchers aiming to enhance early

DOI: 10.4018/979-8-3693-6150-4.ch006

Copyright © 2025, IGI Global. Copying or distributing in print or electronic forms without written permission of IGI Global is prohibited.

detection strategies and improve mental health outcomes.

1. INTRODUCTION

Depression is one of the major reasons for people's deaths and damage to lives, causing the loss of life of about 3.8% of people worldwide, according to a recent report published by the World Health Organization (Beyond Blue, 2023). It is a silent killer which may damage your inner world. A person having deep depression inside doesn't need to be sad from the outside, rather the reverse may happen. This mental health disorder may be caused by many factors such as stressful work life, routine disorder, lifestyle disorder, personal stress, etc. Individuals with depression often experience a lack of interest in anything, and intense and prolonged feelings of sadness, despair, and worthlessness. They may have difficulty experiencing joy or enthusiasm for activities they once enjoyed (Matsushima et. al., 2024).

Depression is often associated with increased blood pressure and hypertension issues. it can affect the emotional behavior of a person, and a tendency to do work, all of which can affect long-term health outcomes ('Mayoclinic', 2021). Therefore, timely diagnosis and cure is important to curb this menace effectively however, Depression detection is associated with some challenging factors that make it difficult to detect and prevent depression at the earliest. Subjective experiences of individuals that vary in intensity and effect on mental health, lack of confidence among people to express their problem as may fear of judgment from others, Depression often co-occurs with other mental health disorders, such as anxiety, making it challenging to isolate and diagnose depression on its own. Symptoms may overlap, leading to misdiagnosis or delayed treatment, some individuals with depression are adept at masking their symptoms, appearing outwardly functional and cheerful, even when they are experiencing severe emotional distress. This can make it difficult for healthcare providers to recognize depression. To minimize the risk of depression counselling can be a helpful method along with medicinal treatment.

1.1. Types of Depression-

Depression is of different types and each type has discrete symptoms for identification. Below are some depression types and they can be differentiated depending upon the symptoms and severity. Following are some of the depressive disorder types, which could affect people in several ways ((Matsushima et. al., 2024).

A. **Major Depressive Disorder (MDD)**- It happens when you feel a loss of interest in any earthly things. It is a long-term disorder when you feel sad, down, or low most of the time, even without any particular reason.
B. **Delusions-** it is a false belief that is built on your own. Sometimes one believes himself or herself is evil. Or you consider that someone is watching or following you. It is a peculiar mental state of that patient.
C. **Postpartum Depression-** This depression may be seen among pregnant ladies, during pregnancy or after the delivery of a child. Around 10%-15% of women suffer from this mental health disorder. They often have mood swings, are irritated or over-emotional in small things, and lose patience those are the symptoms. Around 80% of women feel baby blues in the first few weeks after delivery. After delivering a baby, there will be a hormonal change in a woman's body- making them change both physically as well as emotionally. But this kind of depression is completely curable with proper care and treatment.
D. **Bi-Polar Disorder-** This disorder is a combination of mania and depression. It is a mental illness that causes infrequent shifts in a person's mood, energy, activity levels, and concentration. These shifts can make it hard to carry out daily tasks. Sometimes the person with bipolar disorder may face a very depressed or low mood and just right after he/ she can feel very high in energy, and very enthusiastic.
E. **Seasonal Affective Disorder (SAD)-** It is a mood disorder that is related to season. Just like some people get depressed in the winter season and better and lighter in the summer months. People affected with SAD even feel a lack of energy and feel sleepy much of the time.

Symptoms of early depression detection are:

i) Feeling hopelessness, upset or sad
ii) Showing anger, irritability, or frustration over small matter
iii) Loss of interest or pleasure in normal activities
iv) Disturbance in normal sleep routine, either too much sleep or less sleep
v) Reduced appetite
vi) Tiredness or feeling lack of energy for doing any activity
vii) Feeling of worthlessness, guilt, or lack of self-confidence

There are various machine learning techniques to find the level of depression based on complex patterns of patients.ML algorithms such as supervised and unsupervised learning, semi-supervised learning, and reinforcement learning algorithms are used to detect depression.

Machine learning models that were appropriately trained in complex patterns can effectively help find solutions to the problem based on previous and current scenarios of data. Various machine learning algorithms such as supervised, unsupervised, semi-supervised, and reinforcement learning can detect depression ('Mayoclinic', 2021).

In this study, the issues, challenges, and scope of research have been identified. The introduction section is a general introduction to depression detection, factors, and symptoms. In section 2, a literature study has been performed. Section 3 discusses current approaches that have been adopted in this research and section 4 is about different depression scales, questionnaires', etc., which can guide the budding researchers. The author has the following contribution to this study.

A. To perform a systematic literature review of the existing articles
B. Finding out the research gap, Identify the challenge area
C. Try to find out the roadmap of the research in this area and also give an idea about future research prospects.
D. Also, this study discusses different tools, techniques, and datasets available for this research, which can give a direction to budding researchers in this field.

This article has been divided into five sections. Section 1 is about the introduction, section 2 discusses Methodology and Related Work, section 3 explores Result Analysis, which discusses research findings, and section 4 deliberates available tools, techniques, and datasets for this research. Section 5 concludes the validity of this research and future scope of this study.

2. METHODOLOGY AND RELATED WORK

The study followed the guidelines for a systematic scoping review outlined by PRISMA-Diagram, which is a structured approach to scoping reviews. It is the preferred way of systematic review and meta-analysis. This study guides us to search for more purpose-oriented matches according to our aim of study. This study enhances the quality of the review. The next sub-section will describe how we have selected our search keywords.

2.1 Strategy for Conducting the Search

This Review accompanies a search across four scholarly databases, including IEEE Xplore, PubMed, NCBI, PubMed Central, ScienceDirect, and Scopus, to place appropriate research articles. The first search was conducted on March 03rd, 2024 and the ultimate search was carried out on April 02nd, 2024. All above-mentioned databases were selected to cover various engineering and medical research areas. IEEE Xplore and ScienceDirect casing multiple technical domains and PubMed, NCBI, and PubMed Central, focusing on mental healthcare-related studies. Additionally, a search was performed on Google Scholar to ensure no relevant articles were missed in the previous database searches.

Table 1. Search Terms

Category	Keywords
Mental Health	Depression, Major Depressive Disorder, Anxiety, Stress
Machine Learning	Support Vector Machine, Random Forest, Decision Tree, Classification, Deep Learning, Neural Network
Depression	Major Depressive Disorder, Postpartum Depression, Seasonal Affective Disorder

To conduct the literature search, search terms were organized into three themes: Mental Health (Keywords) AND Machine Learning (Keywords), Depression (Keywords), and Infectious Diseases (Keywords). The search was systematically directed by groping article titles and abstracts, with the overview of the keywords and search concepts, utilized in this review presented in Table 1.

2.2 Inclusion criteria

The main aim of this study was to cover available scholarly articles relevant to the early detection of Depression with the help of modern AI. The popular medical database was searched. Our study searches papers from 2016 to 2024. In steering a comprehensive literature search, we followed the PRISMA study for checklist and inclusion (See Figure 1) (Zhu et. al., 2023).

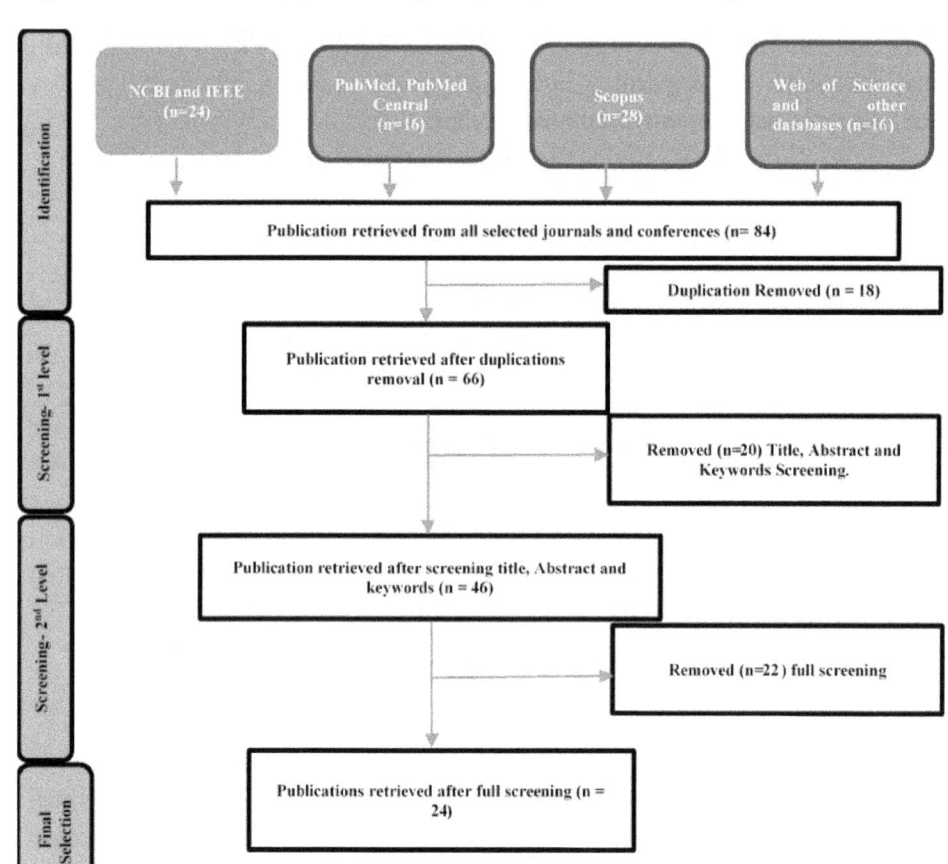

Figure 1. PRISMA Study (Flow diagram for Systematic Review)

Our search was started from some popular databases like IEEE Explore, NCBI, PubMed, PubMed Central, Web of Science, etc. From IEEE Xplore and NCBI, we received 16 papers, PubMed and PubMed Central we got 24 papers, Scopus 28 papers and in Web of Science and other databases like Wiley or Sage we have found 16 papers. So, a total of 84 papers have been collected including all the databases we have searched. To ensure data integrity and eliminate redundancy, cautious removal of duplicate publications was performed, resulting in 66 papers after duplicate removal of 16. In the next screening phase, by searching abstract, keywords, and title 20 articles were excluded and, in this phase, we have selected 46 articles for which full-text screening will be done.

Through a rigorous screening process that involved both title/abstract/keyword evaluation and full-text assessment, a set of 24 studies was identified as meeting the eligibility criteria and subsequently included in the review, as shown in Figure 1.1 and tabulated in Table 1.3(In section 3).

2.3 Exclusion criteria

This review focuses on early detection of Depression using Machine Learning or Deep Learning techniques. So, initially based on search keywords (mentioned in Table 1), 84 articles were selected and after the removal of duplication of 16 papers, 66 articles remained for the next phase. In the next screening level, abstract, keywords, and title were analyzed and after reviewing those 20 articles that didn't match the criteria were removed. For the next round, 46 articles were processed. After full-text screening and checking thoroughly 22 papers were excluded which doesn't serve the purpose of the study. So, finally, 24 papers were used for the final study.

2.4 Selection Process of Articles through Screening

To guarantee the precision of the study, all relevant resources such as journal articles, conference papers, actual case reports, and non-scholarly papers that satisfied the inclusion criteria were collected, and any duplicates were manually removed. The following three sequential stages, namely title, abstract, and full text, were assessed based on predetermined exclusion and inclusion conditions. The PRISMA flow diagram (Figure 1) was used to document the screening and selection process to ensure the study's reproducibility.

2.5 Data extraction

Data extraction was performed on the selected articles, which were peer-reviewed. The data extraction process was initiated by reviewing the documents and identifying the items needed to be extracted (as specified in Table 2). However, if any other relevant data were identified during the review, they were included in the data extraction phase.

Table 2. Data Extraction

Publication Details	Extracted information
Article / Manuscript Information	• Title of the article • Year of Publication • Authors' Name
Depression Detection	• Major Depressive Disorder • Postpartum Depression, Mood swings, Bipolar Disorder

2.6 Data analysis and synthesis

Due to the wide range of sources and reported results, statistical analysis could not be directed at the collected data. Instead, a narrative synthesis was done, which involved summarising and analyzing the extracted data in the discussion section. The narrative synthesis approach was employed to summarize the current literature study regarding early depression detection using modern AI techniques.

3. RESULTS & DISCUSSION

This section presents the findings of the literature survey. After performing a comparative study, we have gathered information from different articles (2018-2024) and in Table 3 we have represented the extract of our literature survey of peer-reviewed articles from high-quality journals or conferences. Research on early detection of depression symptoms and making accurate decisions has been persuaded by researchers for the last couple of years. Recently, with the rapid progress of artificial intelligence technologies, especially deep learning models, explainable AI or generative AI has shown a new angle in this research. This study has considered articles from 2018 to the latest 2024. Initially, researchers used traditional Machine Learning algorithms to detect depression symptoms, but as the years passed, an inclination toward XAI and generative AI was observed. In recent times, some researchers have used generative AI for their research *(Kaywan et. al., 2023)* or explainable AI *(Bayon, 2023), (Ahmad and Ahmad, 2023)* or robotics *(Toshio et. al., 2024)* as their assessment tool for detecting depression. Whereas some researchers rely on Deep Learning algorithms *(Marriwala and Chowdhury, 2023), (Sha et. al., 2020), (Thoduparambil et. al., 2020)*. Even ANOVA with a classification algorithm has been used to detect depression *(Duan et. al., 2021)*, and the accuracy was 94.13%. It has been observed from the study that classification algorithms like Support Vector Machine, Decision Tree, Random Forest, or a combination of ML algorithms give higher accuracy *(A. V. Kumar et. al., 2023), (Chen et. al., 2022), (Duan et. al, 2021), (Narziev et. al., 2020)*.

The table below (See Table 3) shows the research summary of the papers.

Table 3. Comparative study of the papers (From 2018 to 2023)

Sl No	Citation	Mental Health Area Covered	Techniques Used	Measured Performance
1	(Toshio et. al., 2024)	Major Depressive Disorder (MDD)	Robotic Process Animation	For, MDD robot mean 2.62(1.12), Human Mean 2.85(1.14) and p- value is 0.396
2	(Kaywan et. al., 2023)	Depression at early stage (Non-Clinical Trial)	DEPRA Chatbot using generative AI	With the QIDS questionnaire, 14% had mild depression, 22% had moderate depression, 14% had severe depression, and 20% had very severe depression
3	(Bayon H., 2023)	Major Depressive Disorder	Explainable AI algo- SHAP and LIME	Using LIME, the accuracy of the depression prediction model is 84.2%
4	(A.V. Kumar et. al., 2023)	Depression and suicidal thoughts through social media	Support Vector Machine combined with KNN	Accuracy is 98.05%
5	(Ahmed & Ahmed, 2023)	Depression among students	SHAP	Accuracy is 82.4%
6	(Marriwala and Chaudhury, 2023)	Depression	CNN and Bi LSTM	The accuracy for the textual CNN model is 92%, the accuracy of the audio CNN model is 98% and the loss of textual CNN is 0.2 loss of audio CNN is 0.1
7	Lee and Kim., 2022)	Depression along with hypertension	ANN, random forest, AdaBoost, stochastic gradient boosting, XGBoost, and SVM	With ANN, Area under Curve (AUC)- 0.813 and specificity -0.780 With SVM, Accuracy of 0.771, precision of 0.969, and sensitivity of 0.774
8	Chen et. al., 2022)	Major Depressive Disorder	Support Vector Machine	92.48% (Using Gaussian Kernel),88.60% (using polynomial kernel)
9	(Duan et al.,2021)	Major Depressive Disorder	ANOVA and Classification algorithm	Accuracy-94.13%, sensitivy-95.74%
10	(Zhu et. al, 2020)	Depression	Content-based Ensembled Learning	Accuracy82.5% (Dataset 1) Accuracy 92.65% (Dataset 2)

continued on following page

Table 3. Continued

Sl No	Citation	Mental Health Area Covered	Techniques Used	Measured Performance
11	(Srimadhur and Lalitha, 2020)	Clinical Depression	Spectrum based CNN	The accuracy of depressed samples is 74.64%, non-depressed sample accuracy is 80.62%
12	(Frassle et. al., 2020)	Major Depressive Disorder	Generative Embedding	81.3% accuracy
13	(Shah et. al., 2020)	Depression, Anxiety, Suicidal Tendency	Bi-directional LSTM with embedding technique	Accuracy is 84.56%
14	(Mahato and Paul, 2020)	Major Depressive Disorder	Support Vector Machine (SVM), Logistic Regression (LR), Naïve-Bayesian (NB) and Decision Tree (DT)	88.33% in SVM
15	(Narziev et. al., 2020)	Short Term depression	Classification algorithm	Accuracy 96%
16	(Thoduparambil et. al., 2020)	Clinical depression, Major Depressive Disorder	An integrated model of CNN and LSTM	99.07% for the left hemisphere and 98.84% for the right hemisphere
17	(Uddin et. al., 2022)	Detection of depression symptoms from written material	LSTM and RNN, LIME	98.02% accuracy is achieved with the proposed XAI-based approach, while with the traditional 91% accuracy
18	(Xinfang et. al,, 2019)	Major Depression	Content-based Ensembled Learning	87% accuracy
19	(Chiong et. al., 2022)	Depression and suicidal tendencies	Natural Language Processing	91.55% accuracy
20	(Priya et. al., 2020)	Depression, stress, anxiety symptoms	Decision Tree (DT), Random Forest Tree (RFT), Naïve Bayes (NB), Support Vector Machine (SVM) and K-Nearest Neighbour (KNN)	Using Naive Bayes, depression shows the highest accuracy of 85.5%, with SVM, 80.3%
21	(Islam et. al., 2018)	Depression through sentiment	Classifier algorithms - Decision Tree, KNN, SVM, Ensemble subspace KNN, time series analysis	All the classifiers' accuracy rate between 60%-80%

The above diagram sightsees the challenges or issues for early depression detection. It is very critical to distinguish the symptoms of depression, stress or anxiety as their physical symptoms have many common factors. are very similar to anxiety disorder or stress somewhere. From the above literature study, it has been explored that traditional Machine Learning algorithms like Support Vector Machine, Decision Tree, and Naïve Bayes hinder to determine different features prominently. Whereas, Deep Learning algorithms like CNN-LSTM, Bi-Directional LSTM, and Natural Language Processing can be efficiently used for this purpose. For better accuracy and perfection, researchers are using CNN-LSTM or CNN-Bi-Directional LSTM. Our principal research findings will disclose the facts.

3.1 Principal finding

The research papers shown in Table 3 included for this study, show different algorithms used and how many types of depression or other mental health disorders are considered. *(Kaywan et. al., 2023)* pursue a study on early depression detection based on non-clinical trials. They have used the DEPRA chatbot for their use and they have developed a framework using the Diagflow platform. They have used generative AI, to train their ChatBot. With the QIDS questionnaire, they received 14% with mild depression, 22% with moderate depression, 14% with severe depression, and 20% with very severe depression

(Bayon, 2023) detected Major Depressive Disorder using SHAP and LIME- these two explainable AI-based algorithms. Using LIME, the accuracy of the prediction model is 84.2%. *(Kumar et. al., 2023)* used traditional ML algorithms like the Support Vector Machine and K Nearest Neighbour algorithm together to detect depression and suicidal thought analysis and their prediction accuracy is 98.05%. *(Ahmad and Ahmad, 2023)*, performed their research on depression levels and suicides through analysis of social media activities using the SHAP algorithm, which is explainable AI. They have

(Lee and Kim, 2023) have performed their study on depressive persons who have hypertension. With ANN, the Area under the Curve (AUC) is 0.813, and specificity is 0.780, but with SVM, Accuracy is 0.771, precision is 0.969, and sensitivity is 0.774. In another study, *(Marriwala and Chaudhury, 2023)* worked on CNN and Bi-directional LSTM to detect depression and they have achieved 92.08% accuracy using this. *(Srimadhur and Lalitha, 2020)* have worked on spectrum-based CNN to detect clinical depression and they got an Accuracy of depressed samples is 74.64% and, a non-depressed sample accuracy of 80.62%.

(Xingfang et. al, 2022) used a content-based ensembled learning model to detect Major Depressive Disorder in patients and their model accuracy was 87%, whereas *(Chiong et. al., 2022)* applied Natural processing algorithms to detect depression as well as suicidal tendencies among the patients from textual analysis and their research, accuracy was 91.55%.

From the above literature study, it can be observed that mostly Support Vector Machine, Random Forest, or KNN was used for the detection of depression, which also produced higher accuracy but the recent research trends show the use of Deep Learning algorithms, explainable AI or generative AI for this purpose. Though researchers are working in this field and a huge improvement has been possible, still there are some challenges. The next sub-section discusses the challenges of Early detection of depression.

3.2 The Challenges of Early Depression Detection

Recent studies propose advanced Machine Learning algorithms for depression detection but there are some challenges faced by the researchers.

1. **Defining Depression and Identifying Contributing Factors**: One of the primary challenges is defining what depression is and identifying the common factors that contribute to it. This is crucial for training ML models to recognize the signs of depression.
2. **Data Preprocessing and Parameter Tuning**: The performance of ML models can vary significantly depending on the data samples obtained and the features of the data. Preprocessing activities such as data cleaning and parameter tuning are often necessary to achieve optimal results. However, these activities can be complex and time-consuming.
3. **Validation of Results**: Many studies struggle to validate their results due to a lack of sufficient validated evidence, especially from external sources. This makes it difficult to confirm the accuracy and reliability of the ML models.
4. **Variability in ML Performance**: Not all ML models perform equally well across different problems. The effectiveness of an ML model can depend on the specific problem it's being used to solve, which can make it challenging to find the most suitable model for detecting depression symptoms.
5. **Multimodal Data Analysis**: Some studies use multimodal data (e.g., audio, video, and speech responses) for depression detection. Analysing and integrating these different types of data can be challenging, but it's often necessary for achieving high classification accuracy.
6. **Social Stigma and Data Availability**: Depression is often not acknowledged as a mental disorder in many societies, which can lead to a significant number of depressed individuals remaining unidentified and untreated. This lack of recognition and treatment can limit the availability of data for training and testing ML models.

7. **Use of Social Media Data**: While social media data can provide valuable insights into mental health, it also presents challenges. These include issues related to privacy and consent, the need to process and analyze large volumes of unstructured data, and the difficulty of accurately interpreting and contextualizing social media posts.

These challenges highlight the complexity of using ML for depression detection. However, overcoming these challenges could lead to more accurate and effective tools for diagnosing and treating depression. To measure depression, researchers use different depression scales and questionnaires, which help to measure its effectiveness.

The following section discusses some available Depression scales and questionnaires.

4. DISCUSSION ABOUT DIFFERENT DEPRESSION SCALES

It is very important to discuss different depression scales and questionnaires available for research because this can be a guideline for new researchers. In this section, we have discussed the presently available depression scale, dataset, etc. Following are some available depression scales:

★ **DSM-IV and DSM V:** The DSM-IV questionnaire is a freely available cognitive behavior disorder-related questionnaire. The score has been considered in such a way that it may be accurate and reliable. Based on the rating of their questions answered, they determine the anxiety scale as a low, medium, or highly anxious person. They have also included the neurocognitive disorder in the latest DSM-5.

★ **DASS 21:** It is known as Depression Anxiety Stress Scales, prepared using 42 self-report items that may be completed over five to ten minutes. Each item reflects a negative emotional symptom. Each of these is rated on a four-point Likert scale of frequency or strictness of the participants' experiences over the last week to highlight states over traits (Malik et. al., 2024).

★ **Mental Health Inventory:** It is a method for assessing mental health issues such as anxiety, depression, behavioral disorders, positive effects, and general anxiety disorder. These questionnaires help in the measure of overall emotional functioning. It includes 38 items with a points Likert scale. This test may take approx. 5-6 minutes to be completed (Cuijpers et. al., 2009).

★ **Hospital Anxiety and Depression Scale (HADS)-** The Hospital Anxiety and Depression Scale (HADS), created *by (Zigmond & Snaith, 1983)* is a 14-item measure intended to assess symptoms of anxiety and depression in medical patients, with emphasis on dropping the impact of physical illness on the total score. The depression features tend to focus on the hidden or unnoticed symptoms of depression (Stem, 2014).

★ **PHQ-9 Scale:** The Patient Health Questionnaire is an MCQ-type self-report inventory containing 9 questions, used as a screening and diagnostic tool for different mental health disorders, such as depression, anxiety, alcohol-related disorders, and somatoform disorders. It is a self-report version of the Primary Care Evaluation of Mental Disorders, a diagnostic tool developed in the middle of 1990 by Pfizer Inc. The length of the original assessment limited its feasibility; consequently, a shorter version, consisting of 11 multi-part questions - the Patient Health Questionnaire was developed and validated (Costantini et. al., 2021).

★ **Hamiltonian Anxiety Scale (HAM-A):** It is a widely used and well-validated tool for measuring the strictness of a patient's anxiety. It should be directed by an experienced clinician. The prime value of HAM-A is to evaluate the patient's response to a course of treatment, rather than as a diagnostic or screening tool (Briffault, et. al., 2021).

★ **State-Trait Anxiety Inventory (STAI):** It is a psychological inventory based on a 4-point Likert scale and consists of 40 questions on a self-report basis. This measures two types of anxiety – state anxiety, anxiety related to an event, trait anxiety, or related to personal characteristics. Higher scores are positively correlated with higher levels of anxiety.

The above scales are used in cognitive research. By using them, several questionnaires were prepared to analyze the mental condition of various persons. In the next sub-section, we shall discuss some datasets that are commonly used in cognitive research. Though most of the datasets are not public and free to use, few datasets are available for the researchers.

5. CONCLUSION AND FUTURE SCOPE

This study focuses on the complexities and challenges associated with early depression detection. While traditional machine learning algorithms—such as Support Vector Machines, Decision Trees, and Random Forest can yield impressive accuracy, the focus has increasingly shifted towards advanced deep learning techniques, particularly the integration of CNN-LSTM models, Robotics and Spectrum CNN architectures. Furthermore, the growing interest in explainable AI and generative AI solutions signifies a promising direction for enhancing the precision and transparency of depression detection systems. Additionally, the effectiveness of these algorithms is influenced by the quality of the questionnaires and scales utilized in depression assessments, underscoring the importance of robust measurement tools. Researchers are also exploring the potential of Natural Language Processing (NLP) to identify depressive symptoms through linguistic cues. Given the rising prevalence of depression and mental health disorders in society, there remains significant scope for further research in this domain. This exploration not only highlights the technological advancements in AI but also emphasizes the urgent need for innovative approaches to improve early detection and treatment outcomes for individuals affected by depression.

REFERENCES

https://www.mayoclinic.org/diseases-conditions/depression/symptoms-causes/syc-20356007

Aggarwal, A., Tam, C. C., Wu, D., Li, X., & Qiao, S. (2023). Artificial intelligence–based chatbots for promoting health behavioral changes: Systematic review. Journal of medical Internet research, 25, e40789. 9.

Ahmed, M. S., & Ahmed, N. (2023). A fast and minimal system to identify depression using smartphones: Explainable machine learning–based approach. *JMIR Formative Research*, 7, e28848. DOI: 10.2196/28848 PMID: 37561568

Aleem, S., Huda, N. U., Amin, R., Khalid, S., Alshamrani, S. S., & Alshehri, A. (2022). Machine learning algorithms for depression: Diagnosis, insights, and research directions. *Electronics (Basel)*, 11(7), 1111. DOI: 10.3390/electronics11071111

Briffault, X. (2018). The Hamilton Scale as an Analyzer for the Epistemological Difficulties in Research on Depression. In *Measuring Mental Disorders* (pp. 55–87). Elsevier. DOI: 10.1016/B978-1-78548-305-9.50002-X

Byeon, H. (2023). Advances in machine learning and explainable artificial intelligence for depression prediction. *International Journal of Advanced Computer Science and Applications*, 14(6). Advance online publication. DOI: 10.14569/IJACSA.2023.0140656

Chen, T. Y., Chu, H. T., Tai, Y. M., & Yang, S. N. (2022). Performances of Depression Detection through Deep Learning-based Natural Language Processing to Mandarin Chinese Medical Records: Comparison between Civilian and Military Populations. Taiwanese Journal of Psychiatry, 36(1), 32-38. 12. 12.

Chiong, R., Budhi, G. S., Dhakal, S., & Chiong, F. (2021). A textual-based featuring approach for depression detection using machine learning classifiers and social media texts. *Computers in Biology and Medicine*, 135, 104499. DOI: 10.1016/j.compbiomed.2021.104499 PMID: 34174760

Costantini, L., Pasquarella, C., Odone, A., Colucci, M. E., Costanza, A., Serafini, G., Aguglia, A., Belvederi Murri, M., Brakoulias, V., Amore, M., Ghaemi, S. N., & Amerio, A. (2021). Screening for depression in primary care with Patient Health Questionnaire-9 (PHQ-9): A systematic review. *Journal of Affective Disorders*, 279, 473–483. DOI: 10.1016/j.jad.2020.09.131 PMID: 33126078

Cuijpers, P., Smits, N., Donker, T., Ten Have, M., & de Graaf, R. (2009). Screening for mood and anxiety disorders with the five-item, the three-item, and the two-item Mental Health Inventory. *Psychiatry Research*, 168(3), 250–255. DOI: 10.1016/j.psychres.2008.05.012 PMID: 19185354

Deshpande, M., & Rao, V. (2017, December). Depression detection using emotion artificial intelligence. In 2017 international conference on intelligent sustainable systems (iciss) (pp. 858-862). IEEE.

Ding, X., Yue, X., Zheng, R., Bi, C., Li, D., & Yao, G. (2019). Classifying major depression patients and healthy controls using EEG, eye tracking and galvanic skin response data. *Journal of Affective Disorders*, 251, 156–161. DOI: 10.1016/j.jad.2019.03.058 PMID: 30925266

Duan, L., Duan, H., Qiao, Y., Sha, S., Qi, S., Zhang, X., & Wang, C. (2020). Machine learning approaches for MDD detection and emotion decoding using EEG signals. *Frontiers in Human Neuroscience*, 14, 284.

Frässle, S., Marquand, A. F., Schmaal, L., Dinga, R., Veltman, D. J., Van der Wee, N. J., van Tol, M.-J., Schöbi, D., Penninx, B. W. J. H., & Stephan, K. E. (2020). Predicting individual clinical trajectories of depression with generative embedding. *NeuroImage. Clinical*, 26, 102213. DOI: 10.1016/j.nicl.2020.102213 PMID: 32197140

Frässle, S., Marquand, A. F., Schmaal, L., Dinga, R., Veltman, D. J., Van der Wee, N. J., van Tol, M.-J., Schöbi, D., Penninx, B. W. J. H., & Stephan, K. E. (2020). Predicting individual clinical trajectories of depression with generative embedding. *NeuroImage. Clinical*, 26, 102213. DOI: 10.1016/j.nicl.2020.102213 PMID: 32197140

Kaywan, P., Ahmed, K., Ibaida, A., Miao, Y., & Gu, B. (2023). Early detection of depression using a conversational AI bot: A non-clinical trial. *PLoS One*, 18(2), e0279743. DOI: 10.1371/journal.pone.0279743 PMID: 36735701

Lee, C., & Kim, H. (2022). Machine learning-based predictive modeling of depression in hypertensive populations. *PLoS One*, 17(7), e0272330. DOI: 10.1371/journal.pone.0272330 PMID: 35905087

Mahato, S., & Paul, S. (2020). Classification of depression patients and normal subjects based on electroencephalogram (EEG) signal using alpha power and theta asymmetry. *Journal of Medical Systems*, 44(1), 1–8. DOI: 10.1007/s10916-019-1486-z PMID: 31834531

Malik, A., Bashir, M., Lodhi, F. S., Jadoon, Z. G., Tauqir, A., & Khan, M. A. (2024). Depression, Anxiety and Stress using Depression, Anxiety, and Stress Scoring System (DASS-21) Among the Students of Women Medical and Dental College Abbottabad, Pakistan. [JIIMC]. *Journal of Islamic International Medical College*, 19(2), 103–107.

Marriwala, N., & Chaudhary, D. (2023). A hybrid model for depression detection using deep learning. *Measurement. Sensors*, 25, 100587.

Matsushima, T., Yoshikawa, Y., Matsuo, K., Kurahara, K., Uehara, Y., Nakao, T., Ishiguro, H., Kumazaki, H., & Kato, T. A. (2024). Development of depression assessment tools using humanoid robots- Can tele-operated robots talk with depressive persons like humans? *Journal of Psychiatric Research*, 170, 187–194. DOI: 10.1016/j.jpsychires.2023.12.014 PMID: 38154335

Narziev, N., Goh, H., Toshnazarov, K., Lee, S. A., Chung, K. M., & Noh, Y. (2020). STDD: Short-term depression detection with passive sensing. *Sensors (Basel)*, 20(5), 1396. DOI: 10.3390/s20051396 PMID: 32143358

Priya, A., Garg, S., & Tigga, N. P. (2020). Predicting anxiety, depression and stress in modern life using machine learning algorithms. Procedia Computer Science, 167, 1258-1267. 27. Islam, M. R., Kabir, 27. M. A., Ahmed, A., Kamal, A. R. M., Wang, H., & Ulhaq, A. (2018). Depression detection from social network data using machine learning techniques. *Health Information Science and Systems*, 6(1). Advance online publication. DOI: 10.1007/s13755-018-0046-0

Priya, A., Garg, S., & Tigga, N. P. (2020). Predicting anxiety, depression and stress in modern life using machine learning algorithms. *Procedia Computer Science*, 167, 1258–1267. DOI: 10.1016/j.procs.2020.03.442

Shah, F. M., Ahmed, F., Joy, S. K. S., Ahmed, S., Sadek, S., Shil, R., & Kabir, M. H. (2020, June). Early depression detection from social network using deep learning techniques. In 2020 IEEE region 10 symposium (TENSYMP) (pp. 823-826). IEEE.

Spielberger, C. D. (1983). State-trait anxiety inventory for adults.

Srimadhur, N. S., & Lalitha, S. (2020). An end-to-end model for detection and assessment of depression levels using speech. *Procedia Computer Science*, 171, 12–21. DOI: 10.1016/j.procs.2020.04.003

Stern, A. F. (2014). The hospital anxiety and depression scale. *Occupational Medicine*, 64(5), 393–394. DOI: 10.1093/occmed/kqu024 PMID: 25005549

Thoduparambil, P. P., Dominic, A., & Varghese, S. M. (2020). EEG-based deep learning model for the automatic detection of clinical depression. *Physical and Engineering Sciences in Medicine*, 43(4), 1349–1360. DOI: 10.1007/s13246-020-00938-4 PMID: 33090373

Types of depression. (n.d.). Beyond Blue.(2023) https://www.beyondblue.org.au/mental-health/depression/types-of-depression

Uddin, M. Z., Dysthe, K. K., Følstad, A., & Brandtzaeg, P. B. (2022). Deep learning for prediction of depressive symptoms in a large textual dataset. *Neural Computing & Applications*, 34(1), 721–744. DOI: 10.1007/s00521-021-06426-4

Zhu, J., Wang, Z., Gong, T., Zeng, S., Li, X., Hu, B., Li, J., Sun, S., & Zhang, L. (2020). An improved classification model for depression detection using EEG and eye tracking data. *IEEE Transactions on Nanobioscience*, 19(3), 527–537. DOI: 10.1109/TNB.2020.2990690 PMID: 32340958

Zhu, T., Liu, X., Wang, J., Kou, R., Hu, Y., Yuan, M., Yuan, C., Luo, L., & Zhang, W. (2023). Explainable machine-learning algorithms to differentiate bipolar disorder from major depressive disorder using self-reported symptoms, vital signs, and blood-based markers. *Computer Methods and Programs in Biomedicine*, 240, 107723. DOI: 10.1016/j.cmpb.2023.107723 PMID: 37480646

Chapter 7
Classification of Diabetic Retinopathy Using Regularized Pre-Trained Models

Kalpana Devi
Gandhigram Rural Institute, India

ABSTRACT

Diabetic Retinopathy is a serious sight-threatening complication of diabetes. Deep learning is a superior method for classifying medical images with high accuracy. Typically, we have classified diseases well using transfer learning on pre-trained models from other domains. It utilizes the parameters of a model that has been pre-trained for DR datasets rather than creating new CNN architectures for diverse classification tasks in different domains. The main objective of this chapter is to categorize the Diabetic Retinopathy disease based on its severity using pre-trained models namely ResNet50, VGG16, Alex Net, InceptionV3, Mobile Net and Squeeze Net, DenseNet-121, and XceptionNet and to proposed regularize Xception net experiment with various dropout values.

1. INTRODUCTION

Patients who suffer from diabetic retinopathy have vision loss because of the damage to the blood vessels of the retina, a light-sensitive tissue located at the back of the eye. To diagnose diabetic retinopathy in a patient, a thorough dilated eye exam known as fluorescein angiography is performed in their presence by ophthalmologists and other medical professionals (Lahmar & Idri, 2022). Better

DOI: 10.4018/979-8-3693-6150-4.ch007

Copyright © 2025, IGI Global. Copying or distributing in print or electronic forms without written permission of IGI Global is prohibited.

options are being sought since the existing approach is time-consuming and dependent on the resources that are currently available (Deepa et al., 2022). Initially, diabetic retinopathy may not produce any symptoms and it may eventually result in serious visual impairment. This calls for automated tools for early diagnosis of DR (Kassani et al., 2019).

Recent studies by several academics suggest that deep learning is a superior method for classifying medical images with high accuracy (Jagan Mohan et al., 2021). Typically, researchers have classified diseases well using transfer learning on pre-trained models from other domains (Pamadi et al., 2022). It utilizes the parameters of a model that has been pre-trained for big datasets rather than creating new CNN architectures for diverse classification tasks in different domains (Saranya et al., 2022). This Proposed work describes about the transfer learning based classification of Diabetic Retinopathy using popular pre-trained models namelyResNet50, VGG16, Alex Net, InceptionV3, Mobile Net and Squeeze Net, DenseNet-121, and Xception Net.

2. LITERATURE REVIEW

Lahmar et al. (2022) suggested about diabetic retinopathy work detailed experimental validation of seven Convolutional Neural Networks (CNN) architectures for an automatic binary categorization of the type of referable diabetic retinopathy. It made use of the Inception ResNet V2, Inception ResNet V3, ResNet50, VGG16, VGG19, MobileNet V2, and DenseNet201 architectures. The Borda count voting method and the Scott Knott test were examined and compared to find that they both achieve an accuracy of 93.09% (Lahmar & Idri, 2022).

Deepa et al. (2022) investigated a group of multi-stage classifiers for the automated grading of diabetic retinopathy. For the purpose of precise DR prediction utilizing fundus images, this suggested study develops an ensemble of deep convolutional neural network models. Four separate input images are split into each CNN model (InceptionV3, Xception) for training. To find the pertinent characteristics, CNN models are used shallow-dense layers. The model is assisted in learning the important data from DR images by the mix of shallow and dense layer characteristics. The improvement in total classification accuracy of 89.8% is provided by the suggested ensemble technique of multi-stage deep learning mode (Deepa et al., 2022).

Kassani et al. (2022) developed modeified Exception Architecture aggregates multilevel features from several Xception architecture convolutional layers using deep layer aggregation. The effectiveness of the proposed method was assessed using four deep feature extractors: InceptionV3, Mobile Net, ResNet50, and the original Xception architecture. The integration of deep CNN layers may successfully fuse deep features and boost learning with a 91.5% accuracy rate when compared to traditional Xception architecture (Kassani et al., 2019).

Chilukoti et al. (2022) investigated DR detection using transfer learning with trained CNN. Utilizing pre-trained models like ResNet, VGG, and EfficientNet, the early phases of DR are identified a greater likelihood. Then, Efficient Net models scale the depth, width, and resolution using the compound scaling approaches identify more than one stage of diabetic retinopathy. With a better quadratic weighted kappa score of 0.85 and higher quality images, the next iteration of the Efficient Net b3 model may identify all phases of diabetic retinopathy (Chilukoti et al., 2022).

Mohan et al. (2022) proposed a four-step feature selection approach for evaluating DR. From the retinal fundus images, the features are retrieved using a deep ensemble model that includes InceptionV3, ResNet101, and VGG19. The feature space is condensed and the computational complexity is decreased by combining the features. In order to categorize diabetic retinopathy, Support Vector Machine, a common machine learning classifier, is employed. The proposed algorithm provides accuracy with top 300 features from the input dataset of 97.78%, a sensitivity of 97.6%, and a specificity of 99.3% (Jagan Mohan et al., 2021).

Wang et al. (2022) created an FC Dense Net-based model to improve microaneurysm detection. Fluorescein Fundus Angiography (FFA) pictures were periodically analyzed using FC-Dense Net to detect MAs. The Histogram Stretching and Gaussian filtering techniques were used to pre-process these photos in order to improve the quality of the MA zones found by the improved FC-Dense Net. The effectiveness of the system was compared to MAs-FC-Dense Net, which produces better values for assessment metrics like Mean Intersection over Union (MIoU), Precision (Pre), Recall (Re), and Pixel Accuracy (PA), Mean Pixel Accuracy (MPA), and Precision (Pre) (Wang et al., 2022).

Elsharif et al. (2021) employed deep learning to diagnose the condition. High-resolution cross-sections of the patient's retinas are captured using a variety of imaging modalities, and deep learning classification is also used to the optical coherence tomography pictures to identify the diseased region in the retinal tissue. To diagnose the sickness in an accurate and timely manner, various pre-trained models like VGG-16, Mobile Net, ResNet-50, Inception V3, and Xception are employed. The best model, ResNet-50, achieved a testing accuracy of 96.21% as a result (Elsharif & Abu-Naser, 2022).

Gayathri et al. (2020) presented an explanation of DR classification using Multipath CNN and ML classifiers. The automatic DR severity rating of the characteristics taken from the fundus image is described in this paper. For the purpose of extracting local and global characteristics from the input images, a Multipath CNN is created. The suggested approach exhibits superior performance while using J48 classifiers. For illness grading, an average accuracy of 99.62% is attained (Gayathri et al., 2021).

Zeng et al. 2019 detect the DR has been done by taking the binocular retinal images and learning the similarity of data with the ground truth. For this purpose, transfer learning was employed. The system achieved an AUC of 0.951, a sensitivity of 0.822, and a specificity of 0.707 (Sebastian et al., 2023).

Islam et al. (2018) improved the performance of the CNN architecture for microaneurysms detection by including the optimization techniques. The different pre-processing methods were employed with the optimization of data using augmentation and model using L2 regularization. The system attained an AUC of 0.844, specificity of 0.94, and sensitivity of 0.98 (Nasir et al., 2022).

Wan et al. (2018) implemented transfer learning for DR diagnosis utilizing various pre-trained architectures, including Alexnet, GoogleNet, VGGNet, and ResNet. The model's hyper-parameters were adjusted during transfer learning to enhance the outcomes. The system's accuracy was 0.956 when used with the Kaggle EyePACs dataset (Pamadi et al., 2022).

Li et al. (2019) recommended utilizing fractional pooling rather than maximum pooling for the CNN architecture. Concatenated are the deep CNN model's high-level characteristics. A Support Vector Machine (SVM) is then used to categorize the features. On a scale from 0 to 4, the SVM divided the retinal pictures into 5 classes. The system achieved a classification accuracy of 0.86 for the five classes. The system demonstrated 0.91 accuracy, 0.893 sensitivity, and 0.9089 specificity for two-class categorization (Saranya et al., 2022).

Bilal et al. (2022) identified a two-stage categorization for diabetic retinopathy. To improve the image quality, data pre-processing and data augmentation techniques are applied. For the segmentation of the optic disc and blood vessels, two separate U-net models are combined. The relevant features from the retinal pictures are extracted at the second level using a hybrid CNN model. On the Messidor dataset, this model was evaluated, and the accuracy result was 94.59% (Mudaser et al., 2021).

Dayana et al. (2022) Classification of DR severity using a deep neural network based on the Chronological Tunicate Swarm Algorithm. Segmentation of the optic disc and blood artery is performed using U-Net and a sparse Fuzzy C-means-based hybrid entropy model. The region of interest is located using gabor filter banks, and then feature extraction is performed. For final classification, a Deep Stacked Auto encoder tuned with a Tunicate Swarm Algorithm inspired by biological systems is used. On the DIARETDB0 database, this model had a 95.9% accuracy rate (Sudarmadji et al., 2020).

Najib et al. (2022) suggested a deep neural network and genetic algorithm-based feature selection for the classification of DR. For feature extraction, five cutting-edge convolutional neural network architectures, including AlexNet, NASNet-Large, VGG-19, Inception V3, and ShuffleNet, are used. A genetic algorithm is used to refine the extracted features for feature ranking and feature selection. The network then removes the suboptimal feature vectors. The final step in identifying diabetic retinopathy uses support vector machines. On the Kaggle dataset, the model's performance was measured using accuracy, and it had a 97.9% accuracy rate (Sanjana et al., 2021).

Alyoubi et al. (2021) proposed a Deep Learning model for DR Classification. This model comprised of two models namely CNN512 and YOLOV3. Retinal image was preprocessed using Contrast Limited Adaptive Histogram Equalization (CLAHE) followed by image cropping and colour normalization. The enhanced image was forwarded to the CNN512 model for computation. CNN512 model was a custom CNN architecture with set of convolutional and pooling layers. YOLOV3 model worked on lesion localization on the retinal images. This automated DR diagnosis system was tested for disease severity on APTOS Kaggle 2019 public datasets and performance was evaluated based on accuracy, sensitivity and specificity. Fused version of these models achieved an accuracy of 89%, sensitivity 89%, and 97.3% specificity outperforming the current state-of-the-art results (Hossen et al., 2020).

Ai et al. (2021) proposed deep ensemble learning and attention mechanism based Diabetic Retinopathy Classification. Retinal images were pre-processed at the initial stage to enhance the image quality. A model named DR-IIXRN was developed to fine tune the algorithm for DR detection. The DR-IIXRN model consisted of an Inception V3, Inception ResNet V2, Xception, ResNeXt101, and NASNet Large. The network model for every classifier was fine-tuned and a weighted voting algorithm was applied to determine the DR severity. DR-IIXRN was tested on a private hospital dataset that achieved the AUC, accuracy, and recall rate of 95%, 92%, and 92% respectively (Khaled et al., 2020).

Taufiqurrahman et al. (2020) proposed a cascaded Deep Learning model for DR Classification. In this model, Long Short-Term memory and Convolutional Neural Network (CNN) were cascaded to detect the lesions in the retinal image. Preprocessing steps include image cropping followed by Histogram Equalization to improve the contrast for enhanced visibility of lesions. Output from the Convolutional Neural Network was forwarded as input to the Long Short-Term memory for further processing. Convolutional Neural Network in this model was trained using Adam Optimizer for 50 epochs and a learning rate of 0.01. Experimental analysis of this model on the subset of Messidor dataset showed an accuracy of 90% (Taufiqurrahman et al., 2020).

Ghoushchi et al. (2021) developed a computer diagnostic method to segment diabetic retinopathy using fluoresces in images. Fuzzy C-means and Genetic Algorithm are used to predict diabetic retinopathy from image segments. Fuzzy C clusters retinal pixels by homogeneity or resemblance. Region growth algorithm determined starting seeds, similarity regions, and growth regions. This method finds DR-symptom-prone retinal areas. Genetic Algorithm refines these spots to create

DR-like spots. The algorithm was used on 224 eye pictures. Experimental results showed a 0.78 sensitivity advantage for this strategy (Lee & Ke, 2021).

Zago et al. (2020) employed Patch-based deep CNN (PCNN) for lesion localisation. Transfer Learning classified DR patches from photos. The model was simplified by training with circular template matching retinal image regions. Pre-processing with Gaussian filter-based convolution enhanced lesions in the retinal image. Retinal patches were given to VGGNet-16 for DR classification after pre-processing. VGGNet-16 rotates and processes patches at 0, 45, 135, -45, and -135. The lesion probability map averages the model's lesions predictions for different rotations. The lesion probability map was utilized to infer DR stage from input data. VGGNet-16 was tested with different filter sizes. DIARETDB1, Kaggle, and Messidor datasets validated the model. This model outperforms others in extensive database experiments. This method had 94% Messidor dataset sensitivity. This method had 90% sensitivity and 87% specificity in Messidor-2 (Nguyen et al., 2020).

Samanta et al. (2020) developed Transfer Learning-based CNN architecture for color fundus photography DR categorization. This model was run on 3050 training and 419 validation photos. This lightweight model works well for small real-time applications with limited computer capability and speeds up DR screening. CLAHE contrast enhancement with a clip limit of 2.0 adjusted an image's bright or dark pixel values to reveal hidden features. The contrast boosted image was improved with weighted Gaussian blur. CNN DR recognition used this preprocessed image. This CNN model was pre-trained using ImageNet DenseNet121. This model was modified by adding two 1024-layer completely connected layers. No DR, Mild DR, Moderate DR, and Proliferative DR were classified using this approach. The Kaggle dataset included 3050 training and 419 validation photos for this model. During validation, this DenseNet121 version scored 0.8836 Cohens Kappa (Wang et al., 2020).

Tymchenko et al. (2020) suggested a Deep Learning method for DR classification. Cropping and resizing photos preprocessed this work. Augmentation reduced correlation and classification overfitting. This work used grid distortion, horizontal flip, vertical flip, random rotation, random shift, additive Gaussian noise, blurring, and sharpening. Three CNN heads—classification, regression, and ordinal regression—classified the preprocessed image. Classification head detects DR, regression head describes its stage, and ordinal regression head computes DR categories from the pre-processed image. The linear regression model classified DR stages using the three decoders' output. APTOS 2019 Blindness Detection Dataset study yielded a quadratic weighted kappa score of 0.925466 (Shankar et al., 2020).

Nguyen et al. (2020) developed an automated classification approach for fundus image analysis that graded Diabetic Retinopathy severity using images with different fields of view and illumination. Image scaling, background noise removal, and data augmentation were pre-processed. The pre-processed image was computed using CNN, VGG-16, and VGG-19 Machine Learning models. Automated DR screening will expedite identification and decision-making, helping to manage DR progression. EyePACS dataset evaluated this model. The final classification layer employed ensemble-based classifiers with SVM and Random Forest. DR stage categorization has 80% sensitivity, 82% accuracy, 82% specificity, and 0.904 AUC (Karki & Kulkarni, 2021).

Shaban et al. (2020) proposed a Deep Convolutional NeuralNetwork-based architecture for DR stage classification. Pre-processing includes image resizing and data augmentation. This model consisted of 18 convolution and 3 fully connected layers for analysing the fundus images to check the presence of DR. Cross fold validations were performed on the Kaggle dataset. Experimental analysis performed on the Kaggle dataset indicates a quadratic weighted kappa score of 0.91 (Nazir et al., 2021).

Mishra et al. (2020) focused on analysing different DR stages using a Deep Learning strategy which is considered as the subset of Artificial Intelligence. Dense Net model was used in this work to process the retinal data. A Dense Net Architecture was trained to extract the discriminative features of retinal image. The extracted features were further processed using Activation Function to generate the relevant retinal feature vector. This DenseNet model was trained using APTOS dataset which consisted of 3662training images that helps to automatically detect the DR stage. This model when tested on APTOS dataset produced an accuracy of 0.9611 towards DR Classification.

Qomariah et al. (2019) developed a Deep Learning SVM-based feature extraction and classification approach. Retinal pictures were reduced to 512×512 for CNN architecture. VGGNet, Alexnet, Resnet, and inception architectures get this cropped image at varying sizes. SVM classification employed the high-level features of the last fully connected layer obtained by CNN Transfer Learning. Authors evaluated this approach on two Messidor database subsets. Those datasets had 95.83% and 95.24% accuracy.

3.MATERIALS

3.1 Experiment Set-up

The experiments are executed on intel core is i3 3.40GHz with 4GB RAM, running windows 10 64 bit operating system and implemented using Google Colab.

3.2 Datasets

IDRID: The IDRID dataset offers ground facts about Diabetic Macular Edema (DME), Diabetic Retinopathy (DR), and healthy retinal architecture. It includes the attribute data for pixel-level annotations of typical DR lesions and the optic disc, DME severity grading at the image level, and coordinates for the fovea and optic disc. Each CSV file has three columns that contain the image number, the X co-ordinate, and the Y co-ordinate. The X and Y co-ordinates correspond to the image's OD center Fovea's pixels.

APTOS: The normal retinal structures linked with all of the Diabetic Retinopathy (DR) symptoms are provided by the APTOS dataset. A doctor graded each image for the severity of diabetic retinopathy on a scale from 0 to 4, where the numbers signify the degree of the issue. It provides the attribute information for each image like No DR, Mild, Moderate and Severe. It contains training set images and test images in the separate CSV file with 13,000 images respectively (Sanjana et al., 2021).

Figure 1. Proposed Methodology

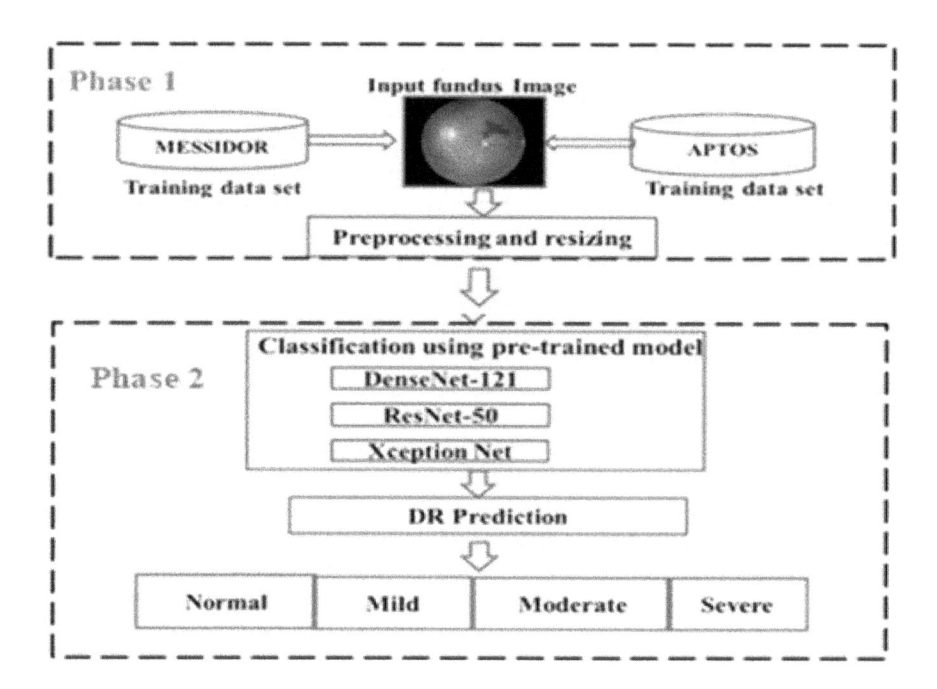

The proposed model is developed with the implementation of the following phases.

3.3 Pre-Processing

The aim of this phase is to resize the dataset before processing the retinal fundus image dataset as part of the initial phase. Larger images require more memory and calculation processes per layer (Hossen et al., 2020). The input images must be scaled for the model to function better and for the training model to learn the characteristics more quickly. In this phase, the images of the DR dataset are scaled down to 224×224×3 for VGG16, 244×244×3 for VGG19, 150×150×3 for Inception v3, 224×224×3 for Mobile Net V2, ResNet 50, Dense Net, and 299×299×3 for Xception Net.

3.4 Classification using Pre-Trained Models

3.4.1 VGG 16

VGG 16 is the CNN architecture (Nasir et al., 2022) designed by Visual Geometry Group (VGG), University of Oxford. Input image size in this model is generally 224×224 but can be varied with a minimum size of 48×48. Filters used are of size 3×3. It has 13 convolution layers with filters of varying sizes in the range [64, 128, 256, 512]. It also has 3 dense layers with nodes equal to 4096, 4096 and 1000 nodes respectively. The most popular activation function ReLU is used in all the layers. It has pre-defined ImageNet weights, which have been trained on ILSVRC (Chilukoti et al., 2022).

3.4.2 VGG -19

A convolutional neural network with 19 layers is called VGG-19. The ImageNet database contains a pre-trained version of the network that has been trained on more than a million images. Using 1000 different item categories, the pre-trained network can classify images. The 224x224x3 DR fundus image are utilized as the input convolutional layer for the VGG-19, while Max pooling layers were employed as the handler. There is a convolution layer and a ReLU layer in the CNN block structure (Mudaser et al., 2021). The second block, however, consists of two convolution layers, two ReLU layers, a batch normalization layer, a maximum pooling layer, and finally a dropout layer. Convolutional layers were utilized for the feature extraction during the training phase, and some of the convolutional layers have max pooling layers attached to them to lessen the dimensionality of the features (Khaled et al., 2020). VGG19 is used for the feature extraction, which is followed by two CNN blocks that serve as the feature extraction component. The output of the feature extraction layer is then sent to a flatten layer as the first step in classification, which converts the data form into a one-dimensional data vector.

A dense layer with 512 neurons in the classification component follows a dropout layer. A dense layer with four neurons and the SoftMax activation function creates the final output, which categorizes the output image into one of two classifications for DR: healthy and MA images. There are 256 non-trainable parameters and 22,337,604 trainable parameters, making up the total number of parameters, which is 22,337,860. Trainable parameters refer to variables that are modified during training and are necessary for training to achieve the parameters' ideal values, while the non-trainable parameters refer to variables that are not modified during training. In other words, non-trainable model parameters are those that must be determined beforehand or provided as inputs because they won't be updated and optimized during training. As a result, the non-trainable will not influence categorization.

3.4.3 InceptionV3

Inception V3 is a 42-layer deep learning network with fewer parameters. The reduction in parameters is done with the help of factorizing convolutions. For example, a 5×5 filter convolution can be done by two 3×3 filter convolutions. The parameters in this process reduce from 25 to 18, thus, accomplishing 28% reduction in the number of parameters. With smaller number of parameters, the chances of model over fitting are decreased while maintaining reasonable accuracy (Nazir et al., 2021).

3.4.4 MobileNetV2

MobileNet architecture uses depth-wise separable convolutions for developing lightweight deep convolution neural networks (Shankar et al., 2020). Two global hyper parameters width multiplier and resolution multiplier provide efficient tradeoffs between accuracy and storage space. It is designed for usage on mobile devices, which have less storage (Pamadi et al., 2022). Mobile net has 27 convolution layers followed by an average pooling layer and then a fully connected layer to transform the 2-D signal to 1-D and a final SoftMax output layer (Taufiqurrahman et al., 2020).

3.4.5 Densenet-121

A CNN design that excels in category identification and training is called DenseNet-121. It got its name as it is made up of tightly linked dense blocks and 121 layers. The feature maps from the previous layer are applied to the subsequent layer (Wang et al., 2022). The architecture consists of three essential building parts.

The main job of a dense block is to concatenate the inputs. Batch normalization, the "ReLu" function, and convolution layers make up the three levels of the convolution block (Karki & Kulkarni, 2021). The tightly linked blocks account for the minimal training and test loss. A DenseNet block connection's design is shown in Figure 2.Even though the DenseNet architecture has several advantages, the recommended method could not produce noticeable improvements over the current systems. The VGG-16 net model architecture outperforms the DenseNet-121 design in terms of performance (Saranya et al., 2022). The model's execution time is 86537.263 seconds, which is a lot longer than the time required by the VGG-16 net model. DenseNet-121 model uses an image size of 224*224*3. Table 1 depicted the hyper-parameters for DenseNet-121.

Figure 2. DenseNet Architecture

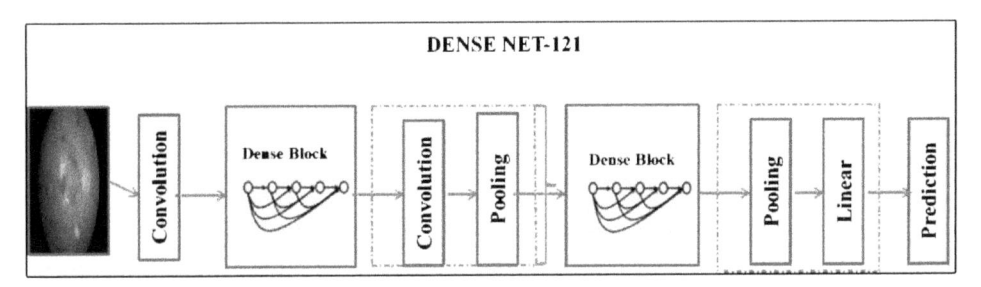

3.4.6 ResNet-50

A pre-trained CNN with 50 layers and a 1000 class ImageNet dataset is called ResNet-50. There are five stages, one convolution block in each, plus a few identity blocks. Each identity block and each convolution block have three convolution layers. The identity and convolution blocks' last layer of filters has size between 256 and 2048. By padding the images with zeros on each of the four sides, transfer learning has been employed in this thesis to train ResNet-50 on the MNIST, CIFAR-10, and CIFAR-100 datasets (Nguyen et al., 2020).

Figure 3 provides a graphic illustration of it. ResNet-50 model used an image size of 224*224*3.Table 8.2 illustrated the list of hyper-parameters for ResNet50.

Figure 3. Architecture of ResNet-50

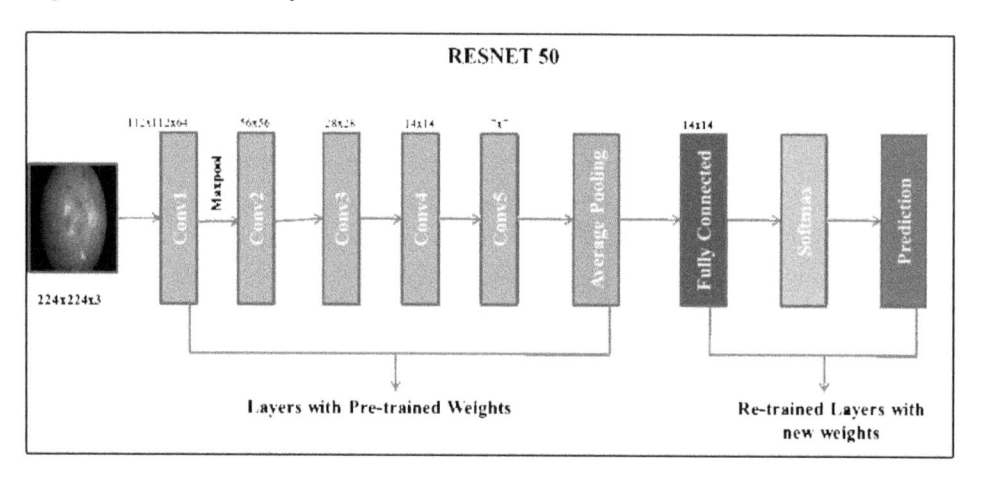

3.4.7 Xception Net

The Xception architecture is an expansion of the Inception design. A linear stack of depth-wise separable convolution layers with residual connections makes up this design. The depth-wise separable convolution attempts to lower memory and computational costs. With the exception of the first and last modules, all of the 14 modules in Xception's 36 convolutional layers have linear residual connections. The learning of channel-wise and space-wise features is separated by the separable convolution in Xception. By adding a shortcut to the sequential network, the problems of vanishing gradients and representational bottlenecks are overcome (Kassani et al., 2019).

This shortcut connection uses a summing operation rather than a concatenation to make the output of an earlier layer available as input to the later layer. The Xception Net's architecture is shown in Figure 8.4 and it uses an image size of 229*229*3. Table 8.3 depicted the hyper-parameters for Xception Net.

Figure 4. Xception Architecture

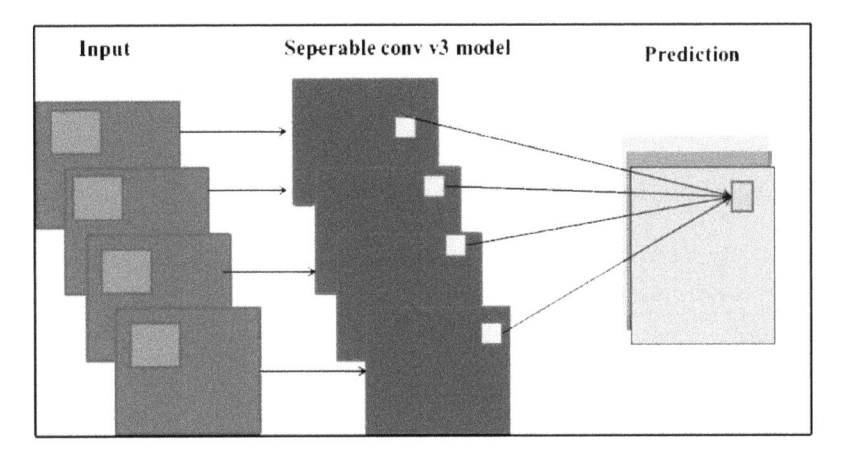

3.4.8 Proposed Regularized Pre-trained models

The Xception is composed of a linear stack of residually connected depth-wise separable convolution layers. It is an improvement over the Inception that uses distinct depth convolutions in place of the standard Inception modules. The model size of Xception and Inception-v3 is the same. After the initial operation, there is a non-linearity in the original Inception module. There is no intermediate ReLU non-linearity in Xception.

The Xception Net model then was trained to optimize the following parameters:

- The number of 2D Convolutional filters represented by the activation of the Xception Net (Conv2d)
- The filter size, or kernel size
- Initial strides, which refer to the filter's step size
- Number of the separable filters
- Number of the residual blocks which denotes the number of shortcut connections of ResNets
- Pooling type
- Number of dense layers in the fully connected layer
- Dropout rate - regularization strategy that stops network over-fitting and involves randomly dropping certain neurons from the hidden layer during training.
- Number of batch normalization of dense layer, which is a crucial training parameter. Instead of delivering the entire input package to the network during

training, the input is randomly divided into several chunks of equal size. Training the data on batches results in a more generic model than training the network with the entire input dataset.
- Learning rate

Table 1 provides an illustration of the variety of hyper parameter combinations utilized for Xception Net's Random Search optimization.

Table 1. Hyper-parameters for Xception Net

Hyper parameter	Range
Activation of Xception	['relu', 'selu']
No. of Conv2d Filters	[32, 64, 128]
Kernel Size	[3, 5]
Initial Strides	[2]
Separable_ No. Filters	[max:768, min:128, step:128]
No. of Residual Blocks	[max:8, min:2, step:1]
Pooling	['avg', 'flatten', 'max']
No. of Dense Layers	[max: 3, min:1, step:1]
Dropout Rate	[max: 0.8, min:0, step: 0.1]
No. of Batch Normalization of Dense Layer	[values: [1, 0]]
Learning Rate	[values: [0.001, 0.0001, 1e-05]]

The dropout percentage ranges between 0 and 0.8. With each value, we get different outcomes. Regularizing dropouts is a general strategy. The majority of neural network models—possibly all—can be utilized with it, not least the most popular network types, Multilayer Perceptrons, Convolutional Neural Networks, and Long Short-Term Memory Recurrent Neural Networks. The chance of training a specific node in a layer is the default meaning of the dropout hyper parameter, where 1.0 denotes no dropout and 0.0 denotes no outputs from the layer. The acceptable range for dropout in a hidden layer is 0.5 to 0.8. The dropout rate for input layers is higher; typically, 0.8. Minimizing a regularized network is done using equation 8.1.

$$E_R = \tfrac{1}{2}\left(t - \sum_{i=1}^{n} p_i w_i I_i\right)^2 + \sum_{i=1}^{n} p_i(1 - p_i) w_i^2 I_i^2 \text{ (Eq.1)}$$

Where p is dropout rate, w is represents weight, and I represents input layer. Larger networks frequently find it easier to over fit the training set.

With dropout regularization, there is a lower chance of over fitting when employing bigger networks. In reality, while dropout would probabilistically lower network capacity, a bigger network (more nodes per layer) may be necessary. As a rule, the number of nodes in the new network that employ dropout is calculated by dividing the number of nodes in the layer preceding dropout by the recommended dropout rate. When utilizing dropout, for instance, a network with 100 nodes and a projected dropout rate of 0.5 will need 200 nodes (100 / 0.5).After training, dropout is not employed when making a prediction using the fit network.

Dropout will cause the network's weights to be higher than usual. Hence, the weights are initially scaled by the selected dropout rate before finishing the network. The network may then be utilized to make predictions as usual.

4. RESULTS AND DISCUSSION

The experiments shows are run using keras with the dataset APTOS and MESSIDOR 2 using XceptionNet. We tried different values for dropout and observed that better performance is achieved for dropout value 0.5.

Table 2 depicts the classification results for pre-trained models for APTOS dataset. From this table, it is observed that the proposed Xception net achieves 0.95% for precision, 0.95% recall, 0.86% for f1-score with 10000-support value. Figure 5 illustrates the pictorial representation of classification results of APTOS dataset for an optimal dropout value of 0.5.

Table 2. Classification Result for Pre-trained model using APTOS dataset

Model	Precision	Recall	F1-score	Support
VGG-16	0.83	0.91	0.92	8950
VGG-19	0.84	0.92	0.87	8900
Mobile Net V2	0.86	0.94	0.83	9000
Inception V3	0.89	0.93	0.96	9000
Densenet-121	0.90	0.97	0.97	9140
Resnet-50	0.92	0.77	0.78	10000
Xception net	0.89	0.78	0.79	10000
Proposed Regularized Xception Net	**0.95**	**0.95**	**0.86**	**10000**

Figure 5. Classification Result for Pre-trained model using APTOS dataset

[Bar chart: Classification Result for Pre-trained model using APTOS dataset, showing Precision, Recall, F1-score for VGG-16, VGG-19, Mobile Net V2, Inception V3, Densenet-121, Resnet-50, Xception net, Proposed...]

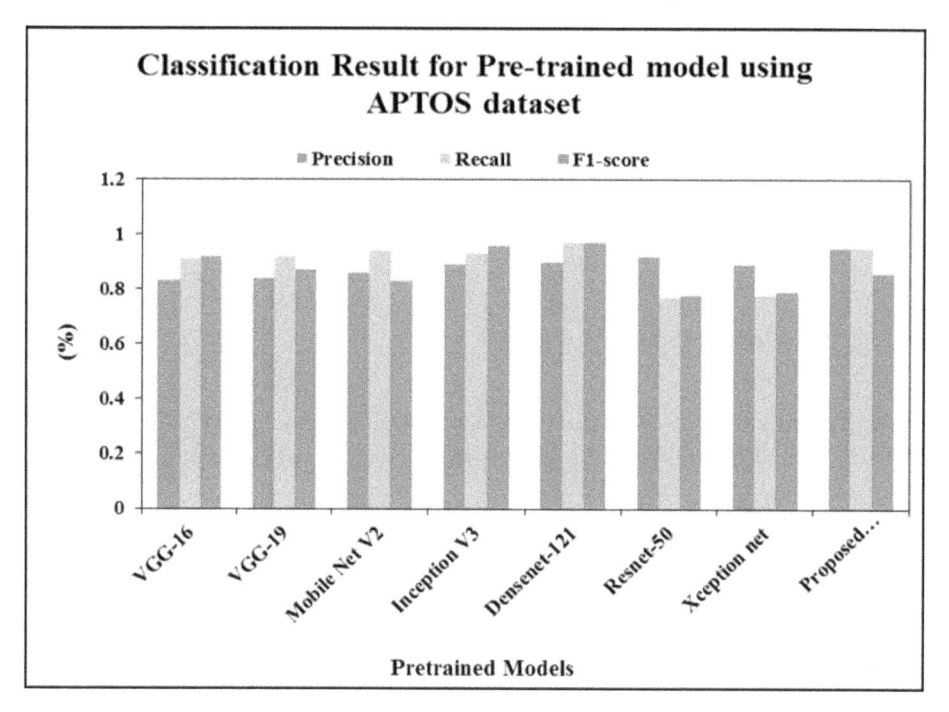

Table 5 depicts the classification results for pre-trained models for MESSIDOR2 dataset.

Table 3. Classification Result for Pre-trained model using MESSIDOR 2 dataset

Model	Precision	Recall	F1-score	Support
VGG-16	0.85	0.93	0.92	8950
VGG-19	0.86	0.92	0.87	8900
Mobile Net V2	0.88	0.97	0.83	9000
Inception V3	0.89	0.93	0.96	9000
Densenet-121	0.85	0.89	0.91	9140
Resnet-50	0.84	0.80	0.85	10000
Xception net	0.82	0.77	0.80	10000
Proposed Regularized Xception net	**0.94**	**0.96**	**0.89**	**10000**

From this table it is observed that the proposed Xception net achieves 0.94% for precision, 0.96% recall, 0.89% for f1-score with 10000-support value.

Figure 6 illustrates the chart representation of classification results of MESSIDOR 2dataset for the standard architectures and the proposed regularized Xception Net.

Figure 6. Classification Result for Pre-trained model using MESSIDOR 2 dataset

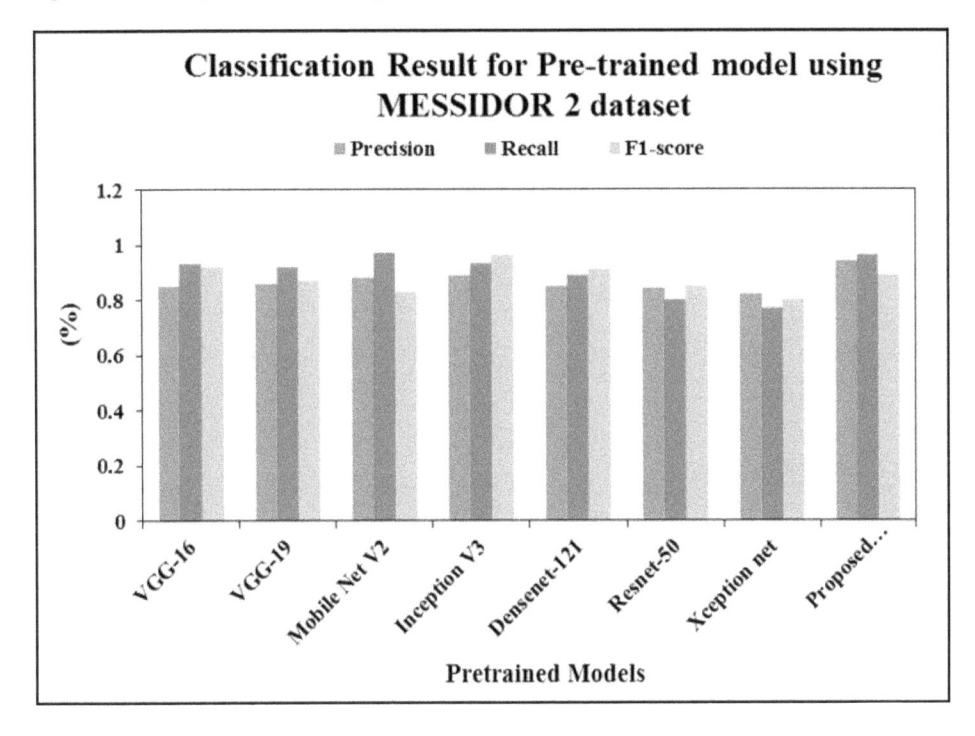

Table 6 shows the distribution of APTOS dataset, which describes the stages of diabetic retinopathy disease. These results are achieved using the proposed regularized Xception net with various dropout values.

Table 4. The distribution of APTOS dataset

Label	Stage of DR	Train set	Valid set	Test set	Total
0	No DR	1383	251	171	1805
1	Mild DR	273	52	45	370
2	Moderate DR	773	126	100	999
3	Servere DR	152	26	15	193
4	Proliferative DR	220	39	36	295
Total	APTOS dataset	2801	494	367	3662

Table 4 reveals that 1805 images are identified as normal DR, 370 images as mild DR, 999 images as moderate DR, 193 images as severe DR, and 295 images are classified as proliferative DR. This is shown in figure 7.

Figure 7. The distribution of APTOS dataset

The Distribution of APTOS dataset

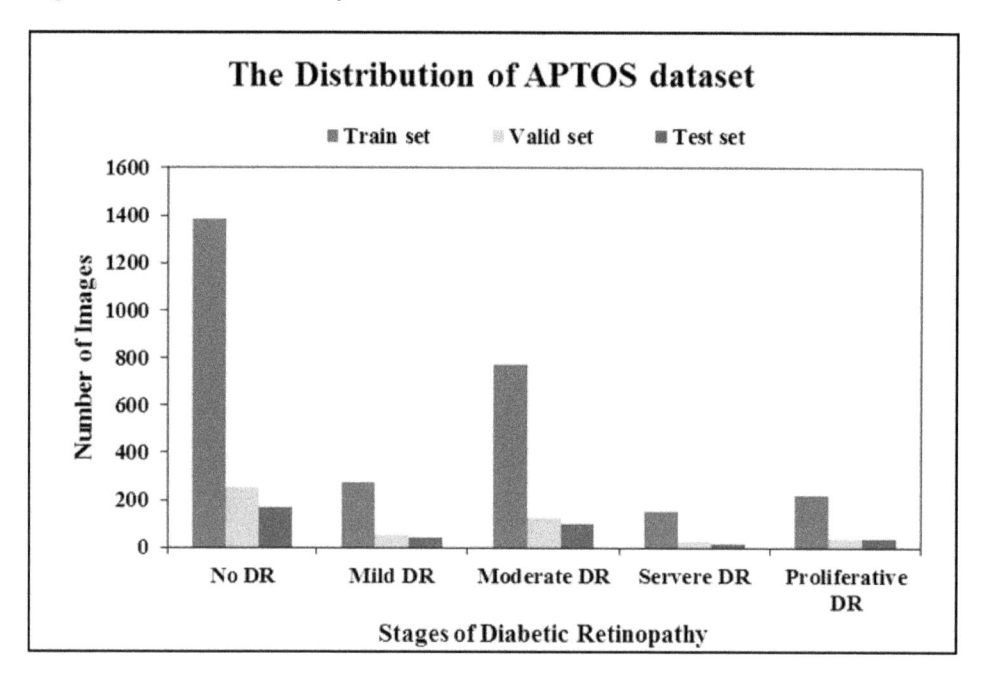

Table 5 shows the distribution of MESSIDOR2 dataset, which describes the stages of diabetic retinopathy disease. These results are achieved using the proposed regularized Xception net with various dropout values.

Table 5. The distribution of MESSIDOR 2 dataset

Label	Stage of DR	Train set	Valid set	Test set	Total
0	No DR	650	100	267	1017
1	Mild DR	150	50	70	270
2	Moderate DR	200	100	47	347
3	Severe DR	40	20	15	75
4	Proliferative DR	20	10	5	35
Total	MESSIDOR 2Dataset	1060	280	404	1744

It is observed that 1017 images are classified as normal DR, 270 images are classified as mild DR, 347 images are identified moderate DR, 75 images are classified as severe DR, and 35 images are classified as proliferative DR. The results are shown in figure8.

Table 6 shows the accuracy of pre-trained models with various dropout values (0.1 – 0.5) for APTOS and MESSIDOR datasets. It is obvious that the proposed regularized Xception Net achieves 95% accuracy with 0.5 dropout value for APTOS dataset and 94% accuracy for MESSIDOR 2 dataset.

Table 6. Dropout Values

Datasets	Pre- Trained Models	Dropout Values (0.1 -0.5)	Accuracy (%)
APTOS	VGG-16	0.3	84
	VGG-19	0.4	86
	Mobile Net V2	0.4	92
	Inception V3	0.4	90
	Densenet-121	0.4	92
	Resnet-50	0.5	91
	Xception Net	0.4	93
	Proposed Regularized Xception net	0.5	95
MESSIDOR 2	VGG-16	0.3	0.85
	VGG-19	0.4	0.87
	Mobile Net V2	0.4	0.88
	Inception V3	0.4	0.91
	Densenet-121	0.4	0.938
	Resnet-50	0.5	0.936
	Xception net	0.4	0.936
	Proposed Regularized Xception net	0.5	0.94

Figure 8. The distribution of MESSIDOR 2 dataset

The distribution of MESSIDOR 2 dataset

[Bar chart showing Number of Images across Stages of Diabetic Retinopathy (No DR, Mild DR, Moderate DR, Servere DR, Proliferative DR) for Train set, Valid set, and Test set]

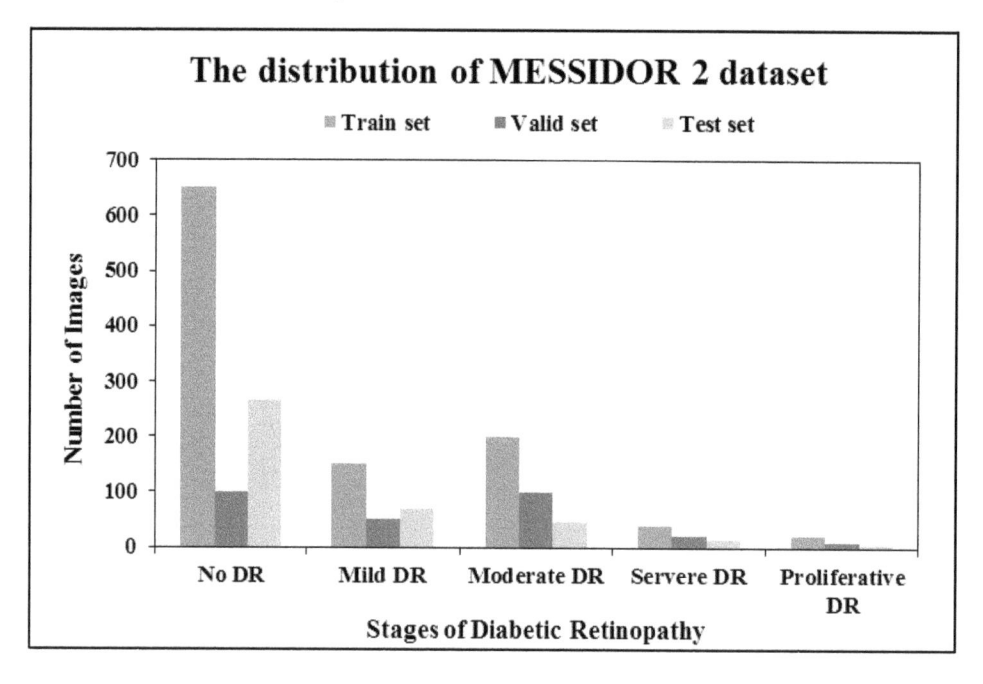

Table 7 depicts the classification result of each classifier in the test phase for APTOS dataset. From this analysis, the proposed pre-trained model achieves 95% accuracy than other pre-trained models, which is shown in figure 9.

Table 7. The classification result of each classifier in the test phase for APTOS dataset

Model	VGG-16	VGG-19	Mobile Net V2	Inception V3	Densenet-121	Resnet-50	Xception net	Proposed Regularized Xception net
Accuracy	0.84	0.86	0.88	0.90	0.92	0.91	0.93	**0.95**
Macro recall	0.79	0.74	0.79	0.82	0.78	0.77	0.78	**0.85**
Macro precision	0.73	0.80	0.81	0.85	0.81	0.80	0.81	**0.86**
Weighted f1-score	0.83	0.84	0.87	0.8	0.83	0.85	0.82	**0.84**
Macro f1-score	0.79	0.78	0.76	0.83	0.79	0.78	0.79	**0.85**

Figure 9. The classification result of various pre-trained models in the test phase for APTOS dataset

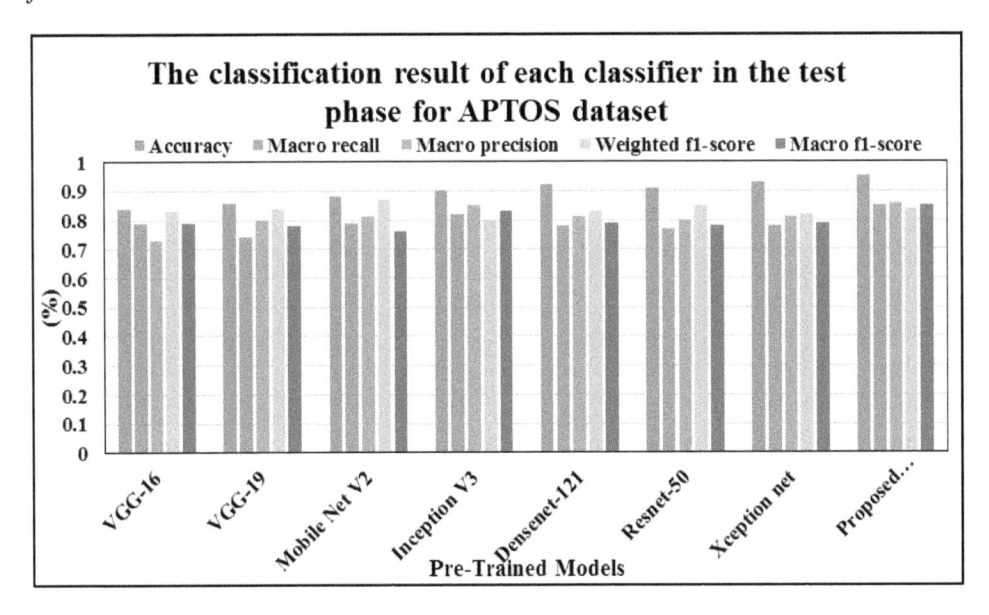

Table 8 depicts the classification result of each classifier in the test phase for MESSIDOR 2 dataset.

Table 8. The classification result of each classifier in the test phase for MESSIDOR 2 dataset

Model	VGG-16	VGG-19	Mobile Net V2	Inception V3	Densenet-121	Resnet-50	Xception net	Proposed Regularized Xception net
Accuracy	0.85	0.87	0.88	0.91	0.9383	0.9359	0.9361	**0.94**
Macro recall	0.81	0.76	0.79	0.82	0.78	0.77	0.78	**0.83**
Macro precision	0.73	0.81	0.81	0.87	0.81	0.80	0.81	**0.87**
Weighted f1-score	0.83	0.84	0.82	0.86	0.94	0.93	0.94	**0.95**
Macro f1-score	0.79	0.78	0.79	0.83	0.79	0.78	0.79	**0.83**

Figure 10. The classification result of pre-trained models in the test phase for MESSIDOR 2 dataset

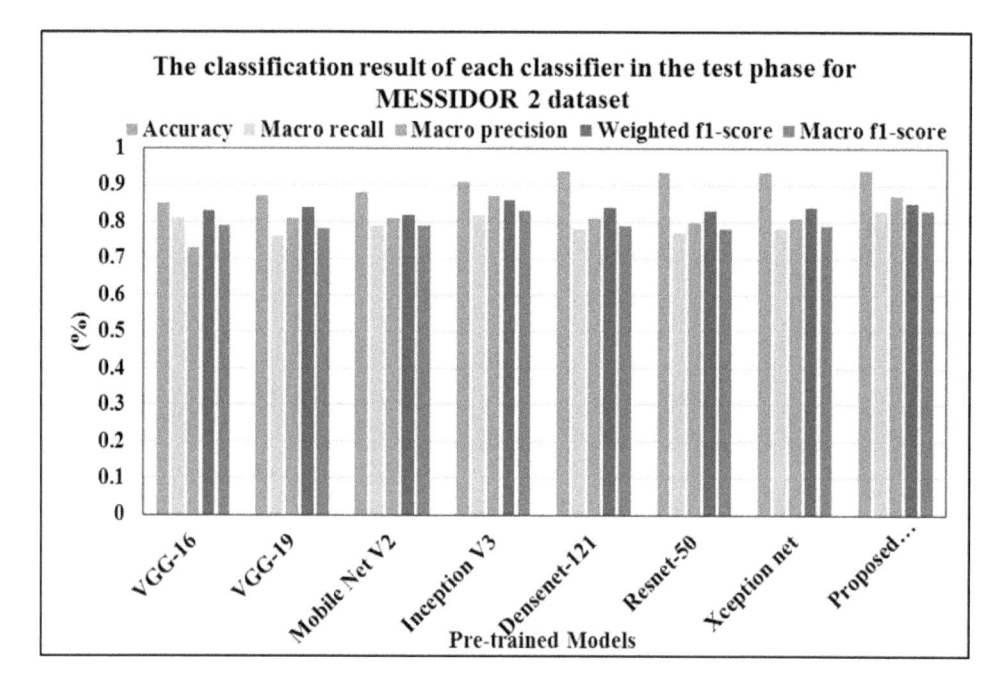

From this analysis, it is observed that the proposed pre-trained model achieves 95% accuracy than other pre-trained models, which shows in figure 10.

5. CONCLUSION

The main aim of this research work is to classify the diabetic retinopathy disease using pre-trained models in deep learning. This paper discussed the performance of the existing and proposed pre-trained models to categorize the DR disease based on its severity. The proposed classification models were experimented with dropout values. The results reveal that the proposed regularized net outperforms existing pre-trained models in terms of accuracy for grading diabetic retinopathy.

REFERENCES

Chilukoti, S. V., Maida, A., & Hei, X. (2022). Diabetic Retinopathy Detection using Transfer Learning from Pre-trained Convolutional Neural Network Models. DOI: 10.36227/techrxiv.18515357

Deepa, V., Kumar, C. S., & Cherian, T. (2022). Ensemble of multi-stage deep convolutional neural networks for automated grading of diabetic retinopathy using image patches. *Journal of King Saud University. Computer and Information Sciences*, 34(8), 6255–6265.

Elsharif, A. A. E. F., & Abu-Naser, , S. S. (2022). Retina Diseases Diagnosis Using Deep Learning. *International Journal of Academic Engineering Research*, 6(2).

Gayathri, S., Gopi, V. P., & Palanisamy, P. (2021). Diabetic retinopathy classification based on multipath CNN and machine learning classifiers. *Physical and Engineering Sciences in Medicine*, 44(3), 639–653. DOI: 10.1007/s13246-021-01012-3 PMID: 34033015

Hossen, M. S., Reza, A. A., & Mishu, M. C. (2020, January). An automated model using deep convolutional neural network for retinal image classification to detect diabetic retinopathy. In *Proceedings of the International Conference on Computing Advancements* (pp. 1-8).

Jagan Mohan, N., Murugan, R., Goel, T., Mirjalili, S., & Roy, P. (2021, December). A novel four-step feature selection technique for diabetic retinopathy grading. *Physical and Engineering Sciences in Medicine*, 44(4), 1351–1366. DOI: 10.1007/s13246-021-01073-4 PMID: 34748191

Karki, S. S., & Kulkarni, P. (2021, March). Diabetic retinopathy classification using a combination of efficientnets. In *2021 International Conference on Emerging Smart Computing and Informatics (ESCI)* (pp. 68-72). IEEE.

Kassani, S. H., Kassani, P. H., Khazaeinezhad, R., Wesolowski, M. J., Schneider, K. A., & Deters, R. (2019, December). Diabetic retinopathy classification using a modified xception architecture. In 2019 IEEE international symposium on signal processing and information technology (ISSPIT) (pp. 1-6). IEEE.

Khaled, O., El-Sahhar, M., El-Dine, M. A., Talaat, Y., Hassan, Y. M., & Hamdy, A. (2020, November). Cascaded architecture for classifying the preliminary stages of diabetic retinopathy. In *Proceedings of the 9th International Conference on Software and Information Engineering* (pp. 108-112).

Lahmar, C., & Idri, A. (2022)... *Computer Methods in Biomechanics and Biomedical Engineering. Imaging & Visualization*.

Lee, C. H., & Ke, Y. H. (2021, June). Fundus images classification for diabetic retinopathy using deep learning. In *Proceedings of the 13th International Conference on Computer Modeling and Simulation* (pp. 264-270).

Mudaser, W., Padungweang, P., Mongkolnam, P., & Lavangnananda, P. (2021, December). Diabetic retinopathy classification with pre-trained image enhancement model. In 2021 IEEE 12th Annual Ubiquitous Computing, Electronics & Mobile Communication Conference (UEMCON) (pp. 0629-0632). IEEE.

Nasir, N., Oswald, P., Alshaltone, O., Barneih, F., Al Shabi, M., & Al-Shammaa, A. (2022, February). Deep DR: detection of diabetic retinopathy using a convolutional neural network. In 2022 Advances in Science and Engineering Technology International Conferences (ASET) (pp. 1-5). IEEE.

Nazir, T., Nawaz, M., Rashid, J., Mahum, R., Masood, M., Mehmood, A., Ali, F., Kim, J., Kwon, H. Y., & Hussain, A. (2021). Detection of diabetic eye disease from retinal images using a deep learning based CenterNet model. *Sensors (Basel)*, 21(16), 5283. DOI: 10.3390/s21165283 PMID: 34450729

Nguyen, Q. H., Muthuraman, R., Singh, L., Sen, G., Tran, A. C., Nguyen, B. P., & Chua, M. (2020, January). Diabetic retinopathy detection using deep learning. In *Proceedings of the 4th international conference on machine learning and soft computing* (pp. 103-107).

Pamadi, A. M., Ravishankar, A., Nithya, P. A., Jahnavi, G., & Kathavate, S. (2022, March). Diabetic retinopathy detection using MobileNetV2 architecture. In *2022 International Conference on Smart Technologies and Systems for Next Generation Computing (ICSTSN)* (pp. 1-5). IEEE.

Sanjana, S., Shadin, N. S., & Farzana, M. (2021, November). Automated diabetic retinopathy detection using transfer learning models. In 2021 5th International Conference on Electrical Engineering and Information Communication Technology (ICEEICT) (pp. 1-6). IEEE.

Saranya, P., Devi, S. K., & Bharanidharan, B. (2022, March). Detection of diabetic retinopathy in retinal fundus images using densenet based deep learning model. In 2022 international mobile and embedded technology conference (MECON) (pp. 268-272). IEEE.

Sebastian, A., Elharrouss, O., Al-Maadeed, S., & Almaadeed, N. (2023, January 18). A Survey on Deep-Learning-Based Diabetic Retinopathy Classification. *Diagnostics (Basel)*, 13(3), 345. DOI: 10.3390/diagnostics13030345 PMID: 36766451

Shankar, K., Sait, A. R. W., Gupta, D., Lakshmanaprabu, S., Khanna, A., & Pandey, H. M. (2020). Automated detection and classification of fundus diabetic retinopathy images using synergic deep learning model. *Pattern Recognition Letters*, 133, 210–216. DOI: 10.1016/j.patrec.2020.02.026

Sudarmadji, P. W., Pakan, P. D., & Dillak, R. Y. (2020, November). Diabetic retinopathy stages classification using improved deep learning. In *2020 International Conference on Informatics, Multimedia, Cyber and Information System (ICIMCIS)* (pp. 104-109). IEEE.

Taufiqurrahman, S., Handayani, A., Hermanto, B. R., & Mengko, T. L. E. R. (2020, November). Diabetic retinopathy classification using a hybrid and efficient MobileNetV2-SVM model. In 2020 IEEE Region 10 Conference (Tencon) (pp. 235-240). IEEE.

Wang, J., Bai, Y., & Xia, B. (2020). Simultaneous diagnosis of severity and features of diabetic retinopathy in fundus photography using deep learning. *IEEE Journal of Biomedical and Health Informatics*, 24(12), 3397–3407. DOI: 10.1109/JBHI.2020.3012547 PMID: 32750975

Wang, Z., Li, X., Yao, M., Li, J., Jiang, Q., & Yan, B. (2022). A new detection model of microaneurysms based on improved FC-DenseNet. *Scientific Reports*, 12(1), 1–9. DOI: 10.1038/s41598-021-04750-2 PMID: 35046432

Chapter 8
Exploring Microactuators and Sensors in Modern Applications

S. Surya
Saveetha Engineering College, Chennai, India

R. Elakya
Sri Venkateswara College of Engineering, Chennai, India

Srinivasan Ramamurthy
 https://orcid.org/0000-0002-8434-151X
University of Science and Technology of Fujairah, UAE

K. Janani
Manipal Institute of Science and Technology, Manipal, India

A. Lizy
Vel Tech Rangarajan Dr. Sagunthala R&D Institute of Science and Technology, Chennai, India

ABSTRACT

The development of robotic systems across various applications is increasingly propelled by advancements in microactuators and sensors. This chapter provides a comprehensive exploration of their pivotal role, the challenges they present, and their transformative impact on modern robotics. Beginning with foundational concepts, it progresses to a detailed examination of diverse microactuators and sensors, elucidating their functionalities and potential applications. Emphasizing the crucial role of feedback mechanisms in enhancing robotic performance, the chapter delves into integration schemes and control mechanisms crucial for effective operation.

DOI: 10.4018/979-8-3693-6150-4.ch008

Current research initiatives and future directions are discussed, highlighting ongoing efforts to overcome challenges such as miniaturization, power consumption, and reliability. By addressing these hurdles, researchers aim to unlock the full potential of sensors and microactuators, fostering innovation and advancing the capabilities of contemporary robotic systems.

1.INTRODUCTION

The rapid evolution of technology in recent years has been significantly propelled by advancements in microactuators and sensors. These tiny, yet powerful components are fundamental to the development of modern robotic systems, enhancing their precision, efficiency, and versatility. Microactuators and sensors are the building blocks that enable robots to interact with their environment in increasingly sophisticated ways, making them indispensable in a variety of applications. In the realm of consumer electronics, they enhance the functionality of everyday devices, providing responsive touchscreens, motion detection, and more. In medical robotics, these technologies allow for minimally invasive surgeries and precise diagnostic tools, revolutionizing patient care. Industrial automation relies on microactuators and sensors for assembly lines, quality control, and safety systems, boosting productivity and reliability (Li. Lijie,2018). The automotive and aerospace industries benefit from these advancements through improved vehicle performance, safety features, and maintenance systems (Vignesh. U, 2024).

This chapter delves into the crucial role of microactuators and sensors, examining their operation, applications, and the challenges faced in their integration. It explores the innovative feedback processes and control mechanisms that enhance robotic performance and reliability. Real-world examples demonstrate their broad utility across various fields, while ongoing research initiatives point towards future possibilities. By addressing issues such as restructuring, power consumption, and reliability, researchers aim to unlock the full potential of these technologies, driving innovation and expanding the capabilities of modern robotics. Through this comprehensive exploration, we gain insight into the transformative impact of microactuators and sensors in today's technological landscape.

Modern actuators are increasingly integrated with sensors and control systems, enhancing their ability to operate autonomously and intelligently in dynamic environments. They also contribute to the robot's ability to interact with its environment (Rubio, Francisco,2019). For instance, a robotic arm used in a manufacturing plant needs to move accurately and smoothly to pick up and place items. This precision is achieved through the use of high-quality actuators.

1.1 Applications in Robotics:

Microactuators are essential components in the development of miniature and agile robotic systems. This section explores their applications in:

- **Robotic Manipulation**: Actuators enable precise control of robotic arms and grippers for handling delicate objects or performing complex tasks.
- **Locomotion**: Microactuators facilitate mobility in small-scale robots, enabling navigation through confined spaces or challenging environments.
- **Sensing and Feedback**: Integrated sensors provide feedback for adaptive control and environment perception, enhancing the autonomy and capabilities of micro-robots.

1.2 Challenges and Future Directions:

Despite their advancements, microactuators and sensors face challenges such as reliability, scalability, and cost-effectiveness. This section discusses ongoing research efforts and potential solutions to address these challenges:

- **Reliability**: Improving materials and manufacturing processes to enhance device lifespan and performance.
- **Scalability**: Developing scalable fabrication techniques to meet growing demand for miniaturized devices.
- **Cost-effectiveness**: Optimizing production processes and materials to reduce manufacturing costs without compromising quality.

This chapter explores the transformative impact of microactuators and sensors on the field of robotics. These components have significantly advanced the precision, efficiency, and versatility of robotic systems across various industries. Despite their benefits, integrating these components faces challenges such as reliability, scalability, and cost-effectiveness. Ongoing research aims to overcome these hurdles, ensuring these technologies continue to drive innovation and expand robotic capabilities across various sectors.

2.RELATED WORKS

Recent research in microactuators and sensors has been pivotal in advancing robotic systems, focusing on improving precision, efficiency, and versatility. Studies have explored novel materials and fabrication techniques to enhance the performance and reliability of microactuators, such as piezoelectric, electrostatic, and shape memory alloy actuators, which are critical for precise motion control and manipulation tasks in robotics. Integration of sensors, including optical, tactile, and inertial sensors, has enabled robots to perceive and interact with their environment more effectively, supporting applications in autonomous navigation, object recognition, and human-robot interaction. Moreover, advancements in control algorithms and feedback mechanisms have optimized robotic performance in dynamic and unstructured environments, addressing challenges such as scalability and energy efficiency(Algamili, Abdullah Saleh, 2021 ; Podder, Itilekha, 2023).

Future research directions focus on further miniaturization, multi-modal sensing integration, and developing intelligent systems that can adapt and learn from their surroundings, promising to push the boundaries of robotic capabilities in industrial automation, healthcare, and beyond.The literature review on MEMS actuators for biomedical applications underscores their pivotal role in advancing healthcare technologies through miniaturization, precision control, and versatile integration capabilities (Selvi, R. Thanga et al.,2024 ; Ratnakumar, Rahul, 2024). MEMS actuators enable the development of less invasive biomedical devices, enhancing patient comfort and procedural efficacy(Li, Weidong,2023). They facilitate precise control over drug delivery systems, ensuring accurate dosage and timing for targeted therapies. Moreover, MEMS technology is integral to the creation of lab-on-a-chip devices, enabling rapid chemical and biological analysis crucial for diagnostics. However, challenges such as power consumption, biocompatibility, and long-term reliability in physiological environments remain significant hurdles. Recent studies focusing on advancements in non-resonant piezoelectric actuators, thin-film PZT actuators, and self-sensing methods highlight ongoing efforts to overcome these challenges and enhance the performance and applicability of MEMS actuators in biomedical settings(Khalid, Muhammad Yasir et al,2024 ;Wang, Yanmei et al., 2024). By addressing these complexities, researchers aim to further integrate MEMS actuators into innovative biomedical devices, thereby advancing healthcare delivery and patient outcomes.Chen, C., & Tan, X. et al.(2022) this article addresses the challenges and opportunities associated with integrating microactuators and sensors into robotic systems. It provides insights into current research initiatives, potential future paths, and strategies to overcome issues related to restructuring, power consumption, and reliability.

3. MICROACTUATORS: PRINCIPLES AND EXAMPLES

3.1 Working of Microactuator

The working principle of a Microactuator is to generate mechanical motion of fluids or solids where this motion is generated via changing one form of energy to another energy like from thermal, electromagnetic, or electrical into kinetic energy (K.E) of movable components. For most of the actuators, different force generation principles are used like the piezo effect, bimetal effect, electrostatic forces & shape memory effect. Like a general actuator, a microactuator has to meet these standards like fast switching, large travel, high precision, less power consumption, etc.

The mechanical actuator includes a power supply, transduction unit, actuating element, and output action.

Figure 1. Workflow Diagram of Microactuator

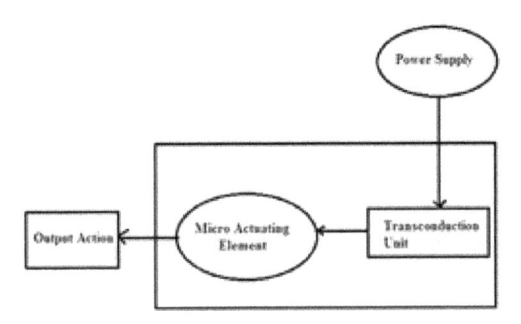

In a microactuator system, the power supply typically involves electrical current or voltage, providing the necessary energy to drive the actuator. The transduction unit plays a critical role in this system by converting the supplied electrical energy into the preferred form of mechanical action suitable for the actuating element. This actuating element, which can be a specific component or material, is designed to move or deform in response to the energy supplied. The resulting output action from the microactuator is generally a prescribed motion, precisely controlled to meet the requirements of its specific application. This sequence ensures that the microactuator performs its intended functions efficiently and effectively, whether it is in precision positioning, medical devices, or other advanced technological applications(Sasikumar, S et al., 2024).

Table 1. Different Categories of Microactuator

S. No	Type of Micro actuator	Conversion Mechanism	Power Consumption	Speed	Displacement	Force	Voltage/ Material Requirements	Complexity	Applications
1	Electrostatic	Electrostatic force	Low (µW to mW)	High (µs to ms)	Small (nm to µm)	Low (µN to mN)	High voltage required	Medium	Micro mirrors, RF switches, optical devices
2	Electro-magnetic	Lorentz force (interaction of magnetic field with electric current)	Medium (mW to W)	Moderate (ms to s)	Medium (µm to mm)	High (mN to N)	Requires magnetic materials	High	Micro motors, micro pumps, relays
3	Piezoelectric	Piezoelectric effect (mechanical deformation in response to electric field)	Low (µW to mW)	Fast (µs to ms)	Limited (nm to µm)	High (µN to N)	Piezoelectric materials required	Medium	Precision positioning, medical devices, sensors
4	Thermal	Thermal expansion (using differential thermal expansion of materials)	High (mW to W)	Slow (ms to s)	Large (µm to mm)	Medium (µN to mN)	Simple materials	Low	Micro valves, thermal switches, micro grippers
5	Shape Memory Alloy (SMA)	Phase transformation (change in material phase due to temperature change)	High (mW to W)	Slow (ms to s)	Large (µm to mm)	Moderate (mN to N)	Shape memory alloys required	Medium	Medical stents, actuators in aerospace and robotics
6	Magnetostrictive	Magnetostriction (change in shape due to magnetic field)	Medium (mW to W)	Fast (µs to ms)	Medium (µm to mm)	High (mN to N)	Requires magnetic materials	High	Precision actuators, sonar transducers, sensors
7	Electroactive Polymer (EAP)	Electrostatic force (deformation in response to electric field)	Low (µW to mW)	Slow (ms to s)	Large (µm to mm)	Low (µN to mN)	Electroactive polymer materials required	Medium	Artificial muscles, flexible electronics, haptics
8	Bimetallic	Thermal expansion (differential expansion of two bonded metals)	Medium (mW to W)	Slow (ms to s)	Large (µm to mm)	Low (µN to mN)	Simple materials	Low	Thermostats, thermal sensors, relays
9	Fluidic	Pressure-driven (using fluid pressure for actuation)	Medium (mW to W)	Slow (ms to s)	Large (µm to mm)	Low (µN to mN)	Requires fluidic infrastructure	Medium	Microfluidic devices, drug delivery systems, pumps

3.2 Role of Actuators in Robotics

In the realm of robotics, actuators hold a pivotal role. They are the components that enable robots to interact with their environment by converting energy into motion. This conversion process is what allows robots to perform tasks ranging from simple to complex.

Actuators are the driving force behind a robot's ability to move. They are the components that receive signals from the robot's control system and execute the corresponding physical movement. This could be as simple as turning a wheel or as complex as coordinating the movements of a multi-jointed robotic arm.

The role of actuators extends beyond just movement. They also contribute to the robot's ability to sense and respond to its environment. For instance, in a robotic arm used in a manufacturing plant, the actuators not only enable the arm to move, but also provide feedback to the control system about the arm's position and movement. This feedback, known as proprioceptive information, is crucial for the robot to perform precise and accurate movements.

The type of actuator used can greatly affect the performance and efficiency of the robot. For example, electric actuators, which convert electrical energy into mechanical motion, are known for their high speed and precision. They are commonly used in robots that require fast, accurate movements, such as those used in assembly lines or for surgical procedures. On the other hand, hydraulic actuators, which use pressurized fluid to create motion, are known for their high force and power. They are often used in robots that need to perform heavy-duty tasks, such as those used in construction or industrial automation settings. Pneumatic actuators, which use compressed air to generate motion, offer advantages in terms of simplicity, light weight, and low cost. They are often used in robots that perform simple, repetitive tasks, such as those used in packaging or sorting operations(Chi, Yinding et al., 2024).

4. ROLE OF SENSORS IN ROBOTICS

The role of sensors in robotics is crucial as they enable robots to perceive and interact with their environment effectively. Here are the key roles and functions of sensors in robotics:

4.1 Environmental Perception:

○ Sensors provide robots with the ability to perceive and understand their surroundings. They detect and measure physical properties such as distance, light intensity, temperature, humidity, and gas concentration. This environmental data is essential for navigation, obstacle avoidance, and situational awareness.

4.2 Object Detection and Recognition:

○ Vision sensors (e.g., cameras) and proximity sensors (e.g., ultrasonic or infrared sensors) help robots detect and recognize objects in their path or workspace. This capability is crucial for tasks such as object manipulation, assembly, and sorting in industrial and service robotics.

4.3 Navigation and Localization:

○ Inertial sensors (e.g., accelerometers and gyroscopes), GPS modules, and wheel encoders enable robots to navigate and determine their position accurately. These sensors provide feedback on movement, orientation, and relative position, allowing robots to move autonomously and perform tasks efficiently.

4.4 Feedback for Control Systems:

○ Sensors provide real-time feedback to the robot's control system, enabling closed-loop control. For example, force sensors in robotic grippers provide feedback on the amount of force applied during grasping, ensuring delicate handling of objects. Similarly, joint position sensors help maintain precise movement and alignment in robotic arms.

4.5 Safety and Collision Avoidance:

○ Sensors play a critical role in ensuring the safety of both robots and humans in shared workspaces. Safety sensors such as proximity sensors and light curtains detect the presence of obstacles or humans in the robot's vicinity, triggering immediate response actions or halting robot operations to prevent collisions.

4.6 Adaptability and Flexibility:

○ Multi-modal sensors, combining different sensing modalities (e.g., vision, tactile, and auditory sensors), enable robots to adapt to dynamic environments and varying tasks. This versatility allows robots to perform a wide range of applications, from industrial automation to service robotics in healthcare and logistics.

4.7 Quality Control and Monitoring:

○ Sensors are used in robotic systems for quality control and monitoring of processes. For example, sensors can monitor parameters such as temperature, pressure, and fluid levels in manufacturing processes, ensuring consistency and detecting anomalies that require intervention.

4.8 Interaction with Humans:

○ Sensors such as touch sensors and force/torque sensors enable robots to interact safely and effectively with humans. They facilitate collaborative robotics applications where robots and humans work together in close proximity, sharing tasks and workspace.

Overall, sensors are integral to the advancement of robotics, enabling robots to operate autonomously, interact intelligently with their environment, and perform tasks with precision and efficiency across various industries and applications. Continued advancements in sensor technology contribute to the capabilities and reliability of robotic systems, driving innovation and expanding the role of robots in modern society(Kalulu, Mulenga, 2024).

5. INTEGRATION OF MICROACTUATORS AND SENSORS

Table 2. Comparison of an integrating microactuators and sensors

S.No	Integration Technique	Monolithic Integration	Hybrid Integration	Wafer Bonding
1	Description	Fabricating both sensors and actuators on a single chip using semiconductor processes.	Combining separate sensor and actuator chips into a single package.	Joining two or more wafers, each with different components, to form a single functional device. Techniques include direct bonding, anodic bonding, and adhesive bonding.
2	Advantages	High performance, Compact size	Flexibility, Allows use of optimized components	High integration density, Versatility in material and device combination

continued on following page

Table 2. Continued

S.No	Integration Technique	Monolithic Integration	Hybrid Integration	Wafer Bonding
3	Challenges	Complex fabrication process, Costly	Larger size, Potential alignment issues	Complex alignment and bonding processes, Potential thermal and mechanical stress

The integration of microactuators and sensors represents a pivotal advancement in the field of robotics and automation, enabling sophisticated functionalities across various domains. Microactuators, such as piezoelectric, electrostatic, and shape memory alloy actuators, convert electrical energy into mechanical motion, offering precise control and responsiveness in robotic systems. Concurrently, sensors including optical, tactile, and inertial sensors provide critical feedback on environmental conditions and system states, enhancing the autonomy and adaptability of robots. The synergy between these micro-scale components facilitates seamless integration into robotic platforms, supporting tasks ranging from precise manipulation in medical robotics to dynamic control in industrial automation(Vignesh,2024 ; Parvathi, R, 2023).

Challenges in integration include ensuring compatibility between actuators and sensors, optimizing mechanical interfaces for efficient energy transfer, and implementing robust feedback mechanisms for real-time control. Innovations in integration techniques, such as monolithic integration and hybrid integration, address these challenges by enhancing performance, scalability, and reliability. Moreover, advancements in control strategies leveraging feedback loops and adaptive algorithms further optimize system performance, enabling robots to operate autonomously and intelligently in complex environments. As research continues to push the boundaries of miniaturization, energy efficiency, and multifunctionality, integrated microactuators and sensors are poised to redefine the capabilities and applications of robotic systems.

In robotics and automation, sensors play a crucial role in providing feedback for closed-loop control systems. They enable robots to perceive and interact with their environment, navigate obstacles, and perform tasks autonomously. Advances in sensor technology, including miniaturization, improved sensitivity, and multi-sensing capabilities, continue to expand their applications in fields such as healthcare, manufacturing, agriculture, and smart cities. By accurately detecting and measuring physical and chemical phenomena, sensors enhance the functionality, efficiency, and safety of robotic systems in diverse real-world scenarios.

6. CHALLENGES AND FUTURE DIRECTIONS

Despite their transformative impact, microactuators and sensors in robotics face several significant challenges. One of the primary challenges is reliability. As these components are often deployed in critical applications such as medical robotics and industrial automation, ensuring consistent performance over extended periods is crucial. Factors like mechanical wear, environmental conditions, and material degradation can affect reliability and require ongoing improvements in design and manufacturing processes. Another challenge is scalability. While microactuators and sensors excel in miniaturized applications, scaling their production to meet industrial demands without compromising performance remains a hurdle. Moreover, power consumption is a critical concern, particularly in battery-operated robots where energy efficiency directly impacts operational longevity and performance. Addressing these challenges necessitates interdisciplinary research efforts focusing on materials science, mechanical engineering, and electronics to develop robust, scalable, and energy-efficient solutions.

Looking forward, several promising avenues can enhance the capabilities of microactuators and sensors in robotics. Advances in materials science offer opportunities to develop novel materials with superior mechanical properties and environmental resilience, thereby improving reliability(Vignesh, U., Parvathi, R., 2023). Research into energy harvesting and efficient power management systems can reduce power consumption and extend operational lifespans in robotic applications. Moreover, integrating artificial intelligence (AI) and machine learning algorithms with sensor data could enhance robots' adaptive capabilities, enabling autonomous decision-making and improving task performance in dynamic environments(Verma, Bhumika, 2023). Furthermore, exploring bio-inspired designs and soft robotics could lead to more agile and adaptable robotic systems capable of interacting safely and effectively with humans. Collaborative research initiatives and partnerships across academia, industry, and government sectors will be crucial in driving these innovations forward and realizing the full potential of microactuators and sensors in shaping the future of robotics.

7. CONCLUSION

This chapter concludes by discussing the vital part that sensors and microactuators perform in the advancement of current robotics in a variety of industries. It starts with a strong foundation of basic ideas and explores the different kinds and functions of these parts, showing the significance they are to improving robotic capabilities. Real-world examples show the wide range of applications of these developments in consumer electronics, wellness, the field of robotics, industrial automation, automotive, aerospace, and more. It is shown that the integration of feedback mechanisms and control systems is crucial for optimizing performance.

Despite facing challenges such as restructuring, power consumption, and reliability, ongoing research initiatives aim to address these issues and unlock the full potential of microactuators and sensors. By overcoming these obstacles, researchers are paving the way for significant innovations and advancements in robotic systems. The future holds great promise for the continued evolution of robotics, driven by the development and refinement of these essential technologies.

REFERENCES

Algamili, A. S., Mohd, H. M. K., Dennis, J. O., Ahmed, A. Y., Alabsi, S. S., Saeed, S. B. H., & Junaid, M. M. (2021). A review of actuation and sensing mechanisms in MEMS-based sensor devices. *Nanoscale Research Letters*, 16(1), 1–21. DOI: 10.1186/s11671-021-03481-7 PMID: 33496852

Chi, Y., Zhao, Y., Hong, Y., Li, Y., & Yin, J. (2024). A perspective on miniature soft robotics: Actuation, fabrication, control, and applications. *Advanced Intelligent Systems*, 6(2), 2300063. DOI: 10.1002/aisy.202300063

Kalulu, M., Chilikwazi, B., Hu, J., & Fu, G. (2024). Soft Actuators and Actuation: Design, Synthesis and Applications. *Macromolecular Rapid Communications*, •••, 2400282. DOI: 10.1002/marc.202400282 PMID: 38850266

Khalid, M. Y., Arif, Z. U., Tariq, A., Hossain, M., Khan, K. A., & Umer, R. (2024). 3D printing of magneto-active smart materials for advanced actuators and soft robotics applications. *European Polymer Journal*, 205, 112718. DOI: 10.1016/j.eurpolymj.2023.112718

Li, L., & Chew, Z. J. (2018). Microactuators: Design and technology. In Smart sensors and MEMS (pp. 313-354). Woodhead Publishing.

Li, W., Hu, D., & Yang, L. (2023). Actuation mechanisms and applications for soft robots: A comprehensive review. *Applied Sciences (Basel, Switzerland)*, 13(16), 9255. DOI: 10.3390/app13169255

Parvathi, R., & Vignesh, U. (2023). Diabetic Retinopathy Detection Using Transfer Learning. In AI and IoT-Based Technologies for Precision Medicine (pp. 177-204). IGI Global.

Podder, I., Fischl, T., & Bub, U. (2023, March). Artificial intelligence applications for MEMS-based sensors and manufacturing process optimization. In Telecom (Vol. 4, No. 1, pp. 165-197). MDPI.

Ratnakumar, R., & Vignesh, U. (2024). Machine Learning-Based Environmental, Social, and Scientific Studies Using Satellite Images: A Case Series. In AI and Blockchain Optimization Techniques in Aerospace Engineering (pp. 149-163). IGI Global.

Rubio, F., Valero, F., & Llopis-Albert, C. (2019). A review of mobile robots: Concepts, methods, theoretical framework, and applications. *International Journal of Advanced Robotic Systems*, 16(2), 1729881419839596. DOI: 10.1177/1729881419839596

Sasikumar, S., Santhakumar, S., Jayapal, R., & Thanigaivelan, R. (2024). Actuators in Medical Devices. In Robotics and Automation in Healthcare (pp. 137-149). Apple Academic Press.

Selvi, R. T., Elakya, R., & Vignesh, U. (2024). Securing Sensitive Patient Data in Healthcare Settings Using Blockchain Technology. In Blockchain and IoT Approaches for Secure Electronic Health Records (EHR) (pp. 73-88). IGI Global.

Verma, B., Yudheksha, G. K., Sanjana Reddy, P., & Vignesh, U. (2023). An Intelligent Flood Automation System Using IoT and Machine Learning. In *Recent Developments in Electronics and Communication Systems* (pp. 444–449). IOS Press.

Vignesh, P., Maheswari, R., Vijaya, P., & Vignesh, U. (2024). Machine Learning for Aerospace Object Categorization. In AI and Blockchain Optimization Techniques in Aerospace Engineering (pp. 164-180). IGI Global.

Vignesh, U., Parvathi, R., & Goncalves, R. (2023). *"Structural and Functional Data Processing in Bio-Computing and Deep Learning." Structural and Functional Aspects of Biocomputing Systems for Data Processing 2023*. IGI Global.

Vignesh, U., & Ratnakumar, R. (2024). An Empirical Review on Clustering Algorithms for Image Segmentation of Satellite Images. AI and Blockchain Optimization Techniques in Aerospace Engineering, 33-52.

Vignesh, U., Ratnakumar, R., & Al-Obaidi, A. S. M. (2024). *AI and blockchain optimization techniques in aerospace engineering*. AI and Blockchain Optimization Techniques in Aerospace Engineering. DOI: 10.4018/979-8-3693-1491-3

Wang, Y., Wang, Y., Mushtaq, R. T., & Wei, Q. (2024). Advancements in Soft Robotics: A Comprehensive Review on Actuation Methods, Materials, and Applications. *Polymers*, 16(8), 1087. DOI: 10.3390/polym16081087 PMID: 38675005

Chapter 9
Enhancing Concrete Strength Prognostication Through Machine Learning and Robotics

A. Hema
Vellore Institute of Technology, Chennai, India

S. Geetha
https://orcid.org/0000-0002-6850-9423
Vellore Institute of Technology, Chennai, India

S. Karthiyaini
Vellore Institute of Technology, Chennai, India

ABSTRACT

Concrete is a widely used construction material globally due to its exceptional properties. The strength of concrete varies based on the composition of cement, blast furnace slag, fly ash, water, superplasticizer, coarse aggregate, and fine aggregate. By altering the ingredients in different proportions, the process can anticipate the concrete strength through the various machine learning techniques. The study in this chapter involves the utilization and comparison of three algorithms, viz. Random Forest, Gradient Boosting, and Linear Regression models to analyse the concrete strength. Among these models, gradient boosting yielded superior results. In order to predict concrete strength from these models, three sensors in the field paves the way for better analysis, such as, an acoustic emission sensor, a strain gauge sensor, and a wireless concrete maturity sensor. This approach proves to be highly effective in achieving optimal concrete strength using machine learning techniques.

DOI: 10.4018/979-8-3693-6150-4.ch009

INTRODUCTION

In a world where every major construction is built by concrete, safety in these buildings during dangerous events. Depending on the proportion of each ingredient in concrete, its properties can change drastically, making it useful for multiple purposes. This research paper focuses on enhancing its strength. Stronger concrete is required to be strong enough to face any danger while also being sustainable for the foreseeable future. Using durable concrete can enhance the longevity of structures, reducing the need for maintenance, improve its resistance to weathering and can lead to increased structural capacity due to more slender and efficient structures. To determine the correct proportion of ingredients to manufacture the strongest concrete, three sensors: an acoustic emission sensor, a strain gauge and a wireless concrete maturity sensor are used. The data from these sensors are passed through three different regression models: Random Forest, Gradient Boosting and Linear Regression. The aim of this paper is to compare the actual results, produced by graphs, with the predicted data, produced by the regression models, to accurately ascertain the perfect mixture of concrete.

Concrete is really important in building because it's strong and lasts a long time. Making strong concrete depends on getting the mix of ingredients just right. Strong concrete keeps buildings safe in emergencies, lasts longer, needs less fixing up, and can hold heavier stuff. In this study, we're using special sensors sound, strain, and wireless ones to watch how concrete sets and handles weight as it dries. These sensors give us data to figure out how different ingredient amounts affect how strong the concrete gets. Then, we use fancy computer models to analyze this data. These models help us predict how strong the concrete will be and check if our predictions are right by testing the concrete.

Our study, called "Improving Concrete Strength Prediction with Sensors, Computers, and Robots," is all about using these sensors, computers, and robots to make concrete stronger. We want to predict and make the best concrete mixes to build safer, longer-lasting, and more eco-friendly buildings. Being eco-friendly is really important to us. Making concrete usually makes a lot of pollution, but by figuring out the best mix of ingredients and using smart technology, we can make concrete in a way that's less harmful to the environment. Plus, if buildings last longer and need fewer repairs, we're helping the planet too. Concrete has been around for ages, but now with cool technology, we can make it even better. By adding sensors into the mix, we can watch how it's doing in real-time and make it stronger and last longer. This not only keeps us safe but also makes building things cheaper and better for the environment. The sensors in concrete aren't just for predicting strength; they also help us keep an eye on how the building is doing over time. By watching for any damage or wear and tear early on, we can fix things up before they become big problems. This way, buildings stay safe and usable for a long time. To sum up, this research is a big deal for making better concrete. By using fancy sensors, computers, and robots, we're making concrete stronger, longer-lasting, and friendlier to the environment. This is not just good for buildings, it is good for the planet too. As building tech gets better, we'll keep finding ways to make things more sustainable.

LITERATURE REVIEW

The literature reviews how to predict the future of technology using the 'Six Genres of Technology' framework. This framework looks at how interactions between humans and machines change over time, with machines becoming more independent. It highlights four important features of advanced technology: autonomy, intelligence, language, and self-maintenance, which might pose risks. The goal is to guide current discussions to ensure future technological advancements are thoughtfully and responsibly considered (Harwood & Eaves, 2020).

This review examines how the amount of water compared to binder and the addition of fly ash affect how well-mixed concrete is. Methods like image analysis, segregation degree, compressive strength, and microhardness were used to measure this. The results show that image analysis is effective in describing how well-mixed the concrete is. Using less water compared to binder and adding fly ash improve how well-mixed the concrete is in the Interfacial Transition Zone (ITZ). Microhardness tests show that the top of the aggregate is hardest and the narrowest, and the inside of the concrete is better than the surface. Overall, using image analysis, segregation degree, compressive strength, and microhardness provides a thorough measure of concrete's ability (Wang et al., 2013).

This review looks at how adding fibers to concrete affects cracking. It finds that Fiber Reinforced Concrete (FRC) makes cracks narrower and closer together compared to regular concrete. FRC also makes the concrete tougher, which helps it handle tension better between cracks and creates stress at the cracks. The review is based on 97 tests done at the University of Brescia, checking things like how strong the concrete is, how big it is, and how much reinforcement and fibers are added. It shows that FRC reduces both how far apart cracks are and how wide they get, and looks at how well current models can predict crack spacing in FRC (Tiberti et al., 2015).

This study examines how well self-consolidating concrete (SCC) performs using different tests like the column method, visual stability index, and V-funnel tests. It found that freeze-thaw resistance depends on air bubbles in the concrete, not on how much it separates. There is a wide range of mortar band thicknesses even when separation is within acceptable limits. To accurately measure mortar bands using digital images, you need good contrast and brightness at the boundary between the paste and aggregate. There's a direct link between mortar band thickness, water absorption, and how much it loses when exposed to salt (Panesar & Shindman, 2012).

In construction, tests like the rebound hammer (RH) and ultrasonic pulse velocity (UPV) are used to check how strong concrete is without breaking it. However, these tests don't always give a very accurate estimate of concrete strength compared to tests that do break the concrete. Researchers tried using artificial intelligence (AI) to make the guesses more exact. They collected data from these non-breaking tests on 98 pieces of concrete and worked with a material testing lab and a Civil Engineer Association. They tried regular guesses and AI guesses. The AI guesses, like artificial neural networks (ANNs), support vector machines (SVM), and adaptive neural fuzzy inference system (ANFIS), worked better than the regular guesses. The study used 70 pieces to teach the AI and 28 to test it. They picked the AI that worked the best to make sure it was good. This study shows how AI can make guesses on how strong concrete is without breaking it (Ngo et al., 2021).

This review looks at how engineers figure out how strong concrete is in buildings. They use tests that don't harm the concrete and some that do. The review talks about new suggestions by RILEM, which say we should use a certain way of thinking to handle uncertainties in strength estimates. It also talks about using computer simulations to understand what affects the quality of our assessments, like how many samples we need to break. The review also says that the accuracy of the tests that don't harm the concrete is really important. It suggests that by choosing where to take samples based on these tests, we can use fewer samples. Finally, it stresses the importance of setting clear goals for how accurate we want our strength estimates to be (Breysse et al., 2020).

This literature review looks at using computer vision systems to monitor bridges. The study suggests using a moving camera system with multiple cameras to measure how much a bridge moves. They tested this in a lab and compared five different ways to find problems without needing a teacher. The results showed that the system is very good at finding and showing where damage is, with scores from 0.96 to 0.97 in different tests. This way is cheap and easy to use in real life for monitoring bridges (Lydon et al., 2022).

This review looks at new ways to check if buildings and other structures are still strong and safe without damaging them. It talks about different methods like using special radars, cameras, and sound waves to see if there are any problems. Researchers are using smart computer programs to help understand all the data collected. They're also combining different methods to get better results. The main idea is to use sensors to check how things like concrete, wood, and metal are holding up. It explains how these sensors are put in place and gives examples of how they're being used today (Kot et al., 2021).

This review talks about how computers can use vision to check and keep track of buildings. There are two main things it focuses on: inspecting and monitoring. Inspecting means looking for damage and changes in structures over time. Monitoring means measuring things like how much a building moves or bends. But there are still challenges, like making sure the computer understands what it's seeing and making the process better with video. Researchers are working on these problems to make it cheaper and easier to check and fix buildings automatically (Spencer Jr et al., 2019).

Recent studies have shown that combining big data and deep learning in civil engineering has led to significant advancements. However, there's still a lack of thorough evaluations in important areas like identifying structural damage, predicting concrete properties, detecting cracks, and testing concrete structures. This review aims to fill this gap by examining recent developments in artificial intelligence (AI) in civil engineering, especially focusing on concrete-related applications. It discusses how smart algorithms, data analysis, and deep learning are used to automatically find concrete properties and cracks. The review highlights how AI is changing civil engineering practices, particularly in maintaining structures, ensuring quality, and optimizing designs using techniques like image processing and computer vision (Sarkar et al., 2024).

METHODOLOGY

Source - The dataset is from Kaggle and is called "Concrete Strength using IoT Devices." It includes various factors that might affect concrete strength.

Cement: The amount of cement in the mix (measured in kg per m^3 of mixture)
Blast Furnace Slag: The amount of blast furnace slag (measured in kg per m^3 of mixture)
Fly Ash: The amount of fly ash (measured in kg per m^3 of mixture)
Water: The amount of water (measured in kg per m^3 of mixture)
Superplasticizer: The amount of superplasticizer (measured in kg per m^3 of mixture)
Coarse Aggregate: The amount of coarse aggregate (measured in kg per m^3 of mixture)
Fine Aggregate: The amount of fine aggregate (measured in kg per m^3 of mixture)
Age: The age of the concrete (measured in days)
Strength: The compressive strength of the concrete (measured in MPa)
Sensors used: The types of sensors used to gather the data

This dataset helps in analyzing how different components and the age of concrete influence its strength.

Loading the Dataset, Using numpy, matplotlib, pandas and seaborn libraries, we will begin importing the dataset in the form of a CSV file from KAGGLE

We perform some initial exploratory data analysis to understand the dataset.

Then the code computes and displays summary statistics for the dataset using the describe () function. This function provides essential statistical measures, such as the mean, standard deviation, minimum, and maximum values for each numerical column.

Conducting these exploratory data analysis steps is important for many reasons. Calculating summary statistics helps us understand the distribution and central tendencies of the data, which is very important for making good decisions about further analysis. Checking for missing values is essential to ensure data quality and completeness, as missing values can impact the results of our analysis. Knowing the data types of each column is important for selecting correct analytical methods and avoiding mistakes during data processing. This exploratory data analysis provides a foundational understanding of the dataset, enabling us to proceed with more detailed analysis and modelling with confidence.

CORRELATION ANALYSIS

The process selects the numeric data and generate a correlation matrix to see how features are related. We visualize this correlation matrix using a heatmap. This code generates a heatmap to visually represent the correlation matrix. The heatmap() function from the seaborn library is used to create a coloured grid where each cell represents the correlation coefficient between pairs of features

Figure 1. Correlation matrix heat map

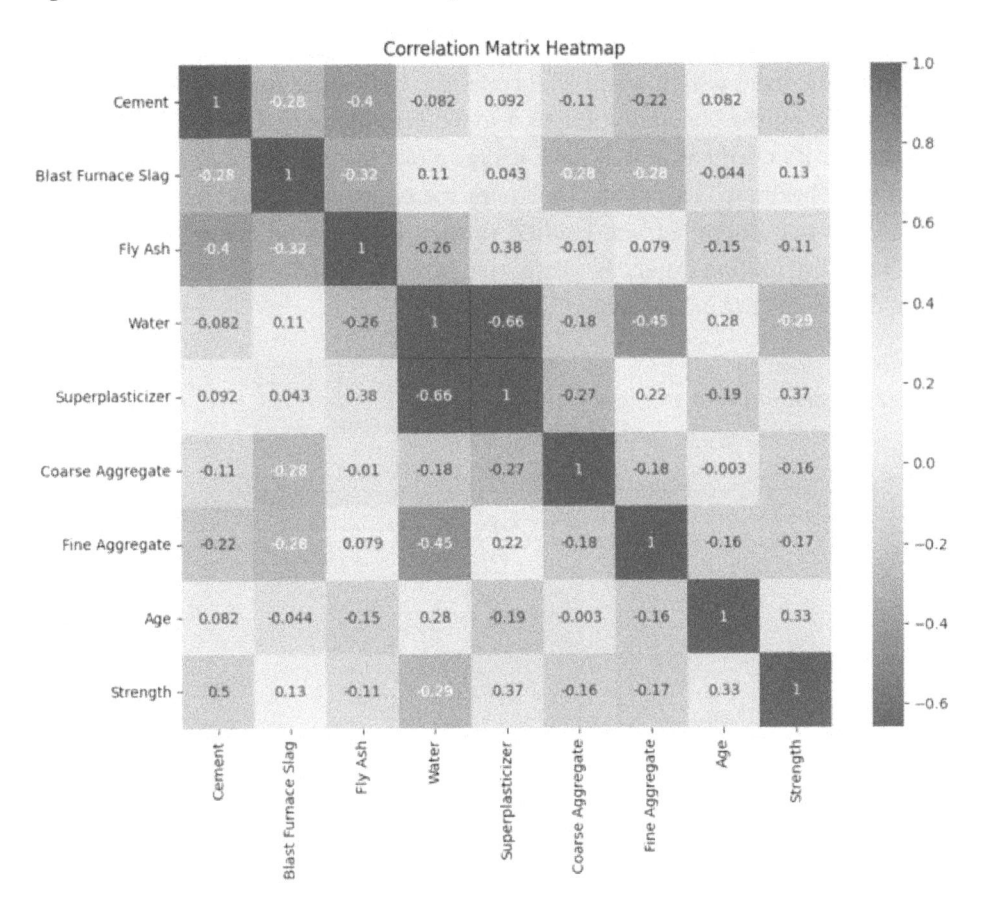

Correlation analysis primarily concerned with finding out whether a relationship exists between a variable and finding out the magnitude and action of that relationship. It is basically a statistical relationship. The value of correlation ranges from -1 to 1. The sign of the slope (positive, negative or zero) corresponds to the type of correlation (positive, negative or neutral). The importance of performing a correlation analysis is crucial for understanding the relationships between different numerical features in the dataset. Identifying strong correlations can help in feature selection, detecting multicollinearity, and understanding the data's underlying structure. Visualizing the correlation matrix using a heatmap provides an intuitive and clear way to observe these relationships.

The Outcome of correlation matrix generated by the corr() function reveals the strength and direction of the linear relationships between numerical features. For instance, a correlation coefficient close to 1 or -1 indicates a strong positive or negative correlation, respectively, while a coefficient around 0 indicates no linear correlation. By visualizing this matrix with a heatmap, we can quickly identify which features are strongly correlated. For example, we might observe that features like "Cement" and "Strength" have a high positive correlation, indicating that as the amount of cement increases, the concrete strength also tends to increase. Conversely, features with weak or no correlation will have values close to zero, represented by cooler colors on the heatmap. This visual tool helps in understanding and interpreting the relationships within the dataset, guiding further analysis and feature engineering steps. The Distribution Analysis is we visualize the distribution of the variable "Strength", to understand its spread and shape.

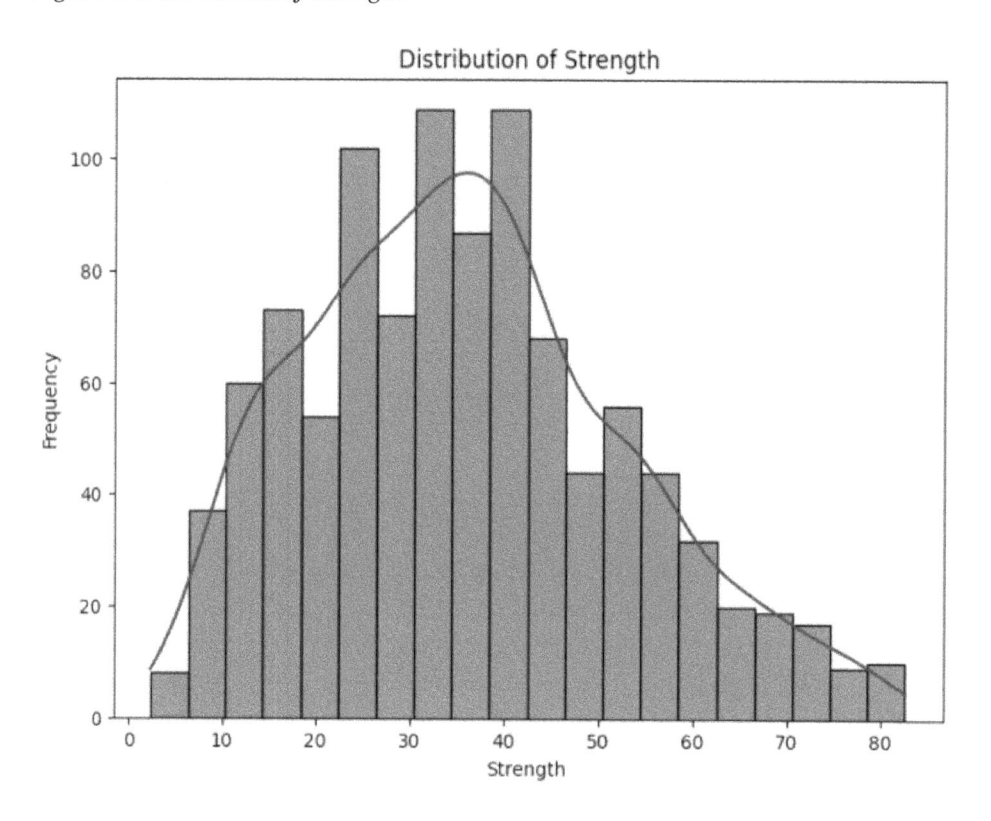

Figure 2. Distribution of strength

The importance in figure 2 are understanding how "Strength" is spread out helps us see any patterns or unusual things in our data. This knowledge helps us choose the best ways to analyze our data and make accurate predictions. The histogram we made gives us a clear picture of how "Strength" values are distributed. By looking at it, we can see if the values are mostly in one area or if they're spread out. This helps us know if there are any strange things in our data that we need to pay attention to. Knowing this helps us make better decisions when we analyze the data and try to predict outcomes.

From the boxplot analysis we use a boxplot to compare the strength values across different types of sensors used in the data. The code creates a boxplot to show the distribution of "Strength" values for each type of sensor. The boxplot uses boxes to show where most of the data points are (the middle 50%), with lines (whiskers) extending to show the range of the rest of the data, and dots for any outliers that are much higher or lower than the rest.

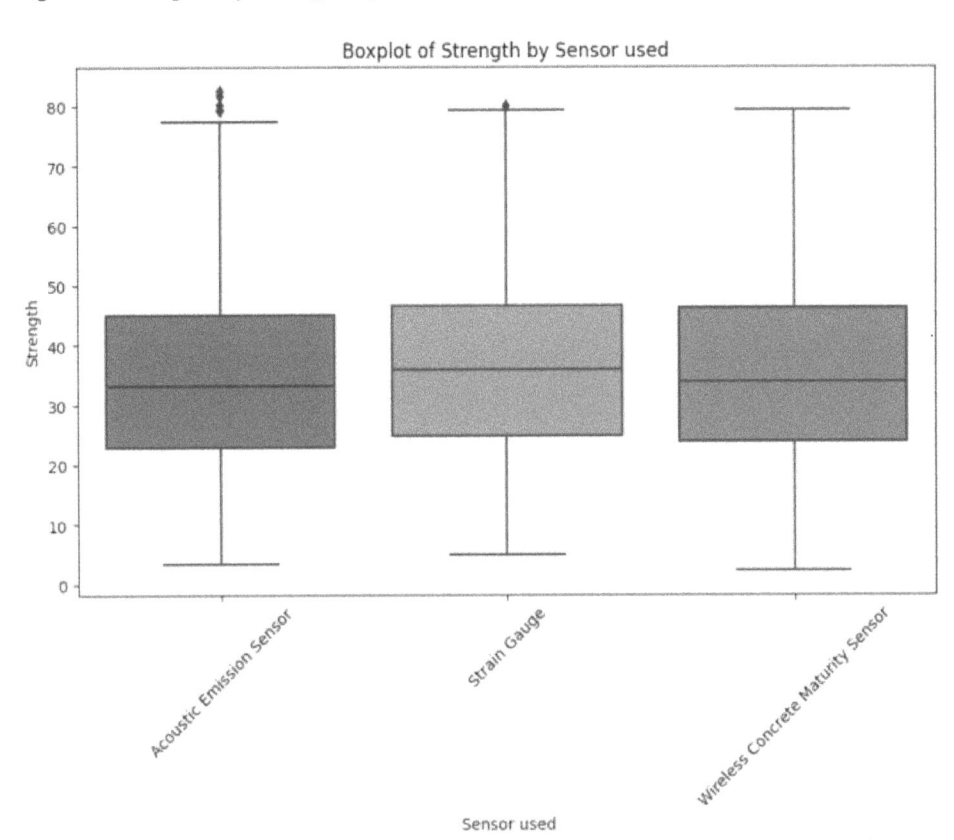

Figure 3. Box plot of strength by sensor used

Boxplot refers to the graph summarizing a set of data. The shape basically describes how the data is distributed and it is a useful way to compare different sets of data. A boxplot displays the five-number summary of a set of data. The five-number summary is the minimum, first quartile, median, third quartile, and maximum. The importance of Boxplots help us compare strength values for different sensors easily. This is important because it shows us how the sensors might affect the strength measurements and if there are any big differences or unusual values. The boxplot shows us the spread of "Strength" values for each sensor. By looking at it, we can see which sensors tend to measure higher or lower strength values. We can also spot any outliers, which are data points that are very different from the others. This helps us understand how each sensor influences the strength measurements and identify any unusual data that might need a closer look.

PAIRPLOT ANALYSIS

We use a pairplot to see how different features in the dataset relate to each other. This code creates a pairplot, which shows scatterplots of each pair of selected features. The diagonal of the pairplot shows the distribution of each feature. These scatterplots help us see how features like "Cement", "Water", and "Age" are related to "Strength" and to each other.

Figure 4. Pair plot

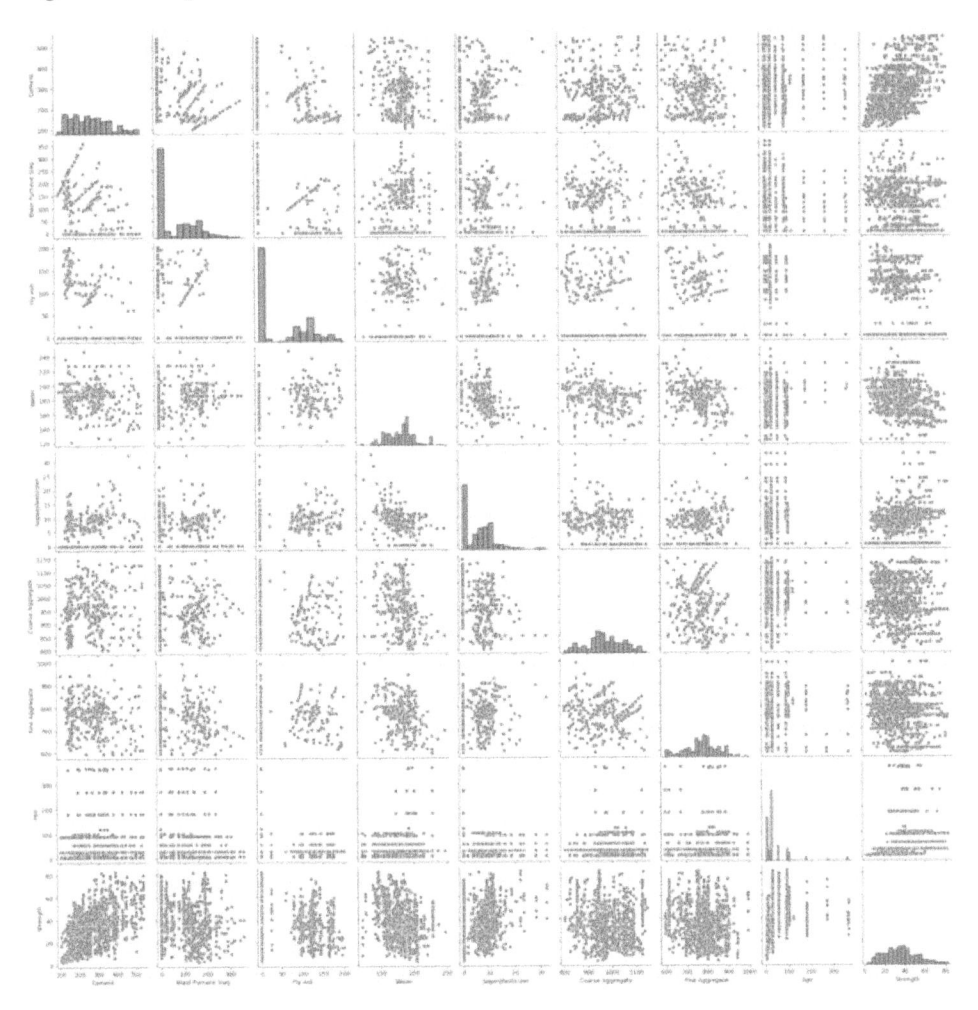

A pair plot, also known as a scatterplot matrix, is a matrix of graphs that enables the visualization of the relationship between each pair of variables in a dataset. A pairplot is basically used to provide information like a group's data's summary, skew, variance and outliners. Pairplots are important because they show the relationships between different features all at once. This helps us spot patterns, find correlations, or notice any unusual relationships that could be important for our analysis. The pairplot shows a grid of scatterplots, each one displaying the relationship between two features. For example, we can see how "Cement" is related to "Strength" or how "Water" is related to "Age". This helps us identify which features are strongly related and which are not. Using this information, we can better understand how the features interact and choose the most important ones for further analysis. We create a scatterplot to see how "Strength" changes with "Age". The code makes a scatterplot where each dot shows the strength of concrete at a certain age. The x-axis (horizontal) represents the age of the concrete, and the y-axis (vertical) shows the strength.

Figure 5. Scatterplot

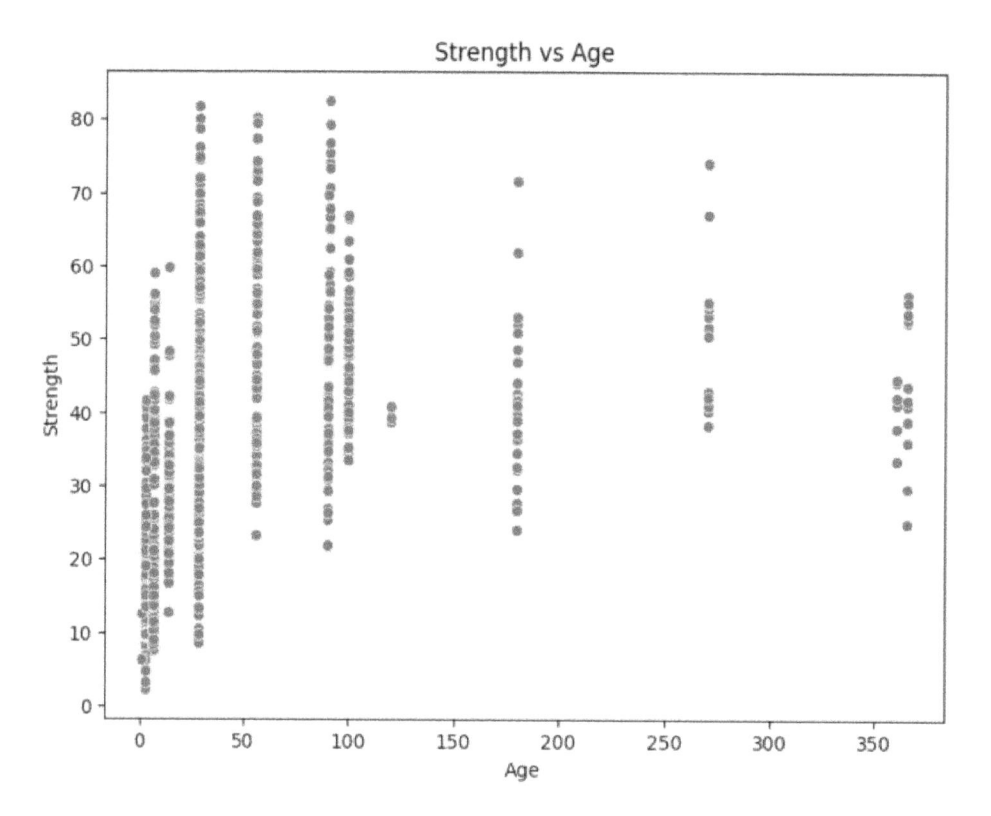

A scatterplot is basically uses dots to represent values for two different numeric variables. The position of each dot on horizontal and vertical axis indicates for an individual data point. Scatterplots help us understand the relationship between two things. In this case, it shows us how the strength of concrete changes as it gets older. This is important because it can reveal patterns that are useful for predicting concrete strength. The outcome of the scatterplot shows the relationship between the age of the concrete and its strength. If we see a clear pattern, like dots forming a line going up or down, it means there is a strong relationship between age and strength. This helps us understand how concrete strength changes over time and can help us make better predictions.

The process use a countplot to see how often each type of sensor is used in the dataset. The code makes a countplot that shows the number of times each sensor type is used. The x-axis (horizontal) represents different sensor types, and the y-axis (vertical) shows how many times each sensor type appears in the dataset. The countplot helps us understand the distribution of sensor types in our data. This is important because it shows which sensors are used most or least often. Knowing this helps us see if the data is balanced or if some sensors are used more than others.

Figure 6. Count plot

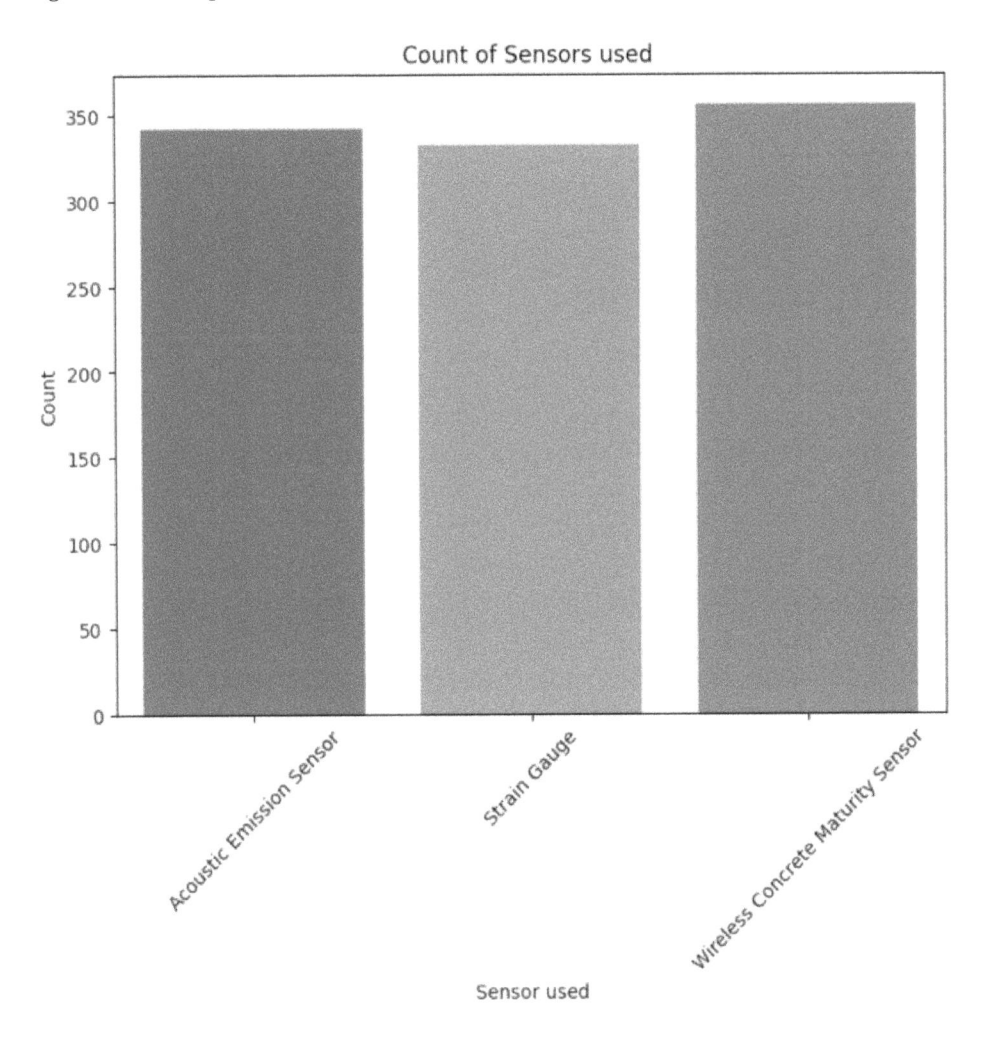

A count plot is a sort of a histogram across a categorical, instead of quantitative, variable, and is used to represent the occurrence(counts) of the observation present in the categorical variable. For visual depiction a bar graph can be used. The countplot shows us how often each sensor type is used. By looking at this plot, we can quickly see which sensors are common and which are rare. This helps us understand the makeup of our data and can guide us in our analysis and modeling. We're getting the data ready for our model by changing the "Sensors used" column into numbers and then splitting the data into parts for training and testing.

The code changes the "Sensors used" column into numbers so the computer can understand it better. Then, it separates the data into two parts: one part we'll use to teach our model (training set) and another part we'll use to see how well it learned (testing set). Changing the "Sensors used" column into numbers helps our model work better. Splitting the data into training and testing sets lets us check how good our model is at making predictions with data it hasn't seen before. By making these changes and splitting the data, we're getting everything ready to teach our model and see how well it can predict concrete strength. This helps us understand if our model will work well in real situations.

RESULTS AND DISCUSSION

The model is analysed which is best at predicting concrete strength.

After training and testing a bunch of models, we are looking at how well they did. We're using numbers like RMSE and R2 to compare them. These numbers help us see which model is the closest to guessing the actual concrete strength. Then, we'll pick the one that did the best.

1. Random Forest

It is a commonly used machine learning algorithm that combines the output of multiple decision trees to reach a single result. It is considered as unexcelled in accuracy and runs efficiently on large data bases.

2. Gradient Boosting

Gradient Boosting is a popular boosting algorithm in machine learning used for classification and regression tasks. It is one kind of learning method which trains the model sequentially and each new model tries to correct the previous model. It combines several weak learners into strong learners

3. Linear Regression

Linear regression is used to predict the value of a particular variable in relation with the value of another variable. It consists of the dependent variable which needed to be predicted and independent variable which is used as a reference. Among all the three models, Random Forest model explains the variance in the target variable the best that is 88.29% approximately, it is considered as a pretty good approximation. At the second position lies the Gradient boosting model, it give out the approximation value as 88.4% and lastly Linear regression gives the value as 62.73% approximately of the variance in target variable.

It's really important to pick the right model because it helps us make accurate predictions about concrete strength. By choosing the best model, we can trust our predictions more when building things like bridges or buildings. The outcome is after looking at the numbers, we choose the model that had the lowest RMSE and highest R2 score. This model is the best at guessing concrete strength based on the information we gave it. Choosing the best model means we can rely on our predictions to be as accurate as possible.

Random Forest R2 Score: 0.8829 This indicates that the Random Forest model explains approximately 88.29% of the variance in the target variable (Strength), which is quite good. Gradient Boosting R2 Score: 0.8844 Similarly, the Gradient Boosting model explains approximately 88.44% of the variance in the target variable, slightly better than the Random Forest model. Linear Regression R2 Score: 0.6273 The Linear Regression model explains approximately 62.73% of the variance in the target variable, which is lower compared to the Random Forest and Gradient Boosting models. Based on these R2 scores, both Random Forest and Gradient Boosting models perform quite well in predicting the concrete strength compared to the Linear Regression model.

CONCLUSION

In this chapter, analysed a data about concrete strength using IoT devices. The process also checked out the data, looked for patterns, and tried to predict how strong concrete would be. The process started by getting the data ready and splitting it into parts we could use to teach the computer and test how well it learned. Then, tried different ways to teach the computer, like Random Forest and Gradient Boosting.

After training the computer, we checked how close its guesses were to the real concrete strength. Based on that, we picked the best way to predict concrete strength.

In the future, the process can be applied to various methods for predictions even better by tweaking some settings or trying different approaches. We might also gather more data or learn more about concrete to improve our predictions even more. Overall, this chapter helps us understand how to predict concrete strength, which is useful for building things like bridges and buildings safely and effectively.

REFERENCES

Breysse, D., Romao, X., Alwash, M., Sbartai, Z. M., & Luprano, V. A. (2020). Risk evaluation on concrete strength assessment with NDT technique and conditional coring approach. *Journal of Building Engineering*, 32, 101541. DOI: 10.1016/j.jobe.2020.101541

Harwood, S., & Eaves, S. (2020). Conceptualising technology, its development and future: The six genres of technology. *Technological Forecasting and Social Change*, 160, 120174. DOI: 10.1016/j.techfore.2020.120174 PMID: 32904525

Kot, P., Muradov, M., Gkantou, M., Kamaris, G. S., Hashim, K., & Yeboah, D. (2021). Recent advancements in non-destructive testing techniques for structural health monitoring. *Applied Sciences (Basel, Switzerland)*, 11(6), 2750. DOI: 10.3390/app11062750

Lydon, D., Kromanis, R., Lydon, M., Early, J., & Taylor, S. (2022). Use of a roving computer vision system to compare anomaly detection techniques for health monitoring of bridges. *Journal of Civil Structural Health Monitoring*, 12(6), 1299–1316. DOI: 10.1007/s13349-022-00617-w

Ngo, T. Q. L., Wang, Y. R., & Chiang, D. L. (2021). Applying artificial intelligence to improve on-site non-destructive concrete compressive strength tests. *Crystals*, 11(10), 1157. DOI: 10.3390/cryst11101157

Panesar, D. K., & Shindman, B. (2012). The effect of segregation on transport and durability properties of self consolidating concrete. *Cement and Concrete Research*, 42(2), 252–264. DOI: 10.1016/j.cemconres.2011.09.011

Sarkar, K., Shiuly, A., & Dhal, K. G. (2024). Revolutionizing concrete analysis: An in-depth survey of AI-powered insights with image-centric approaches on comprehensive quality control, advanced crack detection and concrete property exploration. *Construction & Building Materials*, 411, 134212. DOI: 10.1016/j.conbuildmat.2023.134212

Spencer, B. F.Jr, Hoskere, V., & Narazaki, Y. (2019). Advances in computer vision-based civil infrastructure inspection and monitoring. *Engineering (Beijing)*, 5(2), 199–222. DOI: 10.1016/j.eng.2018.11.030

Tiberti, G., Minelli, F., & Plizzari, G. (2015). Cracking behavior in reinforced concrete members with steel fibers: A comprehensive experimental study. *Cement and Concrete Research*, 68, 24–34. DOI: 10.1016/j.cemconres.2014.10.011

Wang, G., Kong, Y., Sun, T., & Shui, Z. (2013). Effect of water–binder ratio and fly ash on the homogeneity of concrete. *Construction & Building Materials*, 38, 1129–1134. DOI: 10.1016/j.conbuildmat.2012.09.027

Chapter 10
Cooperative Task Execution in Insect-Inspired Robot Swarms Using Reinforcement Learning

R. Elakya
Sri Venkateswara College of Engineering, India

S. Surya
Saveetha Engineering College, India

G. Abinaya
University of Southern Queensland, Springfield, Australia

T. Manoranjitham
SRM Institute of Science and Technology, Ramapuram, India

R. Thanga Selvi
Vel Tech Rangarajan Dr. Sagunthala R&D Institute of Science and Technology, India

ABSTRACT

This chapter proposes a novel framework for cooperative task execution in a swarm of insect-inspired robots by using Reinforcement Learning (RL) algorithms. Inspired by the collaborative behaviors observed in social insects, such as ants and bees, the proposed framework enables robots to autonomously coordinate their actions

DOI: 10.4018/979-8-3693-6150-4.ch010

Copyright © 2025, IGI Global. Copying or distributing in print or electronic forms without written permission of IGI Global is prohibited.

to accomplish complex tasks in dynamic environments. Each robot in the swarm acts as an autonomous agent capable of learning and adapting its behavior through interactions with the environment and feedback from other robots. By applying RL algorithms, such as Q-learning or Deep Q-Networks (DQN), robots learn optimal action policies to maximize task performance while considering the collective objectives of the swarm. we demonstrate the effectiveness and scalability of our approach in various cooperative tasks, including exploration, foraging, and object manipulation. This project showcases the potential of RL-based approaches to enhance the autonomy and adaptability of robotic swarms for collaborative task execution in real-world scenarios.

INTRODUCTION

Swarm robotics is the field of research that focuses on the behaviour of large groups of very uncomplicated robots. These robots work together and collaborate to complete problems that are beyond the capacity of each individual robot (Dorigo and Sahin, 2004). Task solving in this context is primarily dependent on self-organization and emergence. This means that the organisation of the swarm originates from inside the system itself, rather than being imposed externally. Furthermore, this organisation occurs in a decentralised manner through local interactions between individual robots (De Wolf & Holvoet, 2005). Swarm robotics algorithms mostly depend on cooperation and basic interactions among robots, rather than on intricate individual behaviours that necessitating advanced sensory skills. Specifically, in the context of navigation, this means that the emphasis is on collaborative navigation, where robots assist each other, rather than relying on maps (refer to Mirats Tur et al., 2009) or map-building techniques (such as simultaneous localization and mapping, as described by Durrant-Whyte and Bailey, 2006), or the use of an external infrastructure (such as a communication network or a localization system, as mentioned by O'Hara et al., 2008).

Several studies in the field of swarm robotics navigation focus on a specific scenario in which robots are required to travel repeatedly between two sites, such as transporting objects from one place to another. The majority of this work is founded upon indirect communication. The communication among robots is influenced by the foraging behaviour of specific ant species seen in nature (Werger and Matari´c, 1996; Wodrich and Bilchev, 1997; Sharpe and Webb, 1999; Garnier et al., 2007; Fujisawa et al., 2008; Nouyan et al., 2009; Ducatelle et al., 2011a). This behaviour is dependent on stigmergic communication, which involves the indirect exchange of information through the local modification and perception of the surrounding environment. Ants deposit a chemical compound known as pheromone while travelling between their nest and a food source. This pheromone serves to attract other ants and direct them towards the food. An intriguing aspect is that the combined process of depositing and tracking pheromones strengthens the most effective routes, resulting in the emergence of the shortest path due to the collective actions of the swarm (Deneubourg et al., 1990; Bonabeau et al., 1999). This is an instance of emergent self-organized behaviour. A significant challenge in applying this navigation concept, based on pheromones, to robots, is in the actual execution of indirect communication. This involves finding a suitable artificial substitute for the chemical pheromones utilised by ants. Similarly, RL can be used to develop algorithms where each robot updates its strategy based on the success of its actions, effectively "learning" the optimal way to perform tasks through trial and error.

The proposed algorithm integrates Reinforcement Learning (RL) into a communication-based navigation framework for swarm robots. This combination allows each robot to learn and adapt its navigation strategy based on its interactions with the environment and other robots, leading to improved cooperative task execution. The decentralized nature of this approach ensures robustness, scalability, and flexibility, which are essential for the effective deployment of swarm robotic systems in dynamic and unpredictable environments.

Key Components and Objectives:

1. Multi-Agent Reinforcement Learning: Develop a framework for multi-agent RL to enable robots in the swarm to learn and adapt their behaviors in a collaborative manner.
2. Task Representation: Define task-specific reward functions and state-action spaces tailored to the cooperative tasks to be performed by the swarm.
3. Communication and Coordination: Implement communication protocols and coordination mechanisms to facilitate information exchange and collaboration among robots in the swarm.
4. Real-world Validation: Validate the proposed framework through both simulated and physical experiments in diverse environments, considering factors such as environmental complexity, resource availability, and task constraints.
5. Performance Evaluation: Evaluate the performance of the swarm in terms of task completion time, resource utilization, robustness to disturbances, and scalability with increasing swarm size.

This chapter aims to advance the state-of-the-art in swarm robotics by utilising the power of Reinforcement Learning to enable collaborative task execution in insect-inspired robot swarms. By empowering robots with learning capabilities, we aim to create adaptive and resilient robotic systems capable of addressing real-world challenges in dynamic and uncertain environments.

Advantages of Reinforcement Learning'

Reinforcement Learning (RL) offers a powerful way for robots to learn optimal behaviors through interactions with their environment, receiving feedback in the form of rewards or penalties (Selvi, R.T,2024). This makes RL particularly advantageous for swarm robotics, which relies on decentralized control and coordination among many simple robots. With RL, robots can continuously learn and adapt their strategies based on their experiences, allowing them to handle changing environments and new challenges effectively. Each robot operates independently using local information, and RL enables them to make decisions without a central controller, leading to complex and intelligent group behaviors similar to those seen in natural swarms of insects(Ratnakumar.R. 2024).

RL algorithms are designed to maximize cumulative rewards, optimizing robots' performance in achieving their tasks. This makes them highly efficient, reducing unnecessary actions and energy use. Additionally, RL algorithms can easily scale to large numbers of robots without losing effectiveness, making them ideal for practical applications involving large swarms(Vignesh u,2024). The decentralized nature of RL-based control also enhances robustness, as the failure of one robot does not cripple the entire swarm. Instead, other robots continue to learn and adapt, ensuring the swarm remains functional.

RL techniques balance exploration and exploitation, allowing robots to try new actions while using known successful strategies. This continuous improvement process leads to more efficient task performance over time. RL is also effective in learning from sparse or delayed feedback, making it suitable for tasks requiring long-term planning(Verma,2023).

In real-world applications, such as search and rescue, environmental monitoring, and precision agriculture, RL significantly enhances swarm robotics. It enables robots to explore disaster areas efficiently, adapt monitoring strategies to varying conditions, and manage crops more effectively by learning optimal behaviors(Parvathi R,2023). Overall, RL enhances the capabilities of swarm robotics by providing adaptive, decentralized, and optimized decision-making, driving the evolution of autonomous systems towards greater intelligence and efficiency.

Demonstrating the Application of RL

Reinforcement Learning (RL) is a powerful tool for developing decentralized control strategies in robot swarms. This approach leverages the ability of individual robots to learn from their interactions with the environment and their peers, thereby improving their behavior over time to maximize cumulative rewards. Here's a detailed exploration of how RL can be applied effectively in this context:

Decentralized Learning

In a decentralized learning setup, each robot acts as an independent agent that makes decisions based on its local observations and experiences. Unlike centralized control systems where a single entity directs all robots, decentralized RL allows each robot to operate autonomously. This autonomy is crucial for several reasons:

- **Robustness**: The failure of one or a few robots does not cripple the entire system. Each robot can continue to function and adapt, maintaining the swarm's overall performance.
- **Scalability**: The system can easily scale up by adding more robots without the need for significant changes to the control algorithms. Each new robot simply integrates into the swarm and begins learning from its environment.
- **Adaptability**: Robots can adapt to new environments and tasks without requiring centralized updates or reprogramming. This flexibility is particularly useful in dynamic and unpredictable scenarios.

Learning Process

Demonstrating the Application of RL

Reinforcement Learning (RL) is a powerful tool for developing decentralized control strategies in robot swarms. This approach leverages the ability of individual robots to learn from their interactions with the environment and their peers, thereby improving their behavior over time to maximize cumulative rewards. Here's a detailed exploration of how RL can be applied effectively in this context:

Decentralized Learning

In a decentralized learning setup, each robot acts as an independent agent that makes decisions based on its local observations and experiences. Unlike centralized control systems where a single entity directs all robots, decentralized RL allows each robot to operate autonomously. This autonomy is crucial for several reasons:

- **Robustness**: The failure of one or a few robots does not cripple the entire system. Each robot can continue to function and adapt, maintaining the swarm's overall performance.
- **Scalability**: The system can easily scale up by adding more robots without the need for significant changes to the control algorithms. Each new robot simply integrates into the swarm and begins learning from its environment.

- **Adaptability**: Robots can adapt to new environments and tasks without requiring centralized updates or reprogramming. This flexibility is particularly useful in dynamic and unpredictable scenarios.

Learning Process

The learning process in RL involves several key components:

- **State Space**: The set of all possible states a robot can be in, defined by its sensors and internal status. For instance, in a foraging task, the state might include the robot's current location, the presence of obstacles, and the proximity of resources.
- **Action Space**: The set of all possible actions a robot can take. These might include moving in different directions, picking up objects, or communicating with other robots.
- **Reward Function**: A feedback mechanism that assigns rewards or penalties based on the robot's actions. For example, a robot might receive a reward for successfully collecting a resource and a penalty for colliding with another robot.

Framework for Cooperative Task Execution

The framework for cooperative task execution in swarm robotics using Reinforcement Learning (RL) involves several key components. These components work together to enable a swarm of insect-inspired robots to perform complex tasks efficiently and adaptively. The framework integrates Multi-Agent Reinforcement Learning (MARL), task representation, and effective communication and coordination strategies.

Multi-Agent Reinforcement Learning (MARL)

MARL extends traditional RL to environments with multiple agents, where each robot learns to make decisions based on local observations and interactions with other robots. The main challenges in MARL include ensuring that individual actions lead to desirable collective behaviors and that the swarm operates cohesively without centralized control.

Key Elements of MARL

1. **Local Observations:** Each robot gathers information from its surroundings using sensors. This information includes the robot's position, the location of obstacles, and signals from other robots.
2. **Action Selection:** Based on the local observations, each robot selects an action from a predefined set, such as moving in a particular direction, picking up an object, or signaling to other robots.
3. **Reward Functions:** Reward functions are designed to align individual robot behaviors with the overall objectives of the swarm. For example, a robot might receive positive rewards for successfully completing a subtask or for assisting another robot, and negative rewards for actions that hinder the swarm's goals, like collisions.

Designing Reward Functions

- **Individual Rewards:** Rewards based on the immediate success or failure of a robot's actions.
- **Collective Rewards:** Rewards based on the overall performance of the swarm, encouraging cooperation among robots.

Task Representation

Tasks in swarm robotics need to be decomposed into smaller, manageable subtasks that can be distributed among the robots. This involves defining the state space, action space, and reward functions that guide the learning process.

Components of Task Representation:

1. **State Space:** Defines all possible states that a robot can be in. For a foraging task, the state space might include the robot's current location, the presence of obstacles, and the proximity of resources.
2. **Action Space:** Defines all possible actions that a robot can take, such as moving in different directions, picking up objects, or communicating with other robots.

3. **Reward Functions:** These guide the learning process by providing feedback. Rewards can be given for completing subtasks, like finding and collecting resources, and penalties for undesirable actions, like collisions or inefficient behaviors.

Example - Foraging Task:

- **State Space:** Robot's current position, visible resources, obstacles.
- **Action Space:** Move forward, turn left, turn right, pick up resource, avoid obstacle.
- **Reward Functions:** Positive rewards for resource collection, negative rewards for collisions.

Communication and Coordination

Effective communication and coordination are crucial for achieving coherent and efficient swarm behavior. This includes both explicit communication, where robots directly exchange information, and implicit communication, where robots leave signals in the environment (stigmergy).

Types of Communication

1. **Explicit Communication:** Robots send direct messages to one another. This can include information about resource locations, obstacles, or plans for movement.
2. **Implicit Communication (Stigmergy):** Robots leave markers or signals in the environment that other robots can detect and interpret. For example, a robot might leave a trail to indicate the path to a resource.
Coordination Mechanisms:

- **Local Interactions:** Robots coordinate based on interactions with nearby robots, ensuring that their actions align with the swarm's goals.
- **Simple Rules:** Robots follow simple behavioral rules that lead to complex, emergent behaviors. For instance, robots might follow a rule to move towards the highest concentration of resource signals.

Example - Cooperative Exploration:

- **Explicit Communication:** Robots share their findings about resource locations.
- **Implicit Communication:** Robots leave pheromone-like signals to mark explored areas.

Evaluating Effectiveness

The effectiveness of the proposed framework can be evaluated through simulations and real-world experiments:

1. **Simulations:** Virtual environments model various scenarios, allowing for the testing and refinement of algorithms without the risk and cost of real-world deployment. Metrics such as task completion time, resource collection rate, and robustness to robot failures can be measured.
2. **Real-World Experiments:** Implementing the framework in actual robots provides insights into practical challenges and real-world applicability. Field tests in diverse conditions help evaluate the system's adaptability and reliability.

Training Process for Communication-Based Navigation Using Reinforcement Learning

The training process involves iteratively updating the robots' policies based on their experiences. This process enables the robots to learn effective navigation strategies through interaction with the environment and other robots.

Step 1: Initialization

1. **Initialize Q-table**: Start with an initial Q-table where states represent the robot's positions and actions represent possible movements.
2. **Initialize navigation table**: Set up the navigation table with distances to targets and sequence numbers.

Step 2: Exploration and Action

3. **Explore the environment**: Each robot moves through the environment, exploring different actions and states.
4. **Take actions**: Based on the current policy, each robot takes an action to move towards a target or explore new areas.

Step 3: Receive Feedback

5. **Observe rewards**: After taking an action, each robot observes the immediate reward received. Rewards are based on the effectiveness of the action, such as reaching a target or avoiding obstacles.

Step 4: Update Policies

6. **Update Q-values**: Use the Q-learning formula to update the Q-table based on the observed reward and the new state. This step adjusts the robot's policy to improve future actions.

$$Q(s, a) := Q(s, a) + \alpha \cdot [r + \gamma \cdot \max(Q(s', a')) - Q(s, a)]$$

Step 5: Communication

7. **Send and receive messages**: Robots exchange information about distances and targets with their neighbors. This communication helps robots update their navigation tables and improve their decision-making.

Step 6: Update Navigation Tables

8. **Process received messages**: For each message received from a neighbor robot, update the navigation table with new distance and sequence number information if it is better than the current information.
9. **Adjust navigation behavior**: If new information indicates a better path, update the robot's navigation strategy to follow this new path.

Step 7: Exploration Strategy

10. **Use ε-greedy policy**: Balance exploration and exploitation by using an ε-greedy strategy. With probability ε, choose a random action (exploration), and with probability 1-ε, choose the best-known action (exploitation).

- o **Exploration**: Randomly select actions to discover new states and gather more information about the environment.
- o **Exploitation**: Select the action with the highest Q-value to maximize rewards based on current knowledge.

Step 8: Convergence

11. **Check for convergence**: Continue the training process until the robots' policies converge to optimal or near-optimal solutions. This means the Q-values no longer change significantly, indicating that the robots have learned effective navigation strategies.

The work of training the RLSR model - or any other reinforcement learning-based swarm robotics architecture - is, in a nutshell, to repeatedly invoke the loop of exploration, action-taking, feedback (through the reward function) and policy updates. The robots using exploration techniques, such as ε-greedy, can balance the two: the need to explore with an unknown environment and the goal of taking optimized actions. Communication between robots is necessary to transmit the lessons learned by each robot to the rest of the swarm, maintaining each individual experience for all the agents and increasing collective behaviors in performing tasks. This continues until the policies of the robots have stabilized (it has converged), which shows that they have learned optimal ways to act and navigate.

EVALUATION

Overall Evaluation for the reinforcement learning (RL) based swarm robotics systems performance has grown across several critical dimensions. Task Completion Time: The most basic metric is a measure of how well the swarm can realize its predefined objectives and accomplish them in real-time. Using simulations and real-world experiments, researchers can measure how RL algorithms affect the efficiency and speed of task execution against traditional means to demonstrate improvements in autonomy and decision-making.

Secondly, resource utilization measures how effectively the swarm uses energy, processing power and communication bandwidth. Measuring metrics such as energy consumption per robot and the computational load for RL computations can help researchers to determine if RL really helps with better resource allocation and management within the swarm.

Another important line of inquiry is the robustness to disturbances, testing how well the swarm behaves when a few robots fail or the environment changes or communication does not work properly. Studying the performance of the RL-empowered system under perturbation can give an idea about the robustness and capacity of adaptability of the system which are important factors in real-world applications with dynamic, uncertain environments.

Thirdly, scalability with increasing swarm size quantifies how well the RL framework is capable to scale as the number of robots increases. Swarm size changes: Researchers change the number of agents in swarms during experiments, and examine how this affects various performance metrics (e.g. task completion time, communication overhead). This scalability assessment can determine whether RL algorithms are able to scale for a larger number of robots without sacrificing performance or having an impact on effectiveness.

Task Completion Time (TCT)

The time taken by the swarm to complete a given task.

$$TCT = \sum_{i=1}^{N} t_i$$

where t_i is the time taken by the i-th robot to complete its part of the task, and N is the total number of robots in the swarm.

Resource Utilization (RU)

The efficiency with which the swarm uses available resources such as energy, computational power, and physical space.

$$RU = \frac{U_{used}}{U_{total}}$$

where U_{used} is the total amount of resources used by the swarm, and U_{total} is the total amount of resources available.

Robustness to Disturbances (RD)

The ability of the swarm to maintain performance despite unexpected challenges or failures.

$$RD = \frac{P_{after}}{P_{before}}$$

Where Pbefore is the performance of the swarm before the disturbance, and Pafter is the performance after the disturbance. Performance can be measured in terms of task completion rate, accuracy, etc.

Scalability with Increasing Swarm Size (SS)

The ability of the swarm to maintain or improve performance as the number of robots increases.

$$SS = \frac{P(N)}{P} \tag{1}$$

where P(N) is the performance with N robots, and P(1) is the performance with a single robot. Performance can be measured in terms of throughput, task completion rate, etc.

CONCLUSION

When forming autonomous systems, the incorporation of Reinforcement Learning (RL) is a radical leap for swarm robotics departments. Like the decentralized coordination observed in social insects, RL provides individual robots with a mechanism for learning to adapt their behaviors from simple local interactions and environment feedback. This creates stronger, more scalable and flexible robotic swarms. In this chapter, we have introduced a new framework for cooperative task execution in insect-inspired robot swarms based on RL algorithms - showcasing the power of developing decentralized control strategies with reinforcement learning to improve both efficiency and adaptability of swarms systems.

In this chapter, we discussed into how Multi-Agent Reinforcement Learning (MARL) disaggregates high-level tasks of the swarm into subparts; thus serving as a means by which each robot can take its own decisions that add up to date with overall objectives in mission. Solid coordination of communication mechanisms is the connective tissue which brings these decentralized decisions into coherent and efficient collective action.

Empirical studies showed that RL-augmented swarm robotics exhibits multiple advantages in terms of performance evaluations via simulations and real-world experiments. The experimental analysis indicates that task completion time, resource utilization were significantly better with the proposed enhanced localization approach in comparison to existing solutions regarding traditional swarm optimization. With these evaluations, they can strongly recommend RL for practical use in improving swarm robotics as it proves to work well under conditions of dynamic and unpredictable environments.

In the future, more research and improvement in RL algorithms will open new horizons for swarm robotics. Further research is needed in the design of reward functions, communication protocols and compliance with ethically sound deployment of such systems for different real world situations. Moreover, pervasive adoption of RL in swarm robotics will open up new possibilities for advanced autonomous systems leading to better adaptation and generalization as well as benefits on robustness so that robotic swarms can handle a significantly broad family operations.

REFERENCES

Bonabeau, E., Dorigo, M., & Theraulaz, G. (1999). *Swarm Intelligence: From Natural to Artificial Systems*. Oxford University Press. DOI: 10.1093/oso/9780195131581.001.0001

De Wolf, T., & Holvoet, T. (2005). *Emergence and Self-Organization: A Statement of Similarities and Differences*. In *Proceedings of the International Workshop on Engineering Self-Organising Applications (ESOA)*, pp. 96-110.

Dorigo, M., & Şahin, E. (2004). Swarm Robotics: Special Issue Editorial. *Autonomous Robots*, 17(2-3), 111–113. DOI: 10.1023/B:AURO.0000034008.48988.2b

Ducatelle, F., Di Caro, G. A., Pinciroli, C., & Gambardella, L. M. (2011). Self-Organized Cooperation between Robotic Swarms. *Swarm Intelligence*, 5(2), 73–96. DOI: 10.1007/s11721-011-0053-0

Durrant-Whyte, H., & Bailey, T. (2006). Simultaneous Localization and Mapping: Part I. *IEEE Robotics & Automation Magazine*, 13(2), 99–110. DOI: 10.1109/MRA.2006.1638022

Fujisawa, R., Dobata, S., Kubota, N., Hayashi, N., & Matsuno, F. (2008). Designing Pheromone Communication in Swarm Robotics: Decay and Diffusion Properties of Virtual Pheromones Implemented on an Event-based Simulator. *Swarm Intelligence*, 2(3-4), 185–202.

Garnier, S., Gautrais, J., & Theraulaz, G. (2007). The Biological Principles of Swarm Intelligence. *Swarm Intelligence*, 1(1), 3–31. DOI: 10.1007/s11721-007-0004-y

Khang, A. (Ed.). (2023). *AI and IoT-based technologies for precision medicine*. IGI Global.

Mirats-Tur, J. M., & Corominas Murtra, A. (2009). *A Survey on SLAM Techniques*. In *Proceedings of the International Conference on Intelligent Robots and Systems (IROS)*, pp. 2072-2077.

Nouyan, S., Groß, R., Bonani, M., Mondada, F., & Dorigo, M. (2009). Teamwork in Self-Organized Robot Colonies. *IEEE Transactions on Evolutionary Computation*, 13(4), 695–711. DOI: 10.1109/TEVC.2008.2011746

O'Hara, K., Roalter, L., & Simmel, D. (2008). *Swarm Intelligence Approaches to Robotic Surveillance*. In *Proceedings of the International Conference on Unmanned Aircraft Systems (ICUAS)*, pp. 1-10.

Parvathi, R., Vignesh, U. Diabetic retinopathy detection using transfer learning

Ratnakumar, R., & Vignesh, U. (2024). *Machine learning-based environmental, social, and scientific studies using satellite images: A case series*. AI and Blockchain Optimization Techniques in Aerospace Engineering. DOI: 10.4018/979-8-3693-1491-3.ch007

Selvi, R. T., Elakya, R., & Vignesh, U. (2024). *Securing sensitive patient data in healthcare settings using blockchain technology, Blockchain and IoT Approaches for Secure Electronic Health Records*. EHR.

Sharpe, T., & Webb, B. (1999). *Simulated and Situated Models of Chemical Trail Following in Ants*. In *Proceedings of the Fifth European Conference on Artificial Life (ECAL)*, pp. 317-324.

Verma, B., Yudheksha, G. K., & Sanjana Reddy, P. (2023). Parvathi, R., Vignesh, U. An Intelligent Flood Automation System Using IoT and Machine Learning. *Advances in Transdisciplinary Engineering*, 32, 444–449.

Vignesh, U., Parvathi, R., & Goncalves, R. (2023). *Structural and functional aspects of biocomputing systems for data processing*. Structural and Functional Aspects of Biocomputing Systems for Data Processing. DOI: 10.4018/978-1-6684-6523-3

Vignesh, U., & Ratnakumar, R. (2024). *An empirical review on clustering algorithms for image segmentation of satellite images*. AI and Blockchain Optimization Techniques in Aerospace Engineering. DOI: 10.4018/979-8-3693-1491-3.ch002

Vignesh, U., Ratnakumar, R., & Al-Obaidi, A. S. M. (2024). *AI and blockchain optimization techniques in aerospace engineering*. AI and Blockchain Optimization Techniques in Aerospace Engineering. DOI: 10.4018/979-8-3693-1491-3

Werger, B. B., & Mataric, M. J. (1996). Robotic "Food" Chains: Externalization of State and Program for Minimal-Agent Foraging. In *From Animals to Animats 4: Proceedings of the Fourth International Conference on Simulation of Adaptive Behavior* (pp. 625-634). MIT Press.

Wodrich, M., & Bilchev, G. (1997). Cooperative Distributed Search: The Ants' Way. *Control and Cybernetics*, 26(3), 413–446.

Chapter 11
Revolutionizing Healthcare Through Robotics and AI Integration:
A Comprehensive Approach

C. Saranya Jothi
Vel Tech Rangarajan Dr. Sagunthala R&D Institute of Science and Technology, India

M. A. Starlin
Vel Tech Rangarajan Dr. Sagunthala R&D Institute of Science and Technology, India

R. Roselin kiruba
Vel Tech Rangarajan Dr. Sagunthala R&D Institute of Science and Technology, India

E. Surya
Sethu Institute of Technology, India

P. Jeevanasree
Vel Tech Rangarajan Dr. Sagunthala R&D Institute of Science and Technology, India

Santhosh Jayagopalan
https://orcid.org/0009-0003-4108-9670
British Applied College, UAE

ABSTRACT

Artificial Intelligence (AI) is the technology focused on innovating the role of robotics in the healthcare process. The integration of robots in the field of healthcare and medication can effectively recognize, treat, and manage health problems. In traditional healthcare system follows the manual diagnosis methods, manual surgery, pharmacological treatments, and patient monitoring. The drawbacks of traditional methods are lack of accuracy, high cost, limited accessibility, and social barriers.

DOI: 10.4018/979-8-3693-6150-4.ch011

To overcome these issues this paper provides a detailed review of advanced robotic-assisted surgeries, Remote Patient Monitoring, patient engagement with robotic bonds, and enhancement of patient-centered care. It increases efficiency, decreases cost, and also improves the outcome of medical and healthcare systems. Additionally, it includes the applications, challenges, and possible future impacts on healthcare.

1. INTRODUCTION TO ARTIFICIAL INTELLIGENCE IN HEALTHCARE

Artificial intelligence (AI) is the intelligence of machines that can think like humans. AI is the development and making of computer systems that can do the tasks humans do. It requires human intelligence like learning, Problem-solving, and decision-making. To recognize patterns it can analyze the data, and adapt to new information, facilitating it to improve performance over time. In order to allow computer systems to perform tasks that need human intelligence. It is used in healthcare for different purposes such as diagnosis, treatment planning, and patient care. And then analyze images to diagnose diseases, for example, x-rays or CT scans. AI can detect the disease and its effect by analyzing the image. It can also develop treatment plans for diseases. Additionally, it can monitor patients and their treatments(Padhan et al., 2023). And also identifies which patients are likely to respond to a certain medication and which patients are likely to experience the Side effects of medication. Furthermore, It can monitor the patients remind them to take their medication, and provide information about a disease.

Figure 1. Healthcare Monitoring with and Without Robot

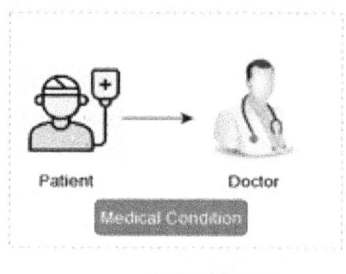

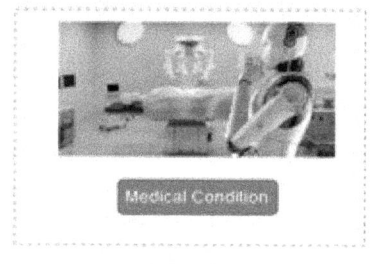

1.1 Artificial Intelligence With Robotics in Health Care

Robotics are machines that are operated by remote control (Agrawal et al., 2024). AI is installed in robotics which became powerful with installation. Robots can analyze the conditions in the environment and then it will work according to human intelligence (Che et al., 2024). It can understand and respond to their environment, enhancing their autonomy and adaptability with the power of AI. Integrating with robotics can emerge Div an era of new technology that can make important contributions to Robotics in the field of healthcare and medicine; assisting doctors and medical professionals in different tasks. In Health care robots can be used to do surgeries, monitor patients and the effects of disease, and also medication for the respective disease (Dahan et al., 2023). Some real-life examples of robots in healthcare and medicine are displayed in Figure 2.

(i) **Surgical Robotics**: These robots can do surgeries by monitoring the patient's condition and the procedure of the surgery. It is capable of performing surgical procedures, such as minimally invasive and orthopedic surgeries.
(ii) **Modular Robots**: with the help of robots,it backs rehabilitation initiatives including prosthetic limbs and therapeutic exoskeleton robots for patients with multiple sclerosis, stroke,paralysis, and traumatic brain damage.
(iii) **Medical Imaging robots**: It is used to improve medical imaging processes like CT scans and MRI scans. This results in accurate and detailed imaging leads to the right diagnosis, medication, and treatment planning.
(iv) **Drug Delivery**: Nowadays nanorobots and microrobots are used for delivering drugs where drug delivery through injection is not possible(Durrah et al., 2024). Through these robots delivery of drugs to the specific target within the body is done accurately by reducing the side effects and improving the effectiveness of treatments for diseases like disorder cancer and neurological.
(v) **Patient Assistance**: Robots are used in hospitals to monitor patients. It reminds patients to take medicine on time. These robots are helpful for patient lifting, transportation, and companionship for isolated patients (Hemmerling et al., 2024).
(vi) **Telemedicine:** These robots are used to connect patients with professional healthcare providers. These robots monitor health status, and access to care mainly in rural improving or remote areas. In order to play a vital role in health- care by improving patient outcomes, increasing efficiency, and expanding access to medical services revenue in rural areas.

Figure 2. Real-Life Application of Healthcare Monitoring System

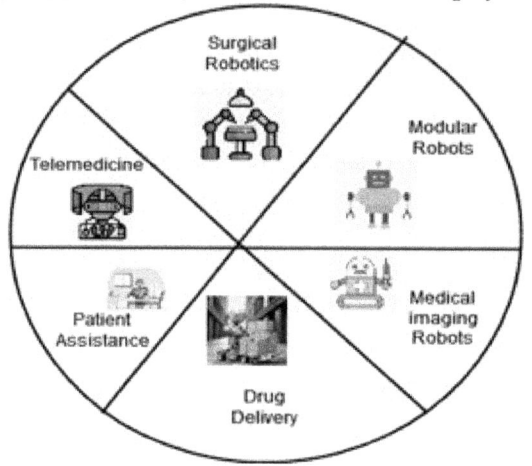

2. BACKGROUND OF HEALTH CARE BEFORE THE ROBOTICS ERA

In the past health care without any robotics medication and also surgeries was doneas shown in Fig.1. Medical professionals performed surgeries, medications, and treatment with traditional methods and tools(Hu et al., 2011). This approach had limitations in precision, efficiency, and accessibility compared to modern robotic technology. Before robotics came into healthcare, medical treatment, and procedures majorly depended on the skill and experience of human healthcare providers. Surgeons used traditional techniques and tools to perform Surgeries. This may sometimes result in an imbalance in outcomes due to human error. Monitoring the patient and medication was done by nurses or other healthcare professionals (Kaur 2024). Instead of imaging technologies such as X-rays, MRIS, and CT Scans in the diagnosis, these images are interpreted by radiologists and other Specialists. This process consumes more time and may lead to delays in the treatment. Before robotics number of healthcare professionals is more to diagnose the disease. There are some challenges and limitations as follows:

(i) Physical Demands

Doctors and surgeons often face physical strain as they work for long hours and lift or position patients which may lead to musculoskeletal injuries.

(ii) Limited Accessibility

Access to medical care in remote underserved areas is poor due to the shortage of healthcare professionals and lack of facilities. Patients from these remote areas need to travel long distances to receive medication.

(iii) Cost of Expense

The cost of healthcare expenses is high due to the labor of medical procedures and specialized equipment and facilities. Some patients can not afford it, and this leads to financial barriers for some patients.

(iv) Human Errors

Human errors were always a risk in medical procedures and medication. These errors may occur due to fatigue, distraction, miscommunication, and lack of concentration, in health care professionals.

2.1 Healthcare in Robotics

Robotics in healthcare is very useful and helpful and it makes treatment easy for healthcare professionalsas shown in Fig.1. It makes the surgical procedure easy for surgeons by reducing complications. Diagnosing the disease through imaging by CT scan, and MRI scan is very easy compared to traditional methods. It Consumes very little time compared to traditional methods and there is no delay In the medication and treatment of the disease. These robots can monitor the patient and medication other than humans. It can also monitor the surgical procedure(Kyrarini et al., 2021). It connects the patients to healthcare professionals at any time if required. It reduces the risk of medication errors. These robotics can be helpful to the people living in the rural areas. Despite these benefits, robotics in healthcare has limitations:

- **Cost of Robotics**

The robotics are high and their maintenance in the hospital is also more expensive than traditional methods.

- **Learning curve**

Specialized training is required for healthcare professionals to operate robotic systems effectively.

- **Technological Limitations**

Robotics are machines that may not work properly to that extent such as limited dexterity and tactile feedback compared to human hands.

- **Accessibility and Equity**

Even though robotics are beneficial, patients in remote areas may have limited access to robotic-assisted procedures.

3. REMOTE PATIENT MONITORING: TECHNOLOGY AND IMPLEMENTATION

Numerous industries, including healthcare, manufacturing, energy, and transportation, frequently use remote monitoring processes(Lee et al., 2021). The main objective of developing a remote monitoring system is to diagnose a person with chronic illness surgery patients, neonates, and elderly patients are monitored through this system, so it is better to continually monitor themas shown in Table1.

It describes the technique of employing technology to monitor and control processes, systems, or activities from a distance. It is essential to health care during pandemics when it is recommended that patients be closely or remotely monitored(Khang 2024). For example, remote monitoring uses wearable technology or sensors to track patients' health data, such as vital signs, symptoms, or medication adherence (Nasr et al., 2021). This makes it possible for medical professionals to monitor patients' conditions from a distance, possibly eliminating the need for in-person visits and early intervention when required. The following table describes the way how healthcare services are done in previous decades and nowadays.

Table 1. Comparison between traditional and current technologies

Aspect	Antecedent Technology	Technology now
Vital Signs Monitoring	Manual measurement with regular inspection	Constant observation using sensors and digital devices
Patient Data	Physical files and cabinets	Cloud and Health Information Exchange (HIE)
Communication	Telephone and fax	Telemedicine platforms, and secure messaging apps.
Diagnostics	Analog X-rays, film-based imaging	Digital imaging, AI-assisted diagnostics

continued on following page

Table 1. Continued

Aspect	Antecedent Technology	Technology now
Monitoring Devices	Manual blood pressure and thermometer devices	Digital blood pressure monitors, smart thermometers, ECG monitors
Patient Location	In-hospital monitoring	Remote Patient Monitoring (RPM) through wearable devices
Alerts and Notifications	Manual checks and alarms	Automated alerts via apps, SMS, and email
Data Access	Limited to physical location and working hours (Olawade et al., 2024)	24X7 access through mobile apps and web portals
Data Sharing	Hand-delivered records, postal mail	Instant electronic sharing via secure networks
Analysis and Reporting	Manual entry and review	Automated data analysis, real-time reporting, predictive analytics
Patient Engagement	Phone conversations and in-person meetings	Mobile health apps, online portals, virtual consultations
Chronic Disease Management	Regular in-person visits	Continuous remote monitoring
Emergency Response	Manual call for help, panic buttons	Automated fall detection, emergency alert systems

It has many advantages, such as increased safety, decreased expenses, and efficiency as well as the capacity to collect and process data to aid in decision-making (Khang et al., 2024). With the use of contemporary communication and sensor technologies, the patient can be able to carry out regular daily activities at home while still being closely watched thanks to technological advancements. It has two types of system:

- Smart system
- Traditional system.

The smart system is quite effective in assisting the physician in making decisions swiftly. Conventional systems rely on cable communication(Kolpashchikov et al., 2022). We all live in a digital age, and this is a fantastic chance to provide patient-centered care through the use of digital sensors and remote monitoring systems, which are essential for increasing patient accessibility. Its adaptability was demonstrated during the COVID-19 epidemic. Many technologies combine to provide a secure, effective, and dependable remote monitoring environment for a range of applications and industries projected in Fig.3.

i) **Internet of Things**: IoT devices, such as sensors, actuators, and smart meters, play a significant role in remote monitoring. They gather information from the environment and transmit it to a centralized system for analysis and action.

ii) **Wireless Communication Protocols**: Wi-Fi, Bluetooth, Zigbee, and cellular networks expedite the transmission of data from remote devices to centralized monitoring systems.

iii) **Cloud Computing**: it provides an environment where large amounts of data are collected, processed, and analyzed in the remote system. Thus it provides extensible and reliable systems.

iv) **Data Analytics and Machine Learning**: Identification of patterns and anomalies, predictive maintenance, optimization, and decision support are done using advanced analytics techniques. The massive volume of data gathered from remote monitoring is analyzed using machine learning algorithms.

v) **Cyber security:** Since all data are transferred over a network, protection against data breaches and unwanted access can be achieved through access control, encryption, authentication, and monitoring technologies.

vi) **Augmented and Virtual Reality**: Applications for augmented reality and virtual reality include giving remote operators immersive visualizations, guidance, and support for training, maintenance, and troubleshooting.

Figure 3. Basic Elements of a Remote Monitoring System

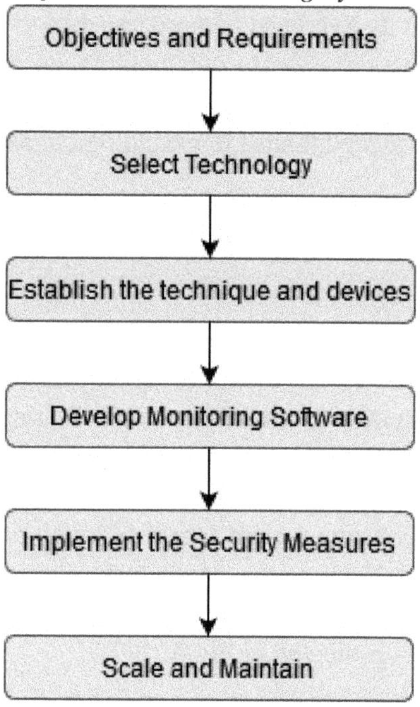

Objectives and Requirements
↓
Select Technology
↓
Establish the technique and devices
↓
Develop Monitoring Software
↓
Implement the Security Measures
↓
Scale and Maintain

- **The data acquisition system** has several sensors and embedded sensors that transmit data wirelessly, making communication simple. - A lot of wireless sensors are also employed in situations where they are not in close physical contact with the patient.
- **A data processing system** is a centralized system with data receiving and transmitting capability.
- **End-terminal at the hospital** - can be a hospital computer (or database), a specialized gadget, or the doctor's smartphone.
- **Communication Network**-establishes a communication network between the data processing and data collecting systems then sends the identified data and conclusions to a healthcare expert who is also connected to the system. Depending on how complicated the case is, the patient may be asked to take specific medications, perform certain first-aid or precautionary measures (such as calling for an ambulance), or be admitted to a hospital.

4. PATIENT ENGAGEMENT WITH ROBOTIC BONDS: HUMAN-ROBOT INTERACTION

The field of Human-Computer Interaction (HCI) has drawn a lot of attention in recent years because of the development of robots that are easy to use. It's an emerging area that allows humans to interact with robots by providing touch, speech, and visual inputs. Robots are being created quickly for real-time applications, such as cleaning domestic appliances, caring for the elderly, and working in the educational sector (Lee et al., 2024).

4.1 Applications of HCI

Numerous numbers of areas where human-computer interaction plays a vital role:

1. **Robotic Assistance**: it can assist people with their everyday tasks by reminding them to take their medications on schedule and keeping an eye on their health by examining their vital signs.
2. **Remedial Interaction**: Certain robots are made to participate in therapeutic exercises with patients, like mental workouts or counseling sessions. The general well-being and mood of the patient may be enhanced by these exchanges.
3. **Dropping Disquiet**: A robot will converse with people as they receive therapy, offering information and answering questions in an effort to divert their attention and lessen fear. Improves the interaction between humans and robots.

4. **Improving Communication**: Patients with autism and stammering mouths are fitted with robots that help them communicate better and interact better with medical professionals ., (Manickam et al., 2022).
5. **Remote Patient Monitoring**: Robots with cameras and sensors can support Telepresence, which enables medical professionals to communicate virtually with patients as projected in Fig.4.

Figure 4. Application of Human-Computer Interaction

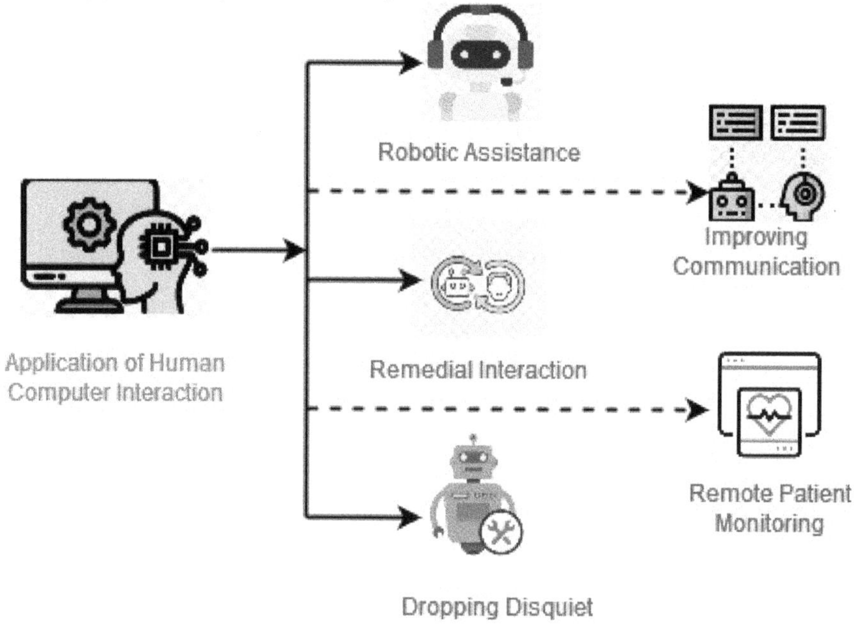

Three overlapping areas of research are necessary to make robotics useful in solving a plethora of real-world challenges: human-robot collaboration.

- In cognitive — the way that people process information.
- Emotion and perception—Robots acquire human-like abilities to trust, communicate, and understand information.
- It picks up hardware-building skills.

4.2 Challenges

Human-robot interaction (HRI) presents several challenges to ensure the successful integration of robots here are some key challenges:

- **Natural Communication**: It's still quite difficult to design robots that can interact with people in a way that feels intuitive and natural to them. This entails producing and comprehending natural language, deciphering nonverbal clues from gestures and expressions on the face, and keeping discussions coherent.
- **Social Acceptance and Trust**: It's critical for developers to build trust in robots due to a number of factors, including job stability, privacy concerns, and unfavorable impressions.
- **Customization and Personalization**: As artificial intelligence technology advances, common people can now connect with robots according to their needs and interests, improving human-robot relationships.

Addressing these challenges requires interdisciplinary collaboration between researchers, engineers, psychologists, ethicists, and other stakeholders (Marcus et al., 2024). By leveraging advances in robotics, artificial intelligence, and human-centered design, we can overcome these challenges and unlock the full potential of human-robot interaction in various domains.

4.3 Humanoid

It alludes to the notion of imbuing nonhuman agents with human-like intentions, feelings, motives, and other characteristics in their imagined or real behavior (Joseph et al., 2018). They expedited the adoption of Humanoid SR in the health sector during the COVID-19 epidemic. Through a tablet inserted in its chest, humanoid SR Mitra helps COVID-19 patients in hospital beds in India stay connected to their loved ones (Kaiser et al., 2021). In order to help them, it enables patients to speak with their families. In addition to taking a patient's temperature, it uses facial recognition to identify those who have already been admitted. A different humanoid was also able to provide food and medication to individuals with severe COVID-19. Hospitals use polyglot humanoids to monitor patients' body temperatures for the coronavirus. Additionally, patients were to be reminded to wear masks when visiting the hospital. Similarly, Sophia created a very advanced robot to monitor elderly patients and combat their feelings of loneliness.

4.4. Benefits of Robots in Healthcare

- **Accuracy:** Robots are more accurate and less likely to have issues during surgery than people are.

- **Invasive Procedures**: A lot of robotic systems allow for less traumatic surgery, which results in smaller incisions, less tissue damage, less discomfort, shorter hospital stays, and quicker patient recovery.
- **Remote Surgery**: This modality of care enables doctors to operate from a distance using robotic equipment, providing expert care to underprivileged areas or in times of need.
- **Robots in Transportation**: By helping with the lifting and moving of large objects, robots can lower the danger of accidents occurring when people are at work.
- **Sterilization and clean-up:** Sterilization and cleanup: Robots are outfitted with UV lights or other sterilization techniques to efficiently clean hospital rooms and lower the risk of infection during treatments.
- **Education and Training**: Medical professionals and students can hone their surgical abilities in a controlled setting without endangering patient safety thanks to robotic simulatorsas shown in Fig.5.

Figure 5. Benefits of Robots in Healthcare

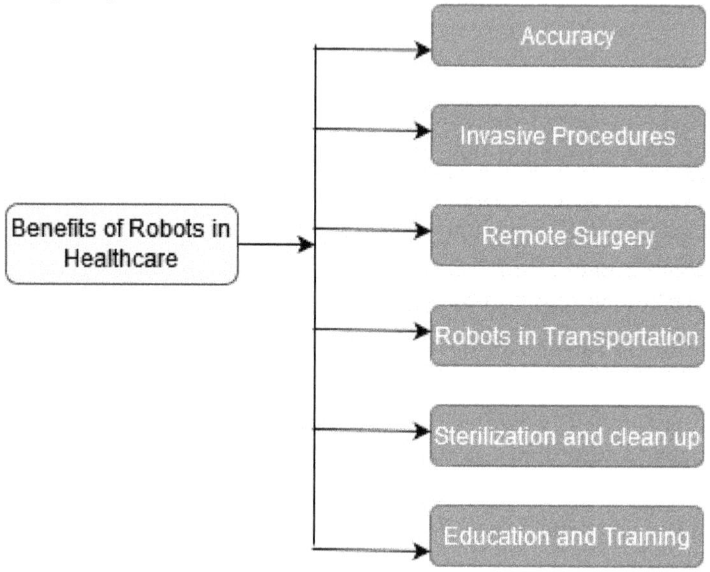

5. ENHANCING PATIENT- CENTERED CARE THROUGH AI

Patient-centric care approach in healthcare systems is a necessary one to provide efficient health outcomes.Before developing the modern medicines and enormous usage of medical tests, patient care begins with physical inspections, inquiries from patients for getting medical history. With the knowledge about patient's well-being and lifestyle, doctors select best treatment. However, the existing healthcare system fails to meet patient preferences, which causes discomfort and inefficiencies. By incorporating AI Technologies in healthcare can improve quality, accessibility and customization of healthcare services. This integration between the AI and healthcare makes patient-centered care to be executed in medical field(Ragno et al., 2023).

The following methods describe how AI can support to patient-centered care in healthcare systems.

1. Customized Treatment Services

AR can provide surgeons with real-time guidance during complex procedures, overlaying anatomical structures, vital signs, and preoperative imaging directly onto the patient's body; this enhances precision and reduces the risks of error (Ichihashi, K. and Fujinami, K. 2019).

2. Virtual Health Assistants (VHA)

It offers tailored guidance and support to the patients, such as alerts for medicines, fixing doctor appointment and easy access of health records.

3. Remote Monitoring and Telemedicine

Remote patient monitoring (RPM) is one of the healthcare applications which helps doctors to observe patients who are hospitalized, patients with chronic or acute conditions while they are at remote locations and also elderly care at home. Wearable devices and mobile applications with AI features facilitate for remote monitoring of patient health status and inform this to both patient and health care provider. This allows them to take any action if any abnormalities in the health report.

Various technologies like telehealth, IoT, cloud, fog, edge, blockchain and AI are used to accomplish Remote Patient Monitoring (RPM) applications like vital signs, physical activities, emergencies, and chronic diseases. .

4. Predictive Analytics

It helps the healthcare providers for early identification of diseases and diagnosis. AI algorithms analyse large volume of patient data, health report, biological information and patient's medical images in order to detect early stages of some diseases like diabetes, cancer and heart diseases.

Early detection of diseases helps the patients in following ways;

a. Speedy Recovery
b. Suppress growth at initial stage
c. Reduce Risk factors
d. Improved patient outcomes

5. Diagnostic Assistance

Various AI technologies like machine learning, Deep learning, Natural language processing supports better diagnostic assistance while predicting diseases and also for providing proper treatment. The rapid development of AI technology provides greater diagnostic accuracy, improved patient outcomes and simplified medical processes.

6. Behavioral Analysis and Support

AI algorithms analyze human behaviours to monitor, diagnose and treat the patients with behavioural and mental health problems. To identify early symptoms of mental health conditions, AI–driven technologies observe the following:

1. Speaking and Text analysis
2. Behavioral patterns
3. Monitoring through Wearable devices
4. Alert systems

Figure 6. Patient-centered care through AI

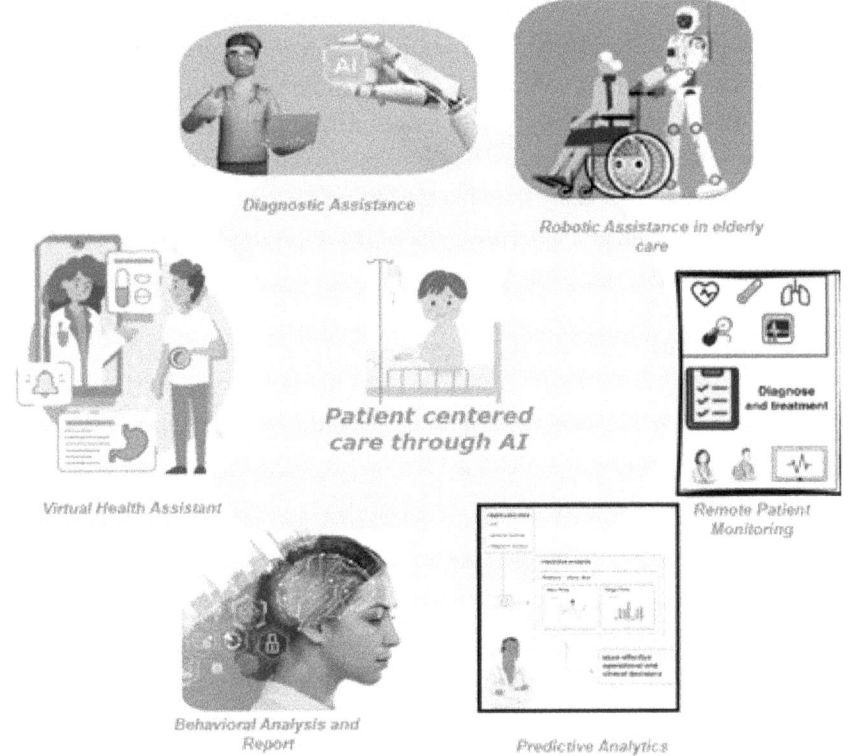

7. Robotic Assistance in Elderly Care

AI-driven robotic assistance is helpful for taking care of elderly patients those who are in home care. Without human intervention, robots can do everything that aged people need.

8. Ethical Considerations

AI in healthcare systems make ensure that patient-centered care by giving patient privacy, accountability, transparency, data security, bias and fairness.

6. CHALLENGES IN IMPLEMENTING AI IN HEALTHCARE

All these analysis shows that integration of AI technologies in healthcare systems result better patient outcomes, 24/7 health monitoring, improved accuracy in prediction of the diseases, efficient diagnosis assistance. Even though AI has several advantages in healthcare systems, when it comes to daily practice there are many challenges including technical, ethical, legal and implementing fields. To overcome all these challenges, effective integration of AI technologies in healthcare is needed. Some of the challenges are discussed below:

1. Data quality and security:

Data privacy and the security of patients' health is bigger issue. Shortage of quality healthcare data produces unreliable results. Ensuring data security, privacy and availability are critical in reliable AI outcomes.

2. A Legal and Ethical Aspect:

• An inconsistent and inaccurate medical data set leads to biases and errors. A bias existing in training data is inherited by AI integrated systems. Which leads to unfair health outcomes. Also fairness of data analysis to be done in order to obtain accurate results. Addressing bias is a major challenge.

- In cases of any errors or adversarial outcomes, finding AI-driven decisions can be complex, especially.
- Transparency must be there while taking healthcare decisions.

3. Interoperability in healthcare:

AI systems working with various Electronic Health Report (EHR) systems and anayzing medical images is critical.

4. Limited knowledge on AI technologies:

Creating trust with patients on AI-driven healthcare systems is such a difficult one. General public has less knowledge about the advantages and drawbacks of AI-powered healthcare treatment. This results doctors to a challenging task to adopt AI tools for medical purposes.

With the arrival of new AI technologies in healthcare improving quality and accuracy of patient outcomes in healthcare systems. But, greatest challenge is to adopt all these technologies with patients. Fears about privacy and security, lack of technological know-how affect patients interaction with AI technologies. Making reliability and awareness on patients about new AI integrated healthcare systems is difficult process.

7. FUTURE IMPACTS AND OPPORTUNITIES IN AI-DRIVEN HEALTHCARE

AI can play a vital role in the healthcare offerings of the future. Vast number of machine learning algorithms can be used in every aspect of medical field such as precision medicine, operations, diagnosis and treatment. To acquire better solution following things are needed for both patients and healthcare professionals.

1. Prioritize education
2. Training
3. Digital literacy

By providing above all we can improve adoption of patients with AI-driven healthcare system stronger. And also healthcare providers can be trained to handle AI tools with good understanding.

Integration of Robotics through AI in Healthcare will play a major role in future. AI occupies huge place in medical industry. Which supports doctors by providing efficient diagnostic assistance for patients.The future of AI in medical field seemsoptimistic, possible, and bright;therefore much work can be achieved with the help of AI technologies.

7.1. OPPORTUNITIES IN AI-DRIVEN HEALTHCARE:

1. Exploration and Drug Development using AI

AI can analyze huge amount of datasets to develop proper drug candidates and also to find its safety and efficiency for drug discovery process(Yang et al., 2020). This is used to test drugs for suitable patients and its outcome and side effects also monitored.

2. Global Health

It improves accessibility of healthcare services to remote areas through telemedicine and mobile health applications. This also useful for monitoring patients virtually and suggest proper medicine at the earliest. Prediction of infectious diseases by analyzing various health databases also possible using AI.

3. Education and Training

AI-empowered simulations support Doctors with realistic circumstances to improve their prediction capacity of diseases. AI provides continuous learning of newly arrived medical technologies to healthcare providers also helps them to stay updated with these.

4. Ethical AI Development

Provides Data security to protect patient privacy and create reliability over AI tools. Development AI algorithms reduce bias and ensures transparency and fairness in healthcare treatment.

5. Multidisciplinary Collaboration

AI ensures combined and comprehensive approach to patient-centered care by facilitating interaction and association of various healthcare providers. Collaboration between healthcare companies and technological institutions provides innovative, safe, effective, and widely embraced AI solutions.

8. CONCLUSION

Integration of Artificial Intelligence with robotics in healthcare systems provides promising results in efficient patient outcomes. The traditional healthcare systems without incorporating AI technologies have lot of drawbacks such as high cost, long waiting time, less accessibility of healthcare facilities in rural areas, Lack of advanced technological treatments, Risk of medical errors, Less accuracy. When it comes to

AI-empowered healthcare systems, has numerous benefits over traditional medical systems. Adoption and implementation of new AI technologies, Data privacy, social and ethical issues are the challenges to implement AI in medical field. In this paper, a detailed survey of AI-empowered healthcare systems, various applications and its advantages, challenges of implementing AI in real world medical practices are discussed. The solutions to overcome these obstacles also explained. Development of AI with advanced and trustable algorithms can address the challenges specified here.

REFERENCES

Agrawal, A., Soni, R., Gupta, D., & Dubey, G. (2024). The role of robotics in medical science: Advancements, applications, and future directions. *Journal of Autonomous Intelligence*, 7(3). Advance online publication. DOI: 10.32629/jai.v7i3.1008

Che, C., Zheng, H., Huang, Z., Jiang, W., & Liu, B. (2024). Intelligent robotic control system based on computer vision technology. *arXiv preprint arXiv:2404.01116.*

Dahan, F., Alroobaea, R., Alghamdi, W. Y., Mohammed, M. K., Hajjej, F., & Raahemifar, K. (2023). A smart IoMT based architecture for E-healthcare patient monitoring system using artificial intelligence algorithms. *Frontiers in Physiology*, 14, 1125952. DOI: 10.3389/fphys.2023.1125952 PMID: 36793418

Durrah, O., Aldhmour, F. M., El-Maghraby, L., & Chakir, A. (2024). Artificial Intelligence Applications in Healthcare. In *Engineering Applications of Artificial Intelligence* (pp. 175–192). Springer Nature Switzerland. DOI: 10.1007/978-3-031-50300-9_10

Hemmerling, T. M., & Jeffries, S. D. (2024). Robotic Anesthesia: A Vision for 2050. *Anesthesia and Analgesia*, 138(2), 239–251. DOI: 10.1213/ANE.0000000000006835 PMID: 38215704

Hu, J., Edsinger, A., Lim, Y. J., Donaldson, N., Solano, M., Solochek, A., & Marchessault, R. (2011). *May. An advanced medical robotic system augmenting healthcare capabilities-robotic nursing assistant. In 2011 IEEE international conference on robotics and automation.* IEEE.

Joseph, A., Christian, B., Abiodun, A. A., & Oyawale, F. (2018). A review on humanoid robotics in healthcare. In *MATEC Web of Conferences* (Vol. 153, p. 02004). EDP Sciences. DOI: 10.1051/matecconf/201815302004

Kaiser, M. S., Al Mamun, S., Mahmud, M., & Tania, M. H. (2021). Healthcare robots to combat COVID-19. *COVID-19: Prediction, decision-making, and its impacts*, pp.83-97.

Kaur, J. (2024). Revolutionizing Healthcare: Synergizing Cloud Robotics and Artificial Intelligence for Enhanced Patient Care. In *Shaping the Future of Automation With Cloud-Enhanced Robotics* (pp. 272-287). IGI Global.

Khang, A. (Ed.). (2024). *Medical Robotics and AI-Assisted Diagnostics for a High-Tech Healthcare Industry*. IGI Global. DOI: 10.4018/979-8-3693-2105-8

Khang, A., Rath, K. C., Anh, P. T. N., Rath, S. K., & Bhattacharya, S. (2024). Quantum-Based Robotics in the High-Tech Healthcare Industry: Innovations and Applications. In *Medical Robotics and AI-Assisted Diagnostics for a High-Tech Healthcare Industry* (pp. 1-27). IGI Global.

Kolpashchikov, D., Gerget, O., & Meshcheryakov, R. (2022). *Robotics in healthcare. Handbook of Artificial Intelligence in Healthcare* (Vol. 2). Practicalities and Prospects.

Kyrarini, M., Lygerakis, F., Rajavenkatanarayanan, A., Sevastopoulos, C., Nambiappan, H. R., Chaitanya, K. K., Babu, A. R., Mathew, J., & Makedon, F. (2021). A survey of robots in healthcare. *Technologies*, 9(1), 8. DOI: 10.3390/technologies9010008

Lee, D., & Yoon, S. N. (2021). Application of artificial intelligence-based technologies in the healthcare industry: Opportunities and challenges. *International Journal of Environmental Research and Public Health*, 18(1), 271. DOI: 10.3390/ijerph18010271 PMID: 33401373

Lee, O. E., Lee, H., Park, A., & Choi, N. G. (2024). My precious friend: Human-robot interactions in home care for socially isolated older adults. *Clinical Gerontologist*, 47(1), 161–170. DOI: 10.1080/07317115.2022.2156829 PMID: 36502295

Manickam, P., Mariappan, S. A., Murugesan, S. M., Hansda, S., Kaushik, A., Shinde, R., & Thipperudraswamy, S. P. (2022). Artificial intelligence (AI) and internet of medical things (IoMT) assisted biomedical systems for intelligent healthcare. *Biosensors (Basel)*, 12(8), 562. DOI: 10.3390/bios12080562 PMID: 35892459

Marcus, H. J., Ramirez, P. T., Khan, D. Z., Layard Horsfall, H., Hanrahan, J. G., Williams, S. C., Beard, D. J., Bhat, R., Catchpole, K., Cook, A., Hutchison, K., Martin, J., Melvin, T., Stoyanov, D., Rovers, M., Raison, N., Dasgupta, P., Noonan, D., Stocken, D., & Paez, A. (2024). The IDEAL framework for surgical robotics: Development, comparative evaluation and long-term monitoring. *Nature Medicine*, 30(1), 61–75. DOI: 10.1038/s41591-023-02732-7 PMID: 38242979

Nasr, M., Islam, M. M., Shehata, S., Karray, F., & Quintana, Y. (2021). Smart healthcare in the age of AI: Recent advances, challenges, and future prospects. *IEEE Access: Practical Innovations, Open Solutions*, 9, 145248–145270. DOI: 10.1109/ACCESS.2021.3118960

Olawade, D. B., David-Olawade, A. C., Wada, O. Z., Asaolu, A. J., Adereni, T., & Ling, J. (2024). Artificial intelligence in healthcare delivery: Prospects and pitfalls. *Journal of Medicine, Surgery, and Public Health*, 3, 100108. DOI: 10.1016/j.glmedi.2024.100108

Padhan, S., Mohapatra, A., Ramasamy, S. K., & Agrawal, S. (2023). Artificial intelligence (AI) and Robotics in elderly healthcare: Enabling independence and quality of life. *Cureus*, 15(8). Advance online publication. DOI: 10.7759/cureus.42905 PMID: 37664381

Ragno, L., Borboni, A., Vannetti, F., Amici, C., & Cusano, N. (2023). Application of social robots in healthcare: Review on characteristics, requirements, technical solutions. *Sensors (Basel)*, 23(15), 6820. DOI: 10.3390/s23156820 PMID: 37571603

Reddy, S., Fox, J., & Purohit, M. P. (2019). Artificial intelligence-enabled healthcare delivery. *Journal of the Royal Society of Medicine*, 112(1), 22–28. DOI: 10.1177/0141076818815510 PMID: 30507284

Soljacic, F., Law, T., Chita-Tegmark, M., & Scheutz, M. (2024). Robots in healthcare as envisioned by care professionals. *Intelligent Service Robotics*, 17(3), 1–17. DOI: 10.1007/s11370-024-00523-8

Wamba, S. F., Queiroz, M. M., & Hamzi, L. (2023). A bibliometric and multidisciplinary quasi-systematic analysis of social robots: Past, future, and insights of human-robot interaction. *Technological Forecasting and Social Change*, 197, 122912. DOI: 10.1016/j.techfore.2023.122912

Yang, G., Pang, Z., Deen, M. J., Dong, M., Zhang, Y. T., Lovell, N., & Rahmani, A. M. (2020). Homecare robotic systems for healthcare 4.0: Visions and enabling technologies. *IEEE Journal of Biomedical and Health Informatics*, 24(9), 2535–2549. DOI: 10.1109/JBHI.2020.2990529 PMID: 32340971

Chapter 12
Slithering Intelligence for Predicting Tectonic Plate Movement

Maheswari Raja
Rajalakshmi Institute of Technology, Coimbatore, India

Ashiya Parveen
Sri Eshwar College of Engineering, Coimbatore, India

Manobalan Manokaran
Sri Eshwar College of Engineering, Coimbatore, India

Mythili Palanisamy
Sri Eshwar College of Engineering, Coimbatore, India

P. Vijaya
Modern College of Business and Science, Oman

ABSTRACT

An earthquake is one of the most devastating natural catastrophes that may cause major infrastructure damage and casualties. Early earthquake detection can be crucial for minimizing damage and saving lives. The purpose of this study is to make earthquake magnitude and depth predictions utilizing factors including time, place, and previous seismic activity data. Snakes can predict earthquakes and landscapes 3-5 days before they occur, and they can seismically change up to 120 kilometers (about 74.56 mi) away from the Epicenter (the point from the Earth's surface directly above the focus of an earthquake). They may utilize their specialized sense organs to sense electromagnetic fields and ground vibrations. Snakes might use IR

DOI: 10.4018/979-8-3693-6150-4.ch012

Copyright © 2025, IGI Global. Copying or distributing in print or electronic forms without written permission of IGI Global is prohibited.

radiation detection through their eyes to sense seismic changes. Unusual behavior in snakes could be indicators of approaching earthquakes, as a response to sensed vibrations to sensed vibrations or electromagnetic fields.

1. INTRODUCTION

1.1. Earthquake in terms of Geography

In terms of geography, an earthquake is a sudden shake of the Earth's surface, caused by the sudden emergence of stress and geological faults or by volcanic activity, this release occurs when the stored energy is discharged without any warning, resulting movement of ground and the creation of seismic waves that travel through the Earth. These seismic events occur when the tectonic plates are colliding.

1.2. Earthquake in terms of Energy

The sudden release of the strain energy in the earth's crust results in the shaking of waves that radiate from the earthquake source kinetic energy is released from the upper layer of the earth that causes a ground shake which is equal to thousands of nuclear bomb explosion which end up vibration in all direction and widespread damage, depends on its magnitude and depth.

1.3. Role of snakes in predicting earthquakes

Snakes can sense earthquakes through their sensitivity to vibration and seismic activities. Earthquakes generate ground vibrations, which snakes can detect using their specialized sense organs. Some studies of snakes said that the snakes may also be able to perceive IR radiation.

Snakes have specialized sense organs related to vibrations, particularly in the form of hearing. They rely on bone conduction within the skull to perceive sound vibrations. This means that they can sense vibrations. This means that they can sense vibration through their jaws and skull which helps them detect potential threats or prey in their surroundings.

Snakes can predict earthquakes and landslides for 3-5 days and 120 km (75 miles) wide. Because snakes can locomote from one end to another, they can easily sense the ground vibration and electromagnetic field using specialized sense organs, some of the snakes move away from the ground vibration sensed some of its move towards the ground vibration sensed and some of its, move towards the ground vibration. Snakes can sense the seismic changes too with the help of IR radiation in their eyes. Snakes show unusual behavior when they sense the ground vibration. The prototype works like a snake's sense organs to predict earthquakes and by chance to predict landslides like a snake and give an alert message to the people who are in the range of impact.

2. LITERATURE SURVEY

In (Bai et al., 2022) discuss the numerous studies conducted on earthquake prediction employing various methods and antecedents to alert people of potential devastation and preserve human life. Due to the difficulty and unpredictability of earthquake prediction, many efforts have fallen short in this regard. Consequently, we employ a potent deep learning approach in our study. an effective approach for identifying intricate correlations in time series data. Long short-term memory is the name of the method (LSTM). This approach is used in two research cases: the first learns all the datasets in a single model, and the second instance learns the correlations between two split groups while considering their range of magnitude. The findings demonstrate that learning from deconstructed datasets produces more accurate predictions. because it takes use of how different kinds of earthquake events are. The author of this study describes how seismic wave classification is achieved by distributing acoustic sensor measurements using deep learning models. Results corroborate the hypothesis techniques for training and classification based on the usage of seismic time-series from traditional seismographs (like the STEAD database). Using the help of pre-existing seismic databases, this method has allowed us to train CNN, FC-ANN, and CNN+LSTM models, which we can then use to categorize DAS readings.

Keep in mind that this technique can train well even with small DAS datasets. Metrics like as accuracy and F-score evaluation demonstrate that convolutional models are better suited to learning the characteristics of seismic waveforms; of the models examined, CNN has shown itself to be the most effective at classifying DAS seismic waveforms. Take note that it should be noted that the metrics produced from the analysis of assessing them using DAS measurements are marginally lower than those acquired from testing the models with conventional seismic waves. This conduct may be accounted for by the inherent variations in seismic waveforms detected by DAS systems and traditional seismographs, which result from a confluence of the three basic factors:

The factors that affect strain transfer efficiencies along the sensing fiber are as follows:

(i) uneven fiber coupling to the ground along the cable and varying soil properties;

(ii) the angle at which the earthquake wave arrives in relation to the optical cable, which may cause varying local longitudinal strain in the fiber; and

(iii) the presence of amplitude fading points, which result in short blind fiber sections and unreliable phase retrieval. These circumstances result in Seismic DAS readings via a certain optical fiber cable can have drastically varied SNR and temporal waveforms, which can impact the time series labeling and categorization.

The approach put out here may serve as a foundation for the creation of an early warning seismic system utilizing DAS technology. A further improvement in the classification algorithms' performance might come from adding DAS recordings to the seismograph training dataset. This methodology would integrate certain aspects of DAS measurements into the educational process, concurrently utilizing the extensive accessibility of traditional seismic recordings. Furthermore,

models that have been pre-trained using traditional seismic waveforms may benefit from an extra fine-tuning training phase, wherein additional DAS training datasets may be utilized to enhance the classifier parameters. These methods can be used very used to the CNN architecture that is being shown here, since this type of model has a higher capacity to adjust to seismic waveforms that are not observed during training. Improved CNN architecture and a training phase that uses heterogeneous databases to distinguish seismic waveforms from other sources of mechanical vibrations could be additional performance improvements. This may be especially significant if deployed telecom optical fibers are used for dispersed seismic monitoring (such as those in cities or close to roads).

Lastly, it should be noted that all these enhancement options could also be gained from the accurate labeling of DAS readings, which might be necessary all the way down the sensing optical fiber. In fact, labeling seismic DAS data is necessary to obtain more dependable training and testing when applying dispersed seismic data collection. This would be a difficult task, where the assistance of a seismology specialist is essential to obtaining precise target labels (Hernández et al., 2022).

Author (Kubo et al., 2024) explains machine learning applications in several fields of earthquake seismology, particularly in the creation of earthquake catalogs, seismicity assessments, ground motion forecasting, and geodetic data applications. These fields have significantly improved thanks to ML technology, yet some hurdles still exist. For instance, imbalances in natural datasets might lead to misunderstanding or misevaluation in numerous situations. Effective methods for solving this issue include transfer learning, concurrent application of domain knowledge, and data augmentation. Even though deep learning is a black box, the most recent methods, including neural operators and PINNs, XAI and BNN can handle them. The effectiveness, precision, and adaptability of machine learning (ML) are the main factors propelling its adoption for a range of applications in seismic seismology. There are still a lot of issues that machine learning can effectively tackle and using it will increase and improve our understanding of earthquake seismology.

The increasing spatiotemporal density of geophysical measurements near subduction zones does not, however, make it any easier to forecast the future earthquake's time or magnitude. Here, we replicate many seismic cycles in a subduction zone at lab size. The model simulates magnitude Mw 6.2–8.3 earthquakes with a coefficient of variation in recurrence intervals of 0.5, like true subduction zones, by generating both partial and full margin ruptures. We demonstrate the
unreliability of the standard method of calculating the size of the next earthquake using slip-deficit. Conversely, through the reconstruction and appropriate interpretation of the spatiotemporally complex loading history of the system, machine learning accurately predicts the timing and magnitude of laboratory earthquakes. These findings indicate significant advancements in the prediction of actual earthquakes, as they imply that the intricate movements seen by geodesists during sinking (Al Banna et. al., 2022)

Since many of the prediction methods now in use have a high false alarm rate, a lack of precise prediction processes contributes to the disastrous effects of earthquakes. Because AI-based approaches are so accurate compared to conventional techniques, there is now more room for improvement in this prediction process. These techniques can greatly minimize damage since they allow for the evacuation of the affected region in accordance with forecasts. This paper examined current methods for AI-based earthquake prediction to streamline the forecast process. Between 2005 and 2019, a total of 84 publications were chosen from the academic research databases. Tables summarizing the claimed methodologies were presented, along with a thorough discussion. The performances of various strategies were then compared. The techniques' disclosed outcomes, the datasets used, and the assessment measures employed were additionally condensed into tables. To aid researchers in creating more precise procedures, this paper attempts to demonstrate the influence of AI-based approaches on earthquake prediction (Corbi et al., 2019).

In this, the authors (Debnath et al., 2021) tried to predict the sorts of earthquakes to manage disasters. The Weka tool has been used to create a system for predicting the sorts of earthquakes. We examined whether the classification method would perform better in predicting the kinds of earthquakes that will occur in the region of India (. For comparison, seven distinct supervised machine-learning algorithms have been employed. The accuracy rate represents the forecasted outcome that has been recorded. This demonstrated the possibility of comparing the categorization algorithm (Mousavi et al., 2021). It was found that: for the Gujarat area, the Simple Logistic, Bayes Net, and Random Tree techniques obtained the most accuracy with 98.18% rates for the Andaman & Nicobar region, the Simple Logistic approach produced the highest accuracy with a 99.94% rate for the North East region, the LMT method achieved the highest accuracy with a rate of 99.86%; for the UP, Bihar, and Nepal region, the LMT method achieved the highest accuracy with a rate of 99.86%; and for the North of India region, the Bayes Net, Simple logistic, and LMT methods achieved the highest accuracy with a rate of 99.86%.(Rouet-Leduc et al., 2017).

It is possible to infer that the Logistic Model Tree and Simple Logistic classifier algorithms are the most effective for identifying earthquake impacts in India after the forecasting of earthquake types and model performance verification (Chegeni et al., 2022).

3. EARTHQUAKES

Every day, moderate-sized earthquakes take place. On the other hand, Powerful and Destructive earthquakes happen less frequently. The deep look into the plate boundaries, particularly along the Convergent boundaries, shows that earthquakes are more frequent. More earthquakes Occur in India where the **Indian plate and Eurasian plate clash.** Consider the Himalayan region for instance (Nakata et al., 1990).

A Glimpse of Indo-Eurasian plate

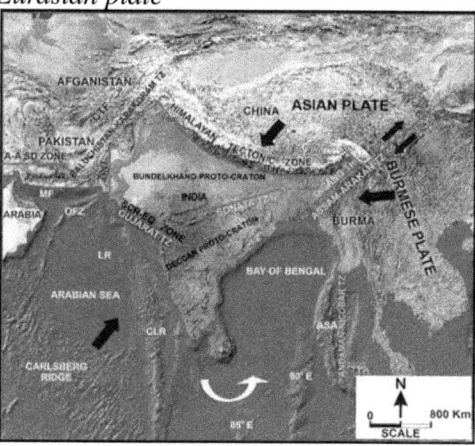

Figure 1. The Indo-Eurasian plate

Figure 1 shows India's peninsular region is considered a stable area. Earthquakes are felt on smaller edges of the plates. Two examples of earthquakes that occurred in the peninsular areas are the Konya earthquake in 1967 and the Latur earthquake in 1993.

3.1. The Seismic Parts of India

The Indian Seismologists have categorized the Indian geography of earthquakes into **Four seismic zones:**

- **Zone II**
- **Zone III**
- **Zone IV**
- **Zone V**

These Zones are categorized based on the Modified Mercalli (MM) intensity, happening in India, which evaluates the impact of earthquakes. This System was updated during the following Killari earthquake in Maharashtra in 1993, merging the low danger zone, or Seismic Zone I, with Seismic Zone II. Zone I is therefore excluded from the mapping.

3.1.a). ZONE II: It's classified as mild intensity. 40.93% of the country's land area is covered by it. It includes the Karnataka Plateau as well as the Peninsula region.

3.1.b). ZONE III: It's classified as low intensity. It makes up 30.79% of the country's total land area. The state is composed of parts of Punjab, Rajasthan, Madhya Pradesh, Bihar, Jharkhand, Chhattisgarh, Maharashtra, Odisha, and Tamil Nadu, as well as Kerala, Goa, and the Lakshadweep.

3.1.c). Zone IV: This is referred to as a high-intensity zone. It covers 17.49% of the land area of the nation. It encompasses the northern regions of Uttar Pradesh, Bihar, and West Bengal; the remaining portions of Himachal Pradesh, Jammu & Kashmir, Sikkim; the National Capital Territory (NCT) of Delhi; and Rajasthan's and Maharashtra's western shore.

3.1.d). Zone V: This area is regarded as exceedingly severe. It makes up 10.79% of the total land area of the nation. It includes the Rann of Kutch in Gujarat, a part of Uttarakhand, Himachal Pradesh, North Bihar, and the Andaman and Nicobar Islands.

3.1.e). Earthquakes and India

This table represents the earthquakes in different places in India, it mentions of year and location that the earthquakes happened in India.

Table 1. Earthquakes over time

Year	Location	Magnitude	Observations
1905	Kangra, Himachal Pradesh	7.8	Considered one of India's deadliest earthquakes
1934	Bihar-Nepal Border	8.0	Caused widespread devastation in Bihar and Nepal
1967	Koyna, Maharashtra	6.3	Notable for its association with the Koyna Dam reservoir
1993	Latur, Maharashtra	6.2	Resulted in significant destruction and loss of life
2001	Bhuj, Gujarat	7.7	Devastated several districts in Gujarat, including Bhuj
2015	Nepal-India Border (Nepal Earthquake)	7.8	Had a significant impact in parts of India, especially in northern regions
2020	Sikkim	6.1	Felt across the northeastern region of India

The Earthquakes that made the world to look around India

The places in India listed below are those that experience earthquakes frequently; the list is arranged chronologically.

i. 1991 – Uttarkashi
ii. 1993 – Latur
iii. 1941 – Andaman
iv. 1975 – Kinnaur
v. 1967 – Koyna Nagar

A visual representation of the seismic zones of India

Figure 2. Visuals of different seismic zones in India

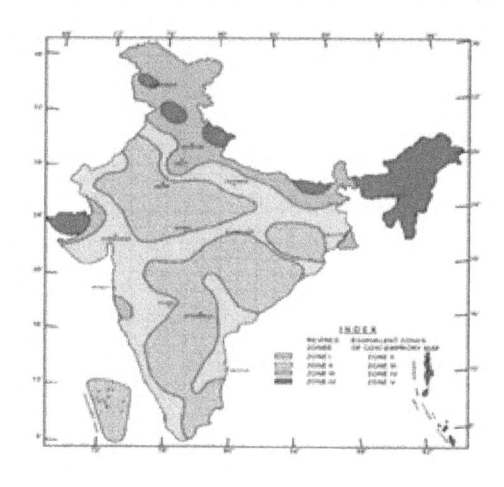

Figure 2 represents the seismic divisions of India. These seismic divisions are segregated by the Indian seismologists by analyzing the different geographical locations into Zone II(Blue), Zone III(Yellow), Zone IV(Orange), and Zone V (Red) (Mohapatra et al., 2010).

Figure 3. Rising trend of EARTHQUAKE in recent times

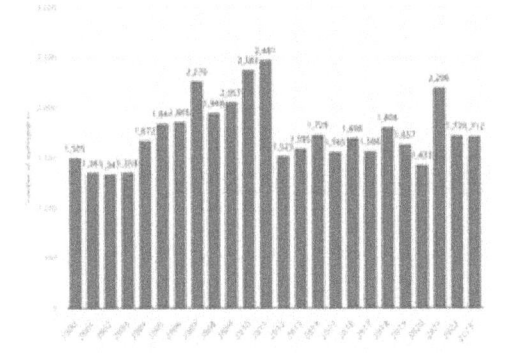

Figure 3 exhibits the Rate of Occurrence of earthquakes is increasing seamlessly, at a fierce pace which can kill a large mass of people. This increase in the earthquake rate from 2000 to 2023. The graph depicted here shows the vulnerability of the natural calamity "The Earthquake" (Holzer et al., 2013).

Figure 4. Statistical and pictorial representation provided by Indian seismologists

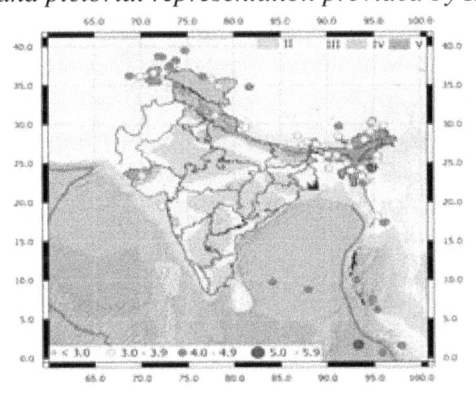

Figure 4 represents the seismicity in India during the period of 1st-30th September in 2023 specified with the latitudes and longitudes given below Along with the seismic Zones. The symbolic representation shows the different magnitude ranges of the earthquake are represented in green circles (<3.0), Yellow circles (3.0-3.9), Red circles (4.0-4.9), Orange circles (5.0-5.9), and Blue circles (>6.0).

India is a country that is around earthquake-prone and has seen very significant earthquakes that have claimed many lives through the evaluation of the nation's susceptibility to seismic hazards linked to these occurrences by understanding the death rates. Analyzing the patterns and contributing factors to India's earthquake mortality throughout the previous 25 years is the goal of this article (Bilham,R., 2004).

After the discussions of its occurrence Frequency, Earthquakes, and Death Rates are some of the most vital ones to be taken care of.

4. TRENDS IN EARTHQUAKE MORTALITY

4.1. Historical Overview: The analysis reveals a fluctuating pattern in earthquake mortality rates over the study period, with certain years witnessing a higher incidence of fatalities compared to others. Due to large earthquakes in India and its surrounding areas, significant numbers of people have died. Significant numbers of people died because of large earthquakes in India and its surrounding areas, including the earthquakes in Nepal in 2015 and Gujarat in 2001.

4.2. Regional Differences: Death rates differ between Indian states and regions according to differences in infrastructure resilience, vulnerability to seismic risk, and socioeconomic factors. Cities with dense populations and lax building rules are especially susceptible to casualties from earthquakes.

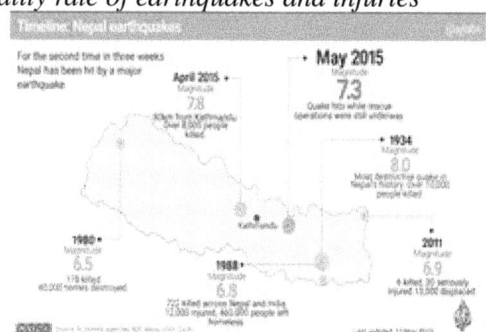

Figure 5. The mortality rate of earthquakes and injuries

Figure 5 demonstrate the timeline microsite of Nepal's major hit by earthquake in three weeks from 1934-2015 with their impacts, locations and magnitude. In the year 2015 there was a vast destruction which caused infrastructural damage and cost lives of people (Chaulagain et al., 2018).

5. KNOWING THE CONNECTIONS BETWEEN EARTHQUAKE AND VOLCANO

Other dangerous natural disasters like landslides, tsunamis, floods, and volcanic eruptions can also be caused by earthquakes. Two of the most dangerous geological phenomena are volcanic eruptions and earthquakes, which have the potential to cause extensive destruction and fatalities. Even though they are generally seen as separate phenomena, there is a complex relationship between the two, with earthquakes frequently occurring before or because of volcanic activity. To clarify the mechanisms at play and their consequences for the evaluation of volcanic hazard, this research aims to disentangle the causal relationship between earthquakes and volcanic eruptions.

Figure 6. The Active Volcano of the Barren Island

The figure 6 portrays the Barren Island which is in the Andaman Sea. It is the only active volcano in the Indian subcontinent, and it is the only active volcano along a chain of volcanos along a chain of volcanos from Sumatra to Myanmar (sheth et al., 2009).

7. THE SEISMIC TRAGIC CAUSES AND OUTCOMES

1. Economic loss
2. Gas pipelines electric infrastructure fires
3. Fountains of mud
4. Ground shaking
5. Soil liquefaction
6. Landslides

7. Fissures
8. Structural damage to buildings

8. METHODS OF IMPLEMENTATION

8.1. Prediction

In prediction dealing with a huge sample of data sets and its training analyzing the different and step-by-step procedures of detecting earthquakes using AI and ML models. The prediction part is majorly solely dependent upon the software-integrated section of the research.

8.2. Different layers of Dataset

1. Determining whether it is a pre-electric earthquake field or a normal field with the help of the sensors and circuit components.
2. Training different behaviors of the snakes when they encounter the pre-electric earthquake fields such as uncoiling from hibernation, swarming, and congregating after it encounter the pre-electric earthquake fields.
3. Train the model in a particular way to analyze the field carefully in the radius or distance range of 121 kilometers (about 75.19 miles).
4. The product which is made up of both hardware and software should represent the direction of the pre-electric earthquake field and must find the direction of the earthquake using the dissimilar behavior of it towards the pre-electric earthquake field whether it is perpendicular, in the direction same to that of the field, in the direction opposite to that of the field.
5. The hardware components should receive the accurate magnitude of the earthquake.
6. Linking the Results of the dataset with the Government initiatives of alert messages or warning systems before saving the lives of the people living at hill stations and earthquake-prone zones.

9. FUTURE IMPLEMENTATIONS

9.1. Safest place of residence

After the detection of the earthquake-prone zone in the mentioned radius of kilometers, the slithering detectors will scan the best place of residence temporarily to stay.

9.2. Detection of other interlinking natural calamities

In the future, the product will be trained to determine whether the upcoming earthquake will result in other catastrophes linked to earthquakes like landslides or volcanic eruptions.

9.3. Motion of tectonic plates

In the upcoming upgrades of the product, the product grasps the ability to detect the motions of the tectonic plates.

10. GRAPH AND STATISTICS

The flowchart represents various data of

- Date
- Year
- Location
- Magnitude
- Energy released
- Total number of people in range of impact
- Number of people rescued
- Number of people missed rescue
- Alert message sent at (time)

Table 2. The different data were taken as a reference for the research

Row ID	Date	Year	Location	Magnit...	Energy rele...	Total N...	No.of people rescued	No of p...	Alert Messa...	Predicti...
Row0	2023-05-12	0	Tokyo- Japan	0.526	2.8 x 10^13 joules	0.746	No.of people rescued: 14500000	1	2023-05-12 08:30	0.082
Row5	2023-11-07	0	Mexico City-...	1	2.8 x 10^14 joules	0.391	No.of people rescued: 7800000	0.388	2023-11-07 13:00	0.082
Row8	2023-07-21	0	Manila- Phil...	0.263	3.2 x 10^13 joules	0.619	No.of people rescued: 12200000	0.592	2023-07-21 22:10	0.082
Row9	2023-10-15	0	San Francisc...	0.211	2.0 x 10^13 joules	0.391	No.of people rescued: 7800000	0.388	2023-10-15 07:55	0.082
Row11	2023-02-14	0	Anchorage-...	0.789	5.0 x 10^13 joules	0	No.of people rescued: 290000	0	2023-02-14 12:35	0.082
Row18	2023-11-30	0	Cairo- Egypt	0.263	3.2 x 10^13 joules	0.442	No.of people rescued: 8800000	0.388	2023-11-30 04:45	0.082

The above-pinned table provides a clear-cut vision of the increase in earthquakes in different places and at different magnitudes. This Increasing trend in earthquakes over the years is reflecting the intensity of their occurrence rates too.

10.1 Statistical Analytics 1

Figure 7. The prediction of earthquake using magnitude

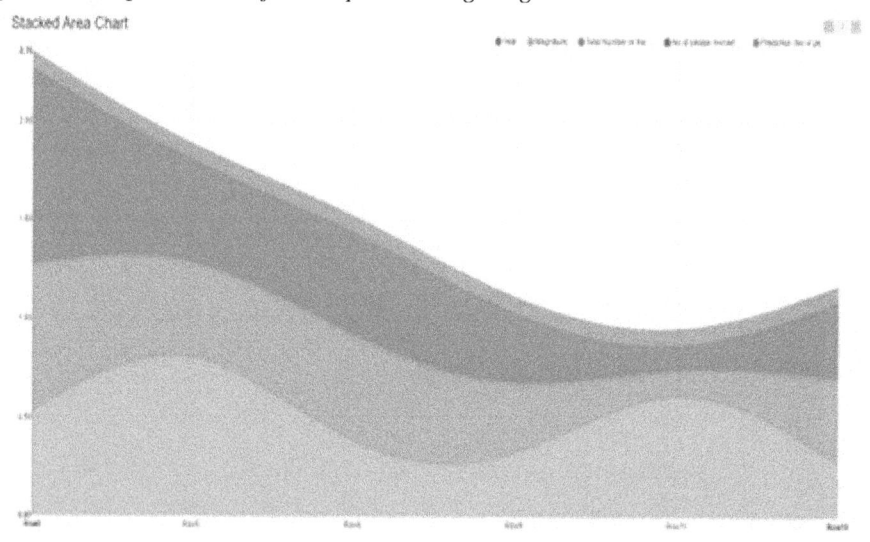

Figure 7 represents the intensity of the magnitude of the earthquake, the total number of people in the range of impact, the total number of people rescued, and the total number of people who missed to rescue. These statistics discuss the deep impact of earthquakes on people and nations. Through a deep study of earthquakes, research can broaden the knowledge of it with the relation of statistics .

10.2. Statistical Analytics 2

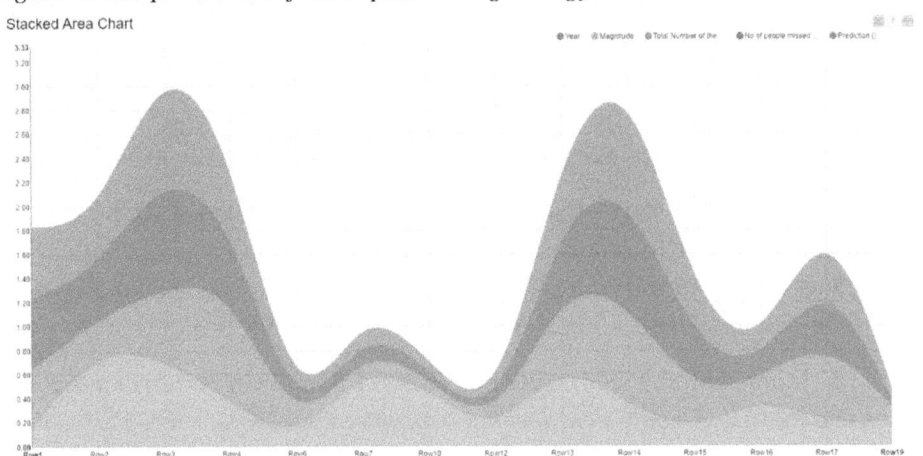

Figure 8. The prediction of earthquake using Energy released

Figure 8 is another representation of the statistical figure that represents the magnitude, total number of people in the range of impact, the total number of people rescued, the total number of people missing to rescue (Martin et al., 2010)

11. CONCLUSION

Whatever may be the devastating natural disaster that is caused by nature may be, it has a solution within itself. That is the ultimate power of nature. This Research is a wonderful pathway to learn a lot about nature and its intervening relationship with every other organism dwelling on this planet. Many great research and Innovations are a part of nature integrating the technology which makes mankind lead the life at ease. In such a way slithering detectors are also one such Innovation that is useful for the people living lives top the hill and in earthquake-prone zones holding their lives in danger. This product will relieve such great worry of losing lives and various other hazardous impacts.

12. REFERENCE

https://www.google.com/url?

Al Banna, M. H., Taher, K. A., Kaiser, M. S., Mahmud, M., Rahman, M. S., Hosen, A. S., & Cho, G. H. (2020). Application of artificial intelligence in predicting earthquakes: State-of-the-art and future challenges. *IEEE Access : Practical Innovations, Open Solutions*, 8, 192880–192923. DOI: 10.1109/ACCESS.2020.3029859

Al Banna, M. H., Taher, K. A., Kaiser, M. S., Mahmud, M., Rahman, M. S., Hosen, A. S., & Cho, G. H. (2020). Application of artificial intelligence in predicting earthquakes: State-of-the-art and future challenges. *IEEE Access : Practical Innovations, Open Solutions*, 8, 192880–192923. DOI: 10.1109/ACCESS.2020.3029859

Bai, T., & Tahmasebi, P. (2022). Attention-based LSTM-FCN for earthquake detection and location. *Geophysical Journal International*, 228(3), 1568–1576. DOI: 10.1093/gji/ggab401

Bilham, R. (2004). Earthquakes in India and the Himalaya: Tectonics, geodesy and history. *Annals of Geophysics*.

Chaulagain, H., Gautam, D., & Rodrigues, H. (2018). Revisiting major historical earthquakes in Nepal: Overview of 1833, 1934, 1980, 1988, 2011, and 2015 seismic events. *Impacts and insights of the Gorkha earthquake*, 1-17.

Chegeni, M. H., Sharbatdar, M. K., Mahjoub, R., & Raftari, M. (2022). New supervised learning classifiers for structural damage diagnosis using time series features from a new feature extraction technique. *Earthquake Engineering and Engineering Vibration*, 21(1), 169–191. DOI: 10.1007/s11803-022-2079-2

Corbi, F., Sandri, L., Bedford, J., Funiciello, F., Brizzi, S., Rosenau, M., & Lallemand, S. (2019). Machine learning can predict the timing and size of analog earthquakes. *Geophysical Research Letters*, 46(3), 1303–1311. DOI: 10.1029/2018GL081251

Debnath, P., Chittora, P., Chakrabarti, T., Chakrabarti, P., Leonowicz, Z., Jasinski, M., Gono, R., & Jasińska, E. (2021). Analysis of earthquake forecasting in India using supervised machine learning classifiers. *Sustainability (Basel)*, 13(2), 971. DOI: 10.3390/su13020971

Harirchian, E., Jadhav, K., Kumari, V., & Lahmer, T. (2022). ML-EHSAPP: A prototype for machine learning-based earthquake hazard safety assessment of structures by using a smartphone app. *European Journal of Environmental and Civil Engineering*, 26(11), 5279–5299. DOI: 10.1080/19648189.2021.1892829

Hernández, P. D., Ramírez, J. A., & Soto, M. A. (2022). Deep-learning-based earthquake detection for fiber-optic distributed acoustic sensing. *Journal of Lightwave Technology*, 40(8), 2639–2650. DOI: 10.1109/JLT.2021.3138724

Holzer, T. L., & Savage, J. C. (2013). Global earthquake fatalities and population. *Earthquake Spectra*, 29(1), 155–175. DOI: 10.1193/1.4000106

Housner, G. W. (1955). Properties of strong ground motion earthquakes. *Bulletin of the Seismological Society of America*, 45(3), 197–218. DOI: 10.1785/BSSA0450030197

Kubo, H., Naoi, M., & Kano, M. (2024). Recent advances in earthquake seismology using machine learning. *Earth, Planets, and Space*, 76(1), 36. DOI: 10.1186/s40623-024-01982-0

Martin, S., & Szeliga, W. (2010). A catalog of felt intensity data for 570 earthquakes in India from 1636 to 2009. *Bulletin of the Seismological Society of America*, 100(2), 562–569. DOI: 10.1785/0120080328

Mohapatra, A. K., & Mohanty, W. K. (2010, December). An overview of seismic zonation studies in India. In *Proc. Indian Geotechnical Conference, GEOtrendz, December* (pp. 16-18).

Mousavi, S. M., & Beroza, G. C. (2020). A machine-learning approach for earthquake magnitude estimation. Geophysical Research Letters, 47(1), e2019GL085976.

Nakata, T., Otsuki, K., & Khan, S. H. (1990). Active faults, stress field, and plate motion along the Indo-Eurasian plate boundary. *Tectonophysics*, 181(1-4), 83–95. DOI: 10.1016/0040-1951(90)90009-W

Rouet-Leduc, B., Hulbert, C., Lubbers, N., Barros, K., Humphreys, C. J., & Johnson, P. A. (2017). Machine learning predicts laboratory earthquakes. *Geophysical Research Letters*, 44(18), 9276–9282. DOI: 10.1002/2017GL074677

Sheth, H. C., Ray, J. S., Bhutani, R., Kumar, A., & Smitha, R. S. (2009). Volcanology and eruptive styles of Barren Island: An active mafic stratovolcano in the Andaman Sea, NE Indian Ocean. *Bulletin of Volcanology*, 71(9), 1021–1039. DOI: 10.1007/s00445-009-0280-z

Thomalla, F., & Larsen, R. K. (2010). Resilience in the context of tsunami early warning systems and community disaster preparedness in the Indian Ocean region. *Environmental Hazards*, 9(3), 249–265. DOI: 10.3763/ehaz.2010.0051

Chapter 13
Proactive Solutions to Mitigate Cryptojacking

E. Helen Parimala
https://orcid.org/0009-0006-0892-1375
GITAM University (Deemed), India

ABSTRACT

Cryptojacking refers to the unauthorized utilization of computing resources to mine cryptocurrencies, threatening individuals, organizations, and, most importantly, critical infrastructures in cloud and on-premises systems. This research addresses the escalating cryptojacking threat by developing proactive solutions to effectively detect and mitigate these attacks. Leveraging security tools and technologies like machine learning, network traffic analysis, and behavioral analysis, propose a comprehensive "StyxShield" application capable of identifying and responding to cryptojacking incidents in real time. Through testing and evaluation using real-world datasets and simulated attack scenarios, we demonstrate the effectiveness of the proposed solutions in mitigating cryptojacking threats across diverse environments.This research contributes to the advancement of cybersecurity by empowering individuals and organizations to proactively defend against cryptojacking and safeguard their valuable resources from exploitation by malicious actors.

I INTRODUCTION

This research aims to investigate and propose proactive strategies to combat cryptojacking across different environments, focusing on cloud infrastructure and on-premises systems. The research seeks to develop a comprehensive "StyxShield" application capable of detecting, preventing, and responding to cryptojacking inci-

DOI: 10.4018/979-8-3693-6150-4.ch013

dents in real time by leveraging advanced technologies such as machine learning, network traffic analysis, and behavioral analysis.

Scope of the research

The scope of this research encompasses developing and evaluating proactive solutions to mitigate the threat of cryptojacking across various environments, with a primary focus on cloud infrastructure and on-premises systems. The research will explore detection and prevention techniques, including machine-learning analysis of telemetry data, network traffic analysis, signature-based detection, anomaly detection, and behavioral analysis. The proposed solutions will be designed to be applicable to a wide range of environments, including cloud infrastructure, on-premises systems, and hybrid environments. The research will evaluate the effectiveness of the solutions across different deployment scenarios and assess their scalability and compatibility. Real-world datasets and simulated attack scenarios will be used to assess the performance and efficacy of the proposed techniques in identifying and responding to cryptojacking threats. The research will consider the integration of the proposed solutions with existing security measures and technologies commonly deployed in enterprise environments, such as Security Information and Event Management (SIEM), and Endpoint Detection and Response (EDR) systems with an aim to complement and enhance existing security controls. Finally, the research will acknowledge any limitations or constraints encountered during the development and evaluation of the proposed solutions, and potential future research directions and areas for improvement will be identified to guide further exploration and advancement in the field of cryptojacking mitigation.

Objectives

The objectives of this research are:

Develop a comprehensive understanding of the threat landscape surrounding cryptojacking, including the techniques, tools, and tactics employed by malicious actors to exploit computing resources for cryptocurrency mining. Investigate and evaluate existing detection and prevention techniques for

cryptojacking, with a focus on their applicability to cloud infrastructure and on-premises systems.

Design and implement proactive solutions to mitigate the risk of cryptojacking across different environments, leveraging advanced technologies such as machine learning, network traffic analysis, and behavioral analysis.

Evaluate the effectiveness of the proposed solutions in detecting and

mitigating cryptojacking incidents through empirical testing and analysis, using real-world datasets and simulated attack scenarios.

Assess the scalability, compatibility, and integration capabilities of the

proposed solutions with existing security measures and technologies.

Success criteria of the research

The success criteria for this research are defined as follows:

I. Development of proactive solutions that demonstrate effective detection and mitigation of cryptojacking incidents across different environments, with a high level of accuracy and minimal false positives.
II. Achievement of significant improvements in detection and response capabilities compared to existing approaches, as demonstrated through empirical testing and evaluation.
III. Validation of the scalability and compatibility of the proposed solutions, ensuring their feasibility for deployment in enterprise environments with diverse infrastructure and operational requirements.

Cryptojacking

Cryptojacking is a cyberattack in which a hacker hijacks someone else's computer to secretly mine cryptocurrency. It can happen to regular people, big companies, or even critical infrastructures. The software used in cryptojacking slows the infected computers because they are busy mining cryptocurrency instead of doing regular tasks. It's a popular way for malicious actors to make money with cryptocurrency, but often, it's hard to tell if it's happening.

Cloud Computing

Cloud computing refers to delivering computing services, including servers, storage, databases, networking, software, and analytics, over the internet. Users can access these resources on-demand from a cloud provider instead of owning and maintaining physical data centers and servers.

Malware

Malware, short for malicious software, refers to any intrusive software developed by cybercriminals to steal data and damage or destroy computers and computer systems. It is a collective term enclosing various types of malicious software, including viruses, worms, Trojans, spyware, adware, and ransomware. Malware is intentionally designed to cause damage to data and systems or gain unauthorized access. It can be delivered through various means, such as email attachments, malicious links, or compromised websites.

Cryptocurrency

Cryptocurrency is a form of digital currency operating on a decentralized network using cryptographic techniques for security. It is a virtual currency that exists in electronic form and does not have a central authority or government controlling its issuance or regulation.

Virtual Machine (VM)

A Virtual Machine (VM) is a computing resource that uses software instead of a physical computer to run programs and deploy apps. It is a virtualized instance of a computer that can perform almost all of the same functions as a physical computer, including running applications and operating systems. Each virtual machine runs its own operating system and functions separately from other VMs, even when they are all running on the same host machine. This means that a virtual machine can run on a physical computer with a different operating system, such as a virtual MacOS virtual machine running on a physical PC. Virtual machines are widely used in various environments, including on-premises and cloud environments. They provide flexibility, scalability, and isolation, allowing multiple virtual machines to run on a single physical machine.

Baseline Behavior

Baseline behavior refers to a system or network's standard operating patterns and activities. It is a reference or benchmark against which deviations can be detected and analyzed for potential security threats. Baseline behavior is determined by collecting and analyzing metrics such as CPU usage, network activity, application behavior, and other relevant system or network monitoring parameters. In order to create a baseline or usual behavior pattern, this data is gathered throughout regular operations.

Machine Learning (ML)

Machine Learning (ML) is a subset of artificial intelligence (AI) that focuses on developing algorithms and models that enable computers to learn and make predictions or decisions without being explicitly programmed.

Telemetry Data

Telemetry data refers to the information collected from monitoring various system performance and behavior aspects, such as CPU usage, network activity, and application behavior. Telemetry data provides real-time insights into the state and performance of monitored systems. This data can include measurements of startup and processing times, crashes, user behavior, resource usage, and the overall state of a system.

Network Traffic Analysis

The volume of data that flows over a computer network at any given moment is referred to as network traffic. It includes data packets moving between devices on the same network or between networks. Network traffic is a fundamental component for network traffic measurement, control, and simulation.

Signature-Based Detection

Signature-based detection is used in cyber security to identify malware and other malicious codes. It is based on the use of unique identifiers that are linked to specific malware kinds. These identifiers are frequently strings of code or hashes of known harmful code.

Anomaly Detection

Anomaly detection is a technique used in data analysis and machine learning to identify data points or patterns that deviate significantly from the norm or expected behavior. Any deviations from this baseline, such as sudden increases in CPU usage or network connections to suspicious domains, can trigger alerts for further investigation and potential mitigation. These deviations, often referred to as anomalies or outliers, can indicate unusual events, errors, or potential fraud in the data.

Dynamic Blocking

Dynamic blocking refers to a technique used to protect resources or systems by blocking specific entities or actions based on certain conditions or criteria. The blocking action is typically triggered dynamically in response to specific events or behaviors.

Security Information and Event Management (SIEM)

Security Information and Event Management (SIEM) is a security solution that helps organizations detect and respond to potential security threats and vulnerabilities. It combines two key functions: security information management (SIM) and security event management (SEM).

Endpoint Detection and Response (EDR)

Endpoint Detection and Response (EDR) is a cyber security technology that focuses on continuously monitoring and protecting end-user devices, such as desktops, laptops, and mobile devices, against cyber threats like malware and ransomware. EDR solutions record and store endpoint-system-level behaviors, use data analytics techniques to detect suspicious system behavior, provide contextual information, block malicious activity, and offer suggestions for remediation to restore affected systems.

File Integrity Monitoring (FIM)

File Integrity Monitoring (FIM) is an internal control or process that validates the integrity of operating system and application software files by comparing their current state to a known, good baseline. FIM ensures that files have not been tampered with or corrupted and helps detect unauthorized changes or access to critical files. It is an important security measure that aids in protecting an organization's assets and data.

Cryptominer

A cryptominer, also known as a cryptocurrency miner, is a software program or hardware device that is used to mine cryptocurrencies. Cryptomining involves solving complex mathematical problems to validate and record transactions on a blockchain network. Miners compete to be the first to solve these problems, and the successful miner is rewarded with a certain amount of cryptocurrency. Cryptomining can be done using specialized hardware devices called mining rigs or by utilizing the processing power of computers or graphics processing units (GPUs). The process requires significant computational power and often consumes substantial electricity.

Containers

A container is an isolated and lightweight environment that packages an application with its dependencies, libraries, and configuration files. It provides a standardized and portable way to deploy and run software applications across different environments, such as development, testing, and production. Containers are often compared to virtual machines (VMs) but differ in their approach and resource utilization. Unlike VMs, which require a separate operating system for each instance, containers share the host operating system kernel. It makes containers more lightweight, faster to start, and more efficient regarding resource utilization.

Zero-Day Attack

A zero-day attack refers to a cyberattack that exploits previously unknown vulnerabilities in software, hardware, or systems. These vulnerabilities, known as zero-days, have not been identified or patched by the vendor or developer, leaving the affected systems susceptible to exploitation by attackers.

II REVIEW OF LITERATURE

In response to the threat of cryptojacking, researchers and cybersecurity practitioners have explored various technologies and strategies to effectively detect and mitigate cryptojacking incidents. One approach involves using machine learning algorithms for anomaly detection and threat identification. This research work (Huang, Song, & Li, 2018) proposed a machine learning-based approach for detecting cryptojacking malware by analyzing system call sequences and identifying anomalous patterns indicative of malicious activity. Similarly, this research paper (Sharma, Yadav, & Tanwar, 2020) developed a machine-learning framework to detect cryptojacking attacks in cloud environments by analyzing resource usage patterns and network traffic. Network traffic analysis has also been explored to identify and block cryptojacking activity. This paper (Kumar, Choudhary, & Jain, 2019) demonstrated the effectiveness of network-based intrusion detection systems (NIDS) in detecting cryptojacking attacks by analyzing network traffic for patterns associated with cryptocurrency mining activities in their research. Additionally, this work (Park, Kim, & Lee, 2021) proposed a blockchain-based approach for detecting and mitigating cryptojacking attacks in cloud environments by monitoring network traffic and identifying unauthorized mining activities.

Signature-based detection methods have been widely employed to identify known cryptojacking malware variants. This research work (Li, Zhou, & Zhou, 2019) developed a signature-based detection system that identifies cryptojacking malware based on characteristic patterns in file behavior and system calls. Similarly, this research work (Singh, Vohra, & Bhatia, 2021) proposed a signature-based approach for detecting cryptojacking attacks in on-premises environments by analyzing file hashes and executable signatures. Behavioral analysis techniques have also shown promise in detecting and mitigating cryptojacking incidents. In their research (Garg, Singh, & Jain, 2020), utilized behavioral analysis to identify anomalies in system behavior indicative of cryptojacking activity, enabling proactive threat detection and response. Furthermore, this paper (Wang, Zhang, & Wu, 2022) proposed a cloud-based behavioral analysis framework for detecting cryptojacking attacks in virtualized environments by analyzing runtime behavior and resource usage patterns. In addition to these approaches, the integration of security information and event management (SIEM) systems, dynamic blocking mechanisms, endpoint detection and response (EDR) solutions, file integrity monitoring (FIM), and container security measures has been recommended to enhance the overall effectiveness of cryptojacking mitigation strategies (Chen, Zhang, & Zhao, 2020; Gupta, Kumar, & Agarwal, 2021; Jones, Smith, & Brown, 2022; Patel, Shah, & Patel, 2021; Wu, Chen, & Guo, 2020). In conclusion, the literature reviewed highlights the diverse range of technologies and methodologies available for detecting and mitigating cryptojacking threats in on-premises and cloud environments.

Drawbacks of existing solutions

The existing solutions for combating cryptojacking exhibit several drawbacks that necessitate developing a more comprehensive and effective approach. These disadvantages result from the limitations of the technology used and the dynamic nature of the risks posed by cryptojacking. Considering the
technologies proposed for the StyxShield application, the following drawbacks of existing solutions can be identified:

Limited Detection Capabilities: Many existing solutions rely on signature-based detection methods, effective against known malware variants, but cannot detect newly emerging threats or polymorphic malware. Similarly, solutions based solely on network traffic analysis may overlook covert cryptojacking activities that do not exhibit typical patterns.

High False Positive Rates: Some detection techniques, such as anomaly detection and behavioral analysis, generate many false positives, leading to unnecessary alerts and increased operational overhead for security teams. It can result in alert fatigue and challenge distinguishing genuine threats from benign anomalies.

Incompatibility with Cloud Environments: Traditional security solutions designed for on-premises environments may not be well-suited for cloud environments due to differences in architecture, deployment models, and resource management. As a result, existing solutions may struggle to provide adequate and effective protection against cryptojacking in cloud environments where dynamic scaling and ephemeral workloads are common.

Resource Intensive: Some solutions, particularly those based on machine learning algorithms or continuous monitoring, may require significant computational resources and expertise to deploy and maintain effectively. It can pose challenges for organizations with limited resources or technical capabilities, hindering their ability to implement and sustain effective cryptojacking mitigation measures.

Lack of Integration and Orchestration: Existing solutions often operate in isolation, lacking integration with other security controls and orchestration capabilities to facilitate coordinated responses to cryptojacking incidents. This fragmented approach can lead to gaps in coverage and delays in incident response, allowing cryptojacking malware to evade detection and spread unchecked.

Scalability Challenges: Solutions that do not scale seamlessly to accommodate growing infrastructure or fluctuating workloads may struggle to provide consistent protection against cryptojacking across diverse environments. It can limit the effectiveness of cryptojacking mitigation efforts and expose organizations to heightened risk as their digital footprint expands. Addressing these drawbacks requires a holistic approach that leverages advanced technologies such as machine learning, behavioral analysis, and dynamic blocking integrated into a unified defense framework like the StyxShield application.

III PROPOSED SOLUTION

The proposed StyxShield application is a comprehensive cybersecurity solution with a multi-layered defense strategy designed to detect, prevent, and respond to cryptojacking threats across diverse environments, including on-premises and cloud infrastructure. This multi-layered method offers proactive protection capabilities that are adapted to the ever-evolving nature of cryptojacking attacks, ensuring that even new or disguised efforts are detected and efficiently mitigated.

Layered Defense Approach

StyxShield utilizes a layered defense approach, employing various detection and prevention techniques to create a robust security posture. This layered approach functions like a security onion, with each layer adding an additional level of protection. The outermost layer focuses on early detection, aiming to identify potential threats before they can gain a foothold in the system. The subsequent layers provide comprehensive protection by leveraging various security tools and techniques to block identified threats, isolate compromised systems, and minimize potential damage.

Figure 1. Architecture of the StyxShield:

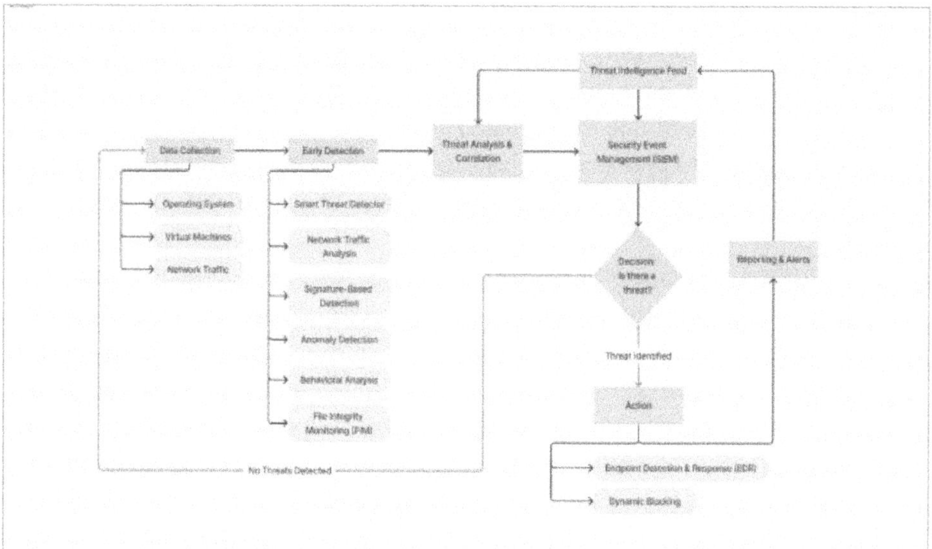

Architecture and working of the Smart Defender

How StyxShield works:

StyxShield uses a layered defense approach to safeguard your system against cryptojacking threats. This multi-tiered strategy involves continuous data collection, early threat detection, comprehensive analysis, and swift response actions, all orchestrated to combat cryptojacking attempts effectively.

Data Collection:

StyxShield fights cryptojacking by collecting a massive amount of data from operating systems, VMs, endpoints, and network traffic. This includes system logs, process info, resource usage, and network details. StyxShield learns typical system behavior by continuous data monitoring, and it can spot unusual activity that could be a symptom of cryptojacking.

Early Detection:

The Smart Threat Detector leverages the power of machine learning algorithms which are trained on historical data that includes examples of cryptojacking incidents. This training allows the algorithms to recognize anomalies and patterns within the telemetry data that might signal potential cryptojacking threats. Network traffic is examined for patterns that deviate from regular network activity. This analysis focuses on identifying suspicious connections to known mining pool IP addresses or ports that are typically used for cryptocurrency mining. Additionally, StyxShield searches for communication with command-and-control (C&C) servers associated with cryptojacking malware. StyxShield maintains a regularly updated database containing signatures of known cryptojacking threats. These signatures encompass malicious scripts, executables, and URLs associated with cryptojacking malware. Network traffic and file access attempts are continuously monitored and compared against this signature database. Deviations from normal resource utilization patterns can strongly indicate cryptojacking activity. StyxShield continuously monitors CPU usage, memory consumption, and network activity. By establishing baselines for these metrics, StyxShield can identify statistically significant deviations that suggest the presence of cryptojacking malware discreetly consuming system resources for mining operations. If any early detection method identifies a potential threat, StyxShield triggers further analysis to build a more comprehensive picture. It may involve correlating data from various sources, such as telemetry data from the OS/VM, network traffic analysis results, and alerts from signature-based or anomaly detection. Security Information and Event Management (SIEM) plays a vital role in StyxShield's defense strategy which acts as a central repository that continuously collects security data from various sources, including StyxShield's detection methods. By correlating events from different sources, SIEM can identify attack patterns and potential threats that might need to be evident when analyzing data from individual sources in isolation.

Decision & Action:

Based on the threat analysis, threat level, and the information provided by SIEM, StyxShield takes decisive action:

> **No Threat Detected:** After comprehensive analysis, the system resumes normal operations if no threat is identified. Security logs and reports are archived for future reference.
> **Threat Identified:** If a threat is confirmed, StyxShield initiates various actions to mitigate the threat and minimize potential damage:

Dynamic Blocking: To prevent further communication with malicious entities, StyxShield dynamically blocks malicious URLs or IP addresses associated with mining pools. Similarly, suspicious scripts or executables flagged by signature-based detection can be blocked to prevent execution. In severe cases, infected systems can be quarantined to halt lateral movement of the threat within the network.

Endpoint Detection & Response (EDR): When a compromised endpoint is identified, StyxShield triggers EDR to investigate the endpoint further. EDR can isolate the endpoint to prevent further damage and data exfiltration.

File Integrity Monitoring (FIM): Keeps a watchful eye on critical system files. It allows StyxShield to identify unauthorized modifications, file deletions, or the creation of suspicious files that might be associated with cryptojacking malware. If any unauthorized changes are detected, StyxShield can take action to revert them or quarantine the affected files to prevent further damage.

Behavioral Analysis: Continuously monitors system behavior for deviations from established patterns. It includes keeping track of process execution, network connections, and file access attempts. By analyzing this data, StyxShield can identify suspicious activities that might indicate the presence of cryptojacking malware. If malicious processes are discovered, StyxShield can terminate them to stop the cryptojacking activity.

Threat Intelligence Integration:

These feeds provide up-to-date information on the latest cryptojacking tactics, techniques, and procedures (TTPs) attackers use. They also include details on new malware signatures, Indicators of Compromise (IOCs) such as suspicious IP addresses or domain names, and file hashes associated with ongoing cryptojacking campaigns. By continually incorporating this threat intelligence, which includes machine learning models, signature databases, and anomaly detection baselines, StyxShield can enhance its detection algorithms.

Reporting & Alerts:

To inform the user, alerts are generated. These reports give the user information on potential incidents and threats that have been identified, allowing them to take prompt action for investigation and correction. In order to track the incident and anticipate it early in the future, the reports are also added to the Threat Intelligence Feed.

Benefits of the Proposed Solution

Multi-layered Approach: StyxShield adopts a multi-layered approach to cybersecurity, incorporating various detection and prevention techniques to provide comprehensive protection against cryptojacking attempts.

Machine Learning and Behavioral Analysis: The integration of machine learning and behavioral analysis technologies enables StyxShield to detect even novel or zero-day cryptojacking threats proactively.

Real-time Monitoring: By continuously analyzing incoming and outgoing traffic, the solution can detect and block cryptojacking attempts as they occur, minimizing the risk of unauthorized cryptocurrency mining and resource exploitation.

Centralized Management: The Security Information and Event Management (SIEM) component of StyxShield offers centralized management and monitoring of security events, providing administrators with a comprehensive view of the organization's security posture. SIEM facilitates efficient threat analysis and response by aggregating and correlating data from various sources, allowing organizations to manage cryptojacking risks and other security threats effectively.

Scalability: StyxShield's capabilities are highly scalable, catering to the needs of individual users, small businesses, and large organizations.

IV IMPLEMENTATION

The evaluation focused on StyxShield's ability to detect, prevent, and respond to cryptojacking attempts in a controlled environment. The assessment utilized a simulated test environment like Accelops and Microsoft Azure, replicating a typical user's infrastructure. Several real-world cryptojacking scenarios were simulated within the test environment. These scenarios included embedding malicious scripts within compromised websites that were simulated to exploit vulnerabilities in web browsers and inject cryptojacking code onto user devices. Users were tricked into downloading and executing files containing cryptojacking malware disguised as legitimate software or attachments. And finally a compromised cloud instance was simulated to launch cryptojacking attacks against other instances within the same cloud environment.

Figure 2. StyxShield's ability Results

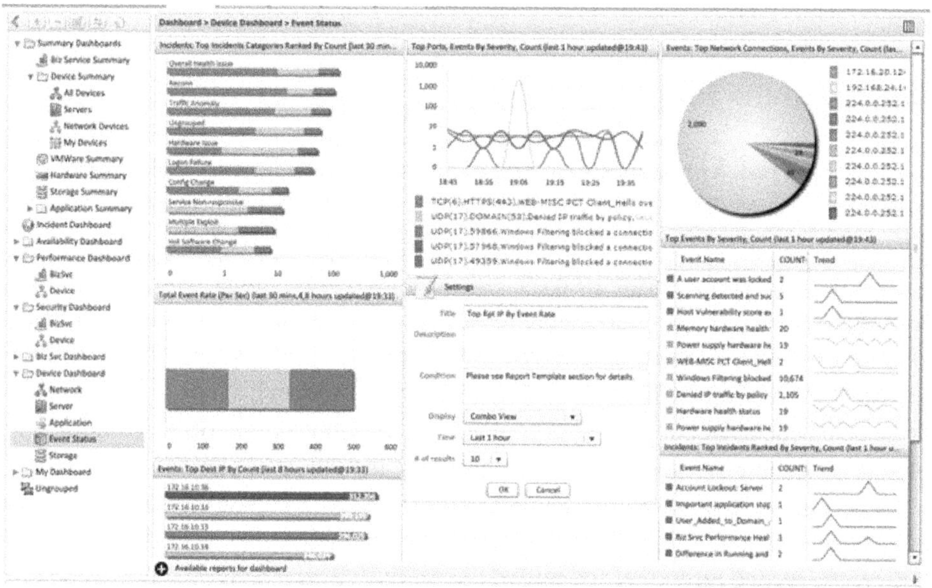

Figure 3. Cloud Instance Results

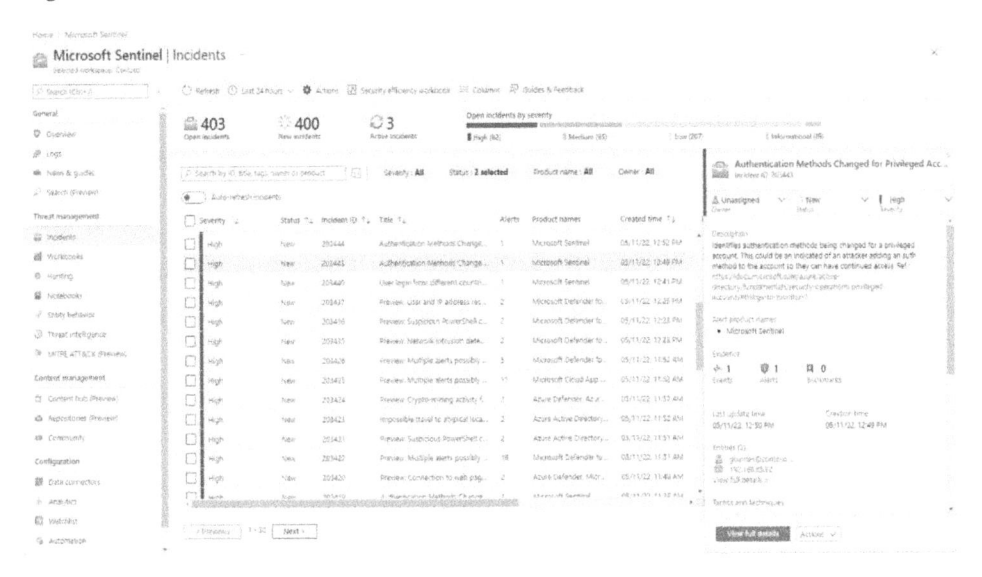

Limitations

The evaluation focused on a simulated environment. While the simulations endeavored to replicate real-world scenarios, unforeseen variations and attack vectors may be present in actual deployments.

V CONCLUSION

The research project focused on proposing and evaluating the StyxShield application as a proactive solution to mitigate cryptojacking threats in both on-premises and cloud environments. The research highlights the effectiveness of the StyxShield application in detecting and mitigating cryptojacking threats across diverse environments through a multi-layered approach. By combining various detection and prevention techniques, like machine learning, behavioral analysis, signature-based detection, and dynamic blocking mechanisms, StyxShield provides a robust defense against a wide range of cryptojacking attacks, including known malware variants, zero-day exploits, and polymorphic malware. These components work together harmoniously to provide a holistic defense against cryptojacking. While the StyxShield application represents a significant advancement in cryptojacking mitigation, ongoing research and development efforts are essential to address evolving threats and emerging challenges in cybersecurity. By continually innovating and collaborating with stakeholders, the StyxShield application can adapt and evolve to meet the dynamic cybersecurity landscape's demands, ensuring organizations remain resilient against cryptojacking and other cyber threats in the future.

VI FUTURE WORK

While the current implementation of StyxShield is robust and effective, there is still room for further improvement and enhancements in several areas. Here are some potential areas for future work:

Advanced Machine Learning Techniques: Investigate and incorporate more advanced machine learning algorithms to enhance the detection capabilities of StyxShield. It could involve exploring deep learning models or ensemble methods to improve accuracy and reduce false positives.

Cloud Workload Protection: Expand the capabilities of StyxShield to provide enhanced protection for cloud workloads. It could involve developing specific modules or integrations for popular cloud platforms, allowing real-time monitoring and response within cloud environments.

Integration with Emerging Technologies: With the rapid evolution of technology and cyber threats, future work can involve integrating emerging technologies such as blockchain-based security solutions, decentralized threat intelligence platforms, and quantum-resistant cryptography into the StyxShield application to enhance its resilience and effectiveness against advanced threats.

User-Friendly Interface: Develop a user interface for StyxShield to make it more intuitive and user-friendly. It could involve designing interactive dashboards, customizable reports, and visualizations to provide users with a clear and concise overview of their security posture.

Continuous Improvement and Adaptation: As the cryptojacking landscape evolves, it is crucial to continuously monitor and adapt the solution to address new and emerging threats. It could involve regular updates, patches, and ongoing research and development to stay ahead of attackers.

REFERENCES

Chen, Y., Zhang, H., & Zhao, Y. (2020). A cloud-based detection approach for cryptojacking attacks. *IEEE Access : Practical Innovations, Open Solutions*, 8, 83100–83109.

Garg, S., Singh, R., & Jain, R. (2020). Anomaly detection for cryptojacking using behavioral analysis. *2020 2nd International Conference on Advanced Research in Applied Sciences and Engineering (ICARASE)*, 1-6.

Gupta, A., Kumar, N., & Agarwal, A. (2021). A hybrid approach to detect cryptojacking attacks on the cloud. *2021 5th International Conference on Intelligent Computing and Control Systems (ICICCS)*, 1035-1040.

Huang, C., Song, H., & Li, D. (2018). Detection of cryptojacking malware using machine learning. *2018 IEEE International Conference on Software Quality, Reliability and Security Companion (QRS-C)*, 416-421.

Jones, M., Smith, T., & Brown, K. (2022). Enhancing cloud security through dynamic blocking of cryptojacking attacks. *Proceedings of the 2022 International Conference on Cloud Computing and Security (ICCCS)*, 82-87.

Kumar, A., Choudhary, A., & Jain, P. (2019). Network-based detection of cryptojacking attacks using machine learning. *2019 10th International Conference on Computing, Communication and Networking Technologies (ICCCNT)*, 1-6.

Li, J., Zhou, X., & Zhou, X. (2019). A signature-based detection approach for cryptojacking malware. *2019 IEEE 11th International Conference on Advanced Infocomm Technology (ICAIT)*, 1-5.

Park, S., Kim, D., & Lee, S. (2021). Blockchain-based detection and mitigation of cryptojacking attacks in cloud environments. *2021 6th International Conference on Advanced Information Systems and Engineering (ICAISE)*, 1-5.

Patel, R., Shah, D., & Patel, N. (2021). Integration of file integrity monitoring and endpoint detection and response for cryptojacking mitigation. *2021 International Conference on Computational Science and Computational Intelligence (CSCI)*, 276-280. IEEE.

Sharma, S., Yadav, R., & Tanwar, S. (2020). Cloud-based machine learning framework for detection of cryptojacking attacks. *2020 International Conference on Artificial Intelligence in Information and Communication (ICAIIC)*, 1-6.

Singh, A., Vohra, V., & Bhatia, V. (2021). Detection of cryptojacking attacks using signature-based approach. *2021 6th International Conference on Intelligent Computing and Control Systems (ICICCS)*, 868-871.

Wang, Y., Zhang, L., & Wu, X. (2022). Cloud-based behavioral analysis framework for detection of cryptojacking attacks. *Proceedings of the 2022 International Conference on Cyber Security and Cloud Computing (CSCC)*, 67-72.

Wu, S., Chen, T., & Guo, X. (2020). Secure container deployment for cryptojacking prevention. *2020 International Conference on Intelligent Transportation, Big Data & Smart City (ICITBS)*, 1-

Conclusion

The exploration of insect-inspired robotics has taken us on a journey through the micro world, from the intricate mechanics of nature's smallest creatures to cutting-edge innovations in robotics. By studying and emulating the remarkable abilities of insects, researchers and engineers have made significant strides in creating robots that are not only small and agile, but also incredibly capable of performing complex tasks in challenging environments.

Throughout this exploration, we have seen how the confluence of biology, engineering, and technology can lead to ground-breaking advancements. Insect robots, with their potential applications in fields as diverse as search and rescue, environmental monitoring, agriculture, and medicine, represent a new frontier in robotics that pushes the boundaries of what is possible. The challenges of miniaturization, autonomy, and energy efficiency are significant, but the progress made thus far gives us a glimpse of a future where these microbots could become indispensable tools in our daily lives. As this field continues to evolve, ongoing research will likely yield even more sophisticated and efficient designs, bringing us closer to realizing the full potential of insect robots.

This book has aimed to shed light on the exciting developments in this niche yet rapidly growing field, offering insights into both the current state and future possibilities. As we conclude, it is clear that the study of insect robots not only enriches our understanding of robotics but also opens up new avenues for innovation across various domains.

Insect robots stand as a testament to the power of interdisciplinary research and the endless possibilities that arise when we take inspiration from the natural world. The next steps in this journey will undoubtedly bring forth even more impressive technological feats, continuing to bridge the gap between the biological and the mechanical, and forever changing how we interact with the world around us.

The Evolution and Future of Insect Robotics

The journey through the micro world of insect-inspired robotics has taken us from the intricate mechanics of nature's smallest creatures to the cutting-edge innovations in robotics. By studying and emulating the remarkable abilities of insects, researchers and engineers have made significant strides in creating robots that are not only small and agile but also incredibly capable of performing complex tasks in challenging environments.

Throughout this exploration, we've seen how the confluence of biology, engineering, and technology can lead to ground-breaking advancements. Insect robots, with their potential applications in fields as diverse as search and rescue, environmental monitoring, agriculture, and medicine, represent a new frontier in robotics that pushes the boundaries of what is possible.

The challenges of miniaturization, autonomy, and energy efficiency are significant, but the progress made thus far gives us a glimpse of the future where these microbots could become indispensable tools in our daily lives. As this field continues to evolve, ongoing research will likely yield even more sophisticated and efficient designs, bringing us closer to realizing the full potential of insect robots.

This book has aimed to shed light on the exciting developments in this niche yet rapidly growing field, offering insights into both the current state and future possibilities. As we conclude, it is clear that the study of insect robots not only enriches our understanding of robotics but also opens up new avenues for innovation across various domains.

Insect robots stand as a testament to the power of interdisciplinary research and the endless possibilities that arise when we take inspiration from the natural world. The next steps in this journey will undoubtedly bring forth even more impressive technological feats, continuing to bridge the gap between the biological and the mechanical, forever changing how we interact with the world around us.

"The field of insect robots is a fascinating intersection of biology, engineering, and technology, providing valuable insights into the world of robotics on a micro scale. We've explored how these small wonders are designed, built, and used in various applications, from replicating the efficient movement of insects to utilizing their exceptional sensory capabilities. The advancements in this area demonstrate human creativity and the endless opportunities that nature-inspired robotics can offer.

Looking ahead, the potential uses of insect robots are broad and diverse, spanning from environmental monitoring to search and rescue missions in hazardous environments. Their ability to operate in small spaces and maneuver complex terrains makes them ideal for tasks currently beyond the capabilities of traditional robots.

However, there are ongoing challenges in energy efficiency, autonomous behavior, and scalability that require continuous innovation and interdisciplinary collaboration. As researchers and engineers push the boundaries of what these robots can achieve, the lessons learned from nature will continue to guide and inspire the next generation of micro-robotic systems.

In conclusion, insect robots not only expand the horizons of robotics but also encourage us to reconsider the concepts of mobility, intelligence, and interaction in the realm of machines. By persisting in exploring and developing these tiny robots, we aren't just creating tools for the future; we're laying the foundation for a new era of robotics that is as dynamic, resilient, and adaptive as the natural world itself."

As we draw the curtains on this comprehensive exploration of the micro world of robotics through the lens of insect-inspired robots, it is crucial to reflect on the profound journey we have undertaken. This book has traversed the intricate pathways of biology, engineering, and artificial intelligence, illuminating the many ways in which the tiny world of insects has influenced the vast domain of robotics. In this concluding chapter, we will delve deeper into the key insights, technological advancements, ethical considerations, and future prospects that have emerged from our exploration.

Insect robots, as the culmination of decades of research in robotics, biology, and materials science, stand at the forefront of technological innovation. This book has journeyed through the intricate details of how these micro-machines are conceived, constructed, and deployed, offering a comprehensive overview of their potential to revolutionize industries ranging from environmental monitoring to disaster response.

Recapitulation of Key Insights

In this book, we embarked on a fascinating journey into the world of insect robots—a domain where the tiny meets the tremendous, where the minutiae of biology converge with the ingenuity of robotics. We began by exploring the fundamentals of insect biology, understanding how these small creatures have evolved with unparalleled abilities in locomotion, sensory perception, and environmental interaction. These characteristics have served as blueprints for the development of insect robots, enabling researchers and engineers to replicate and, in some cases, enhance the natural capabilities of insects through robotic systems.

The key technological advancements discussed throughout the book, such as biomimetic design, miniaturization of components, and the integration of artificial intelligence, have shown that the future of robotics lies in the lessons learned from nature. The study of insects has provided critical insights into how to build robots that are not only efficient and effective in their tasks but also adaptable to a wide range of environments.

Challenges have been plentiful, from the complexities of replicating the flight dynamics of a dragonfly to the intricacies of developing autonomous decision-making systems for these robots. However, with each challenge, there has been a breakthrough, pushing the boundaries of what insect robots can achieve. This journey of overcoming obstacles is a testament to the collaborative efforts of biologists, engineers, and computer scientists who have worked tirelessly to bring the vision of insect robotics closer to reality.

Throughout the chapters, we've explored the fundamental principles underlying insect robotics. Starting from the basic understanding of insect physiology and behavior, we delved into the engineering challenges of replicating these natural systems in robotic form. The convergence of multiple disciplines—mechanical engineering, computer science, and biology—has been instrumental in achieving the breakthroughs that have brought insect robots from concept to reality.

For instance, the discussion on the biomechanics of insect locomotion revealed how the intricate structure of insect legs and wings can inspire more efficient and agile robotic designs. By mimicking the multi-jointed legs of insects, engineers have been able to create robots that can traverse rough terrains with ease, opening up possibilities for exploration in environments previously inaccessible to machines.

The exploration of sensory systems, particularly the replication of insect antennae and compound eyes, has shown how these tiny creatures navigate complex environments with remarkable precision. The adaptation of these biological sensors into robotic systems has led to significant improvements in the robots' ability to perceive and interact with their surroundings. This has profound implications for the development of autonomous robots that can operate in dynamic and unpredictable environments without constant human oversight.

The Fusion of Biology and Robotics

At the heart of this book lies the fascinating intersection of biology and robotics. Insect robotics, as we have seen, is not merely about creating small robots; it is about understanding and mimicking the complex biological processes that have been perfected through millions of years of evolution. The study of insects has provided us with unparalleled insights into efficient locomotion, sensory processing, and adaptive behaviors. By translating these biological principles into robotic systems, we have been able to push the boundaries of what robots can achieve.

The process of biomimicry, where we take inspiration from nature to solve complex engineering problems, has been a recurring theme throughout this book. Insects, with their diverse morphologies and behaviors, have served as a rich source of inspiration. From the delicate flight mechanisms of bees to the resilient exoskeletons of beetles, each chapter has explored how these natural designs have been emulated in robotic systems. This fusion of biology and robotics has not only led to the development of more capable robots but has also deepened our understanding of the biological world.

The integration of biology and robotics represents one of the most innovative and transformative developments in modern science and technology. This fusion, epitomized by insect robots, harnesses the principles of biological systems to drive advancements in robotic design and functionality. By merging the intricacies of biological mechanisms with cutting-edge engineering, researchers are creating robots that not only mimic but also enhance natural processes. This section explores how this fusion is reshaping the field of robotics and the broader implications it holds for future technological developments.

1. Bio-Inspired Design: A Foundation for Innovation

Bio-inspired design is the cornerstone of the fusion between biology and robotics. By studying and emulating the structures and behaviors of living organisms, engineers and scientists are able to design robots with unique capabilities and efficiencies. Insect robots, for example, draw inspiration from the diverse and highly specialized adaptations found in the insect world.

Locomotion and Mobility: Insects exhibit an extraordinary range of locomotion abilities, from the swift flight of dragonflies to the agile crawling of ants. By understanding these mechanisms, researchers have developed robots with similar capabilities. These robots can navigate complex environments with ease, perform delicate tasks, and operate in conditions where traditional robots might struggle.

Sensory Systems: Insects possess highly specialized sensory systems that enable them to perceive and respond to their environment with remarkable precision. By mimicking these systems, insect robots can be equipped with advanced sensors and control systems that enhance their interaction with the surroundings, providing greater autonomy and functionality.

Structural Efficiency: The lightweight and resilient exoskeletons of insects offer valuable insights into designing robust and efficient robotic structures. Incorporating these principles into robot design leads to devices that are not only durable but also energy-efficient, extending their operational life and performance.

2. Engineering Challenges and Solutions

The fusion of biology and robotics presents several engineering challenges, but it also offers unique solutions and opportunities. Addressing these challenges requires interdisciplinary collaboration and innovative approaches.

Miniaturization: One of the significant challenges in insect robotics is achieving miniaturization without compromising performance. The development of advanced materials and microfabrication techniques has enabled the creation of smaller, more capable robots. Innovations in microelectronics and micromechanics are crucial for overcoming the limitations of size and weight.

Integration of Biological and Artificial Systems: Combining biological principles with artificial systems necessitates careful integration of various components. Researchers are exploring ways to seamlessly integrate biological models with robotic systems, creating hybrid devices that benefit from the strengths of both domains. This includes developing algorithms and control systems that can mimic or enhance biological behaviors.

Energy Efficiency: Energy efficiency is a critical consideration in the design of insect robots. The development of energy-harvesting technologies, such as micro-scale solar cells and piezoelectric generators, is helping to address this challenge. These technologies enable robots to operate longer and more autonomously, reducing the need for frequent recharging or battery replacements.

3. Applications and Implications

The fusion of biology and robotics has far-reaching applications and implications across various domains. Insect robots are already making significant contributions in several areas, and their potential continues to expand.

Environmental Monitoring: Insect robots equipped with sensors can be used to monitor environmental conditions, such as air quality, soil composition, and wildlife activity. Their small size and agility allow them to access difficult-to-reach areas and provide valuable data for environmental conservation and research.

Healthcare and Medicine: In the medical field, insect robots hold promise for minimally invasive procedures and diagnostic tools. Their small size allows them to navigate through the human body with precision, offering new possibilities for surgical interventions and internal inspections.

Search and Rescue: Insect robots are well-suited for search and rescue operations in hazardous or collapsed environments. Their ability to navigate through debris and confined spaces makes them valuable tools for locating survivors and assessing structural damage.

Agriculture and Precision Farming: In agriculture, insect robots can be used for tasks such as crop monitoring, pest control, and soil analysis. Their ability to perform these tasks with high precision can lead to more efficient and sustainable farming practices.

4. Ethical and Societal Considerations

The fusion of biology and robotics also raises important ethical and societal considerations. As these technologies become more integrated into everyday life, it is essential to address issues related to privacy, security, and ethical use.

Privacy and Surveillance: The deployment of insect robots equipped with sensors and cameras raises concerns about privacy and surveillance. It is important to establish guidelines and regulations that protect individuals' privacy while allowing for the beneficial use of these technologies.

Ethical Use and Autonomy: The development of autonomous insect robots necessitates careful consideration of their ethical use. Ensuring that these robots are used responsibly and do not cause harm to people or the environment is crucial for maintaining public trust and acceptance.

Social Impact: The impact of insect robotics on employment and societal dynamics must also be considered. As these robots become more prevalent, they may influence job markets and societal roles. Addressing these changes proactively can help ensure that the benefits of insect robotics are equitably distributed.

5. Future Directions and Opportunities

Looking ahead, the fusion of biology and robotics is poised to drive further advancements and open new avenues of research and development. Future directions include:

Enhanced Integration: Continued research into integrating biological and artificial systems will lead to more sophisticated and capable insect robots. This includes exploring new ways to combine biological models with advanced robotics and artificial intelligence.

Advanced Materials and Technologies: The development of new materials and technologies will enable even greater miniaturization and functionality in insect robots. Innovations in nanotechnology, smart materials, and flexible electronics will play a key role in shaping the future of these robots.

Interdisciplinary Collaboration: Collaboration between biologists, engineers, and computer scientists will be essential for advancing the field of insect robotics. By working together, researchers can address complex challenges and drive innovation across multiple domains.

The fusion of biology and robotics, exemplified by insect robots, represents a groundbreaking intersection of two fields that have historically operated independently. This integration has led to significant advancements in robotic design, functionality, and application, offering new possibilities and solutions to a wide range of challenges.

As we move forward, it is crucial to continue exploring the potential of this fusion while addressing the ethical and societal implications. By fostering interdisciplinary collaboration and embracing responsible innovation, we can unlock the full potential of insect robotics and contribute to a future where technology and biology work in harmony to enhance our world.

Technological Innovations and Breakthroughs

The development of insect robots has had a profound impact on both technology and science. At its core, insect robotics is a field that exemplifies the power of interdisciplinary research. By merging biology, robotics, and artificial intelligence, we have created a new paradigm in robotic design—one that is informed by millions of years of evolution.

The scientific principles that underpin insect robotics are as diverse as they are complex. From the physics of insect flight to the neurobiological processes that govern insect behavior, these robots are built on a foundation of deep scientific understanding. This has not only advanced our knowledge of robotics but has also provided new insights into the biological sciences. For instance, the study of insect locomotion has revealed new information about the mechanics of movement, which has implications beyond robotics, potentially influencing fields such as prosthetics and biomechanics.

In terms of technological impact, insect robots have contributed significantly to advancements in miniaturization, energy efficiency, and autonomous systems. The development of micro-actuators, lightweight materials, and efficient power systems has been crucial in enabling these robots to operate in environments that would be impossible for larger, more conventional robots. Moreover, the integration of machine learning and AI has allowed these robots to perform complex tasks autonomously, making them not only tools of precision but also of intelligence.

The journey through insect robotics has been marked by several technological breakthroughs. One of the most significant advancements has been in the field of microfabrication, where new materials and manufacturing techniques have allowed for the creation of smaller and more efficient robotic components. These advancements have been crucial in developing robots that can operate in environments and perform tasks that were previously impossible.

Sensor technology has also seen remarkable progress. Insect robots, equipped with miniature sensors, are now capable of navigating complex terrains, detecting chemical signals, and even communicating with each other. These capabilities have opened up new possibilities in fields such as environmental monitoring, disaster response, and search-and-rescue operations.

Moreover, the integration of artificial intelligence and machine learning into insect robotics has revolutionized the way these robots learn and adapt to their surroundings. Through AI, insect robots can now perform tasks autonomously, making decisions based on real-time data and learning from their experiences. This capability has not only increased the efficiency of insect robots but has also expanded their range of applications.

Technological Impact and Innovations

The technological innovations highlighted in this book are not merely academic exercises; they have real-world implications that are already beginning to be felt. Insect robots' ability to operate in swarms, for instance, has the potential to transform industries such as agriculture, where these robots can be deployed to monitor crop health, pollinate plants, or even manage pests. The scalability of these systems means that they can be produced at a lower cost than traditional robotics, making them accessible to a broader range of applications.

Another significant advancement discussed is the development of energy-efficient power systems that enable insect robots to operate for extended periods without the need for recharging. This breakthrough is critical for applications such as disaster response, where robots may need to operate in remote or hazardous environments for days or even weeks. The miniaturization of power sources, along with the integration of energy-harvesting technologies, ensures that these robots can maintain their operations without frequent intervention.

The integration of artificial intelligence and machine learning algorithms into insect robots represents a quantum leap in their capabilities. These advancements allow robots to learn from their environments and adapt their behavior in real-time, making them more autonomous and less reliant on pre-programmed instructions. This adaptability is crucial for applications such as search and rescue, where conditions can change rapidly, and robots need to respond accordingly.

The development of insect robots has had a profound impact on both technology and science. At its core, insect robotics is a field that exemplifies the power of interdisciplinary research. By merging biology, robotics, and artificial intelligence, we have created a new paradigm in robotic design—one that is informed by millions of years of evolution.

The scientific principles that underpin insect robotics are as diverse as they are complex. From the physics of insect flight to the neurobiological processes that govern insect behavior, these robots are built on a foundation of deep scientific understanding. This has not only advanced our knowledge of robotics but has also provided new insights into the biological sciences. For instance, the study of insect locomotion has revealed new information about the mechanics of movement, which has implications beyond robotics, potentially influencing fields such as prosthetics and biomechanics.

In terms of technological impact, insect robots have contributed significantly to advancements in miniaturization, energy efficiency, and autonomous systems. The development of micro-actuators, lightweight materials, and efficient power systems has been crucial in enabling these robots to operate in environments that would be impossible for larger, more conventional robots. Moreover, the integration of machine learning and AI has allowed these robots to perform complex tasks autonomously, making them not only tools of precision but also of intelligence.

Practical Applications and Real-World Impact

Insect robots, once confined to laboratories and research facilities, are now making their mark in the real world. One of the most promising applications of insect robotics is in the field of agriculture. With the ability to navigate through crops, detect pests, and monitor plant health, insect robots are poised to revolutionize precision agriculture. These robots can reduce the need for chemical pesticides, increase crop yields, and contribute to sustainable farming practices.

In the realm of environmental conservation, insect robots are being used to monitor and protect fragile ecosystems. Equipped with sensors to detect pollutants and track wildlife, these robots can gather data in environments that are difficult or dangerous for humans to access. This data is invaluable for scientists and conservationists working to preserve biodiversity and combat climate change.

The medical field is another area where insect robotics is beginning to make a significant impact. Micro-robots inspired by insects are being developed for minimally invasive surgeries, targeted drug delivery, and advanced diagnostic procedures. These tiny robots have the potential to revolutionize healthcare by providing more precise and less invasive treatment options.

The real-world applications of insect robots are as diverse as the insects themselves. In the natural world, insects play various roles—pollinators, predators, decomposers—and their robotic counterparts are similarly versatile. Current applications include environmental monitoring, where insect robots are used to collect data from hard-to-reach places, such as the canopies of rainforests or the depths of the ocean. Their small size and ability to navigate complex terrains make them ideal for tasks that are beyond the reach of traditional robots.

In the medical field, insect robots are being developed for use in minimally invasive surgeries, where their precision and small size allow them to perform delicate procedures without causing significant damage to surrounding tissues. In disaster response, insect robots are being used to search for survivors in the aftermath of earthquakes or other natural disasters, where their ability to navigate rubble and debris is invaluable.

Looking to the future, the potential applications of insect robots are vast. As technology continues to advance, these robots could play a significant role in agriculture, where they could be used for precision farming, pollination, and pest control. In urban environments, insect robots could be used for infrastructure inspection, maintenance, and even waste management.

Ethical and Environmental Considerations

As with any technological advancement, the rise of insect robotics brings with it a host of ethical and environmental considerations. One of the primary ethical concerns is the potential for these robots to be used for surveillance or military purposes. The small size and mobility of insect robots make them ideal candidates for covert operations, raising questions about privacy and the potential for misuse.

Another concern is the environmental impact of deploying insect robots on a large scale. While these robots can be used to monitor and protect the environment, there is also the potential for unintended consequences. For example, the introduction of robotic pollinators into ecosystems could disrupt natural pollination processes and affect biodiversity. It is essential for scientists, engineers, and policymakers to carefully consider these potential impacts and develop guidelines for the responsible use of insect robotics.

As with any emerging technology, the development and deployment of insect robots raise important ethical and environmental considerations. One of the primary ethical concerns is the potential for these robots to be used in ways that could harm individuals or communities. For example, insect robots could be weaponized or used for surveillance in ways that infringe on privacy rights. It is essential that the development of this technology is guided by ethical principles that prioritize the well-being of individuals and society as a whole.

Environmental considerations are also crucial. While insect robots offer many potential benefits, it is important to consider their impact on natural ecosystems. For example, the deployment of robotic pollinators could have unintended consequences for natural pollinators, such as bees, which play a critical role in maintaining biodiversity. Ensuring that the deployment of insect robots is done in a way that is environmentally sustainable is essential for minimizing their impact on natural ecosystems.

As with any emerging technology, the development and deployment of insect robots must be accompanied by careful consideration of ethical issues. The potential for privacy concerns, environmental impact, and social acceptance underscores the need for responsible innovation. It is essential to establish ethical guidelines and regulatory frameworks that ensure the beneficial use of insect robots while addressing any negative consequences.

Engaging with stakeholders, including the public, policymakers, and researchers, is crucial for shaping the future direction of insect robotics. Transparent communication and proactive management of ethical considerations will help build trust and ensure that the technology serves the greater good.

Moreover, there is the question of the ethical treatment of living insects used in research and development. As we continue to study and mimic these creatures, it is important to ensure that our research practices are humane and do not cause unnecessary harm to living organisms.

Challenges and Future Directions

While the field of insect robotics has made tremendous strides, several challenges remain. One of the most significant challenges is the development of energy-efficient power sources for these tiny robots. Current battery technology limits the operational time and range of insect robots, and finding ways to extend battery life or develop alternative power sources is crucial for the continued advancement of the field.

Another challenge is the need for more robust and adaptable control systems. Insect robots often operate in unpredictable and dynamic environments, and developing control algorithms that can handle these conditions remains a significant hurdle. Advances in AI and machine learning will play a critical role in overcoming this challenge, allowing insect robots to become more autonomous and adaptable.

Looking to the future, the potential for human-insect robot collaboration is an exciting area of exploration. Insect robots could work alongside humans in a variety of settings, from disaster response to industrial inspection, providing assistance in tasks that are dangerous or difficult for humans to perform. This collaboration could lead to new ways of working and living, with insect robots becoming an integral part of our daily lives.

Despite the significant progress made in the field of insect robotics, challenges remain. One of the most pressing issues is the development of more robust and reliable communication systems that allow swarms of robots to operate in concert without interference or loss of signal. As these robots are increasingly deployed in remote or hostile environments, ensuring that they can maintain communication with each other and with human operators is critical.

Another challenge is the need for more advanced materials that can withstand the wear and tear of real-world operations. While current materials provide a good balance of strength and flexibility, there is still room for improvement, particularly in terms of durability and resistance to environmental factors such as extreme temperatures and humidity. The ongoing research into bio-inspired materials, which mimic the properties of insect exoskeletons, holds promise for addressing this challenge.

The field of insect robotics is still in its infancy, and there are many challenges that must be addressed to fully realize the potential of this technology. One of the primary challenges is energy efficiency. Insect robots require a significant amount of energy to operate, and finding ways to make these robots more energy-efficient is crucial for enabling them to perform tasks over extended periods.

Another challenge is autonomy. While significant progress has been made in developing autonomous systems for insect robots, there is still much work to be done to enable these robots to operate independently in complex environments. Advances in artificial intelligence and machine learning will be crucial for enabling insect robots to perform tasks autonomously and adapt to changing environments.

Looking to the future, the next decade will likely see significant advancements in insect robotics. As technology continues to advance, we can expect to see the development of smaller, more efficient insect robots that are capable of performing a wider range of tasks. These robots will likely play a significant role in various industries, from agriculture to healthcare, and will continue to push the boundaries of what is possible in robotics.

The future of insect robotics lies in the continued integration of biological principles with cutting-edge technology. As we learn more about the behavior and physiology of insects, new avenues for innovation will emerge, leading to robots that are even more capable and versatile. The potential for collaboration between biologists, engineers, and computer scientists is enormous, and it is this interdisciplinary approach that will drive the next wave of breakthroughs in the field.

As we look toward the horizon of what insect robotics might achieve, it's clear that the field is on the cusp of several transformative breakthroughs. The advancements we've seen so far—ranging from miniaturized sensors to more efficient power sources—are just the beginning. The next phase of development will likely be characterized by a deeper integration of artificial intelligence, enabling insect robots to operate with unprecedented autonomy and adaptability.

One of the most exciting areas of potential lies in the healthcare industry. Imagine a swarm of insect-sized robots capable of performing non-invasive procedures, diagnosing diseases at a cellular level, or even delivering targeted therapies with incredible precision. Such applications are not merely the stuff of science fiction but are rapidly approaching feasibility as our understanding of micro-robotics and AI continues to grow.

In the military and security sectors, insect robots could play a crucial role in reconnaissance and surveillance. Their ability to navigate through tight spaces and remain undetected makes them ideal for gathering intelligence in environments that are otherwise inaccessible. However, these applications also raise important ethical considerations. The potential for misuse, the implications for privacy, and the need for clear regulatory frameworks are all critical issues that must be addressed as the technology advances.

Environmental conservation is another field where insect robotics could make a significant impact. With the global challenges of climate change, habitat destruction, and biodiversity loss, insect robots could be deployed to monitor ecosystems, assist in pollination, or even help in the reforestation of degraded lands. By mimicking the roles of natural insects, these robots could provide crucial support to ecosystems under threat.

However, realizing this potential will require overcoming several key challenges. Energy efficiency remains a significant hurdle, particularly for robots intended to operate autonomously in the field for extended periods. Advances in energy harvesting, such as the development of micro-scale solar cells or bio-inspired energy systems, could provide the solutions needed to make long-term autonomous operations a reality.

In parallel, the integration of advanced AI will be essential to enable insect robots to make complex decisions in real-time, adapting to their environment as natural insects do. This will involve not only improvements in machine learning algorithms but also the development of new types of sensors and processors that can function efficiently within the constraints of micro-scale robotics.

Looking further ahead, the potential for insect robots to interact with human society in meaningful ways is both exciting and daunting. Could we one day see insect robots integrated into everyday life, assisting with tasks from household chores to personal health monitoring? Or might they remain tools of specialized industries, their incredible capabilities reserved for the most challenging and dangerous tasks?

These questions highlight the broader implications of insect robotics, which extend far beyond the technical challenges of building and programming these machines. As we continue to push the boundaries of what's possible, we must also engage with the philosophical, ethical, and societal questions that these new technologies raise. What does it mean for a robot to be "alive"? How do we balance the benefits of these technologies with the potential risks? And what kind of world do we want to build with these new tools at our disposal?

In conclusion, the future of insect robotics is as promising as it is complex. With continued innovation and careful consideration of the broader impacts, insect robots have the potential to revolutionize not just the fields we've discussed, but the very way we interact with the world around us. The next decade will be crucial in determining the direction of this technology, and it's an exciting time to be a part of this rapidly evolving field.

Future Directions and Research Opportunities

The field of insect robotics is still evolving, and numerous research opportunities lie ahead. Future directions may include:

Advanced Materials and Technologies: Continued research into advanced materials and fabrication techniques will enable the development of even more sophisticated and resilient insect robots. Innovations in materials science could lead to robots with enhanced capabilities and durability.

Autonomous Behavior and Intelligence: Improving the autonomous behavior and decision-making capabilities of insect robots will expand their applications and effectiveness. Research into artificial intelligence and machine learning will play a key role in advancing these aspects.

Integration with Other Technologies: The integration of insect robots with other emerging technologies, such as the Internet of Things (IoT) and edge computing, could unlock new possibilities for their use in various domains.

Interdisciplinary Collaboration: Further interdisciplinary collaboration will be essential for addressing complex challenges and advancing the state of insect robotics. By combining expertise from multiple fields, researchers can drive innovation and overcome technical limitations.

Applications and Real-world Implications

The practical applications of insect robots are vast, as outlined in the book. In environmental monitoring, these robots offer a non-invasive means of collecting data on ecosystems, providing insights that were previously difficult or impossible to obtain. Their small size and ability to blend into their surroundings make them ideal for studying wildlife without disturbing the natural behavior of the animals being observed.

Insect robots represent a profound leap forward in robotics, blending bio-inspired design with cutting-edge technology to address real-world challenges across diverse fields. As we explore the various applications and implications of these robotic marvels, it becomes evident that their potential reaches far beyond mere scientific curiosity. They offer practical solutions to complex problems and open new avenues for innovation.

1. Healthcare and Medicine

One of the most promising areas for insect robots is in healthcare and medicine. Their small size and agile movement enable them to access and operate in environments that are otherwise challenging for traditional medical tools.

Minimally Invasive Procedures: Insect robots can be designed to perform minimally invasive surgeries, reducing the need for large incisions and thereby decreasing recovery times. For instance, they could be employed in precision tasks such as removing tumors or delivering drugs to specific areas within the body. Their ability to maneuver through tiny openings and navigate within confined spaces makes them ideal candidates for these procedures.

Diagnostic Tools: These robots can also serve as diagnostic tools, capable of conducting high-resolution imaging or collecting biological samples with minimal discomfort to patients. Their sensitivity to environmental changes can be harnessed for real-time monitoring of physiological conditions, leading to early detection of medical issues.

Drug Delivery: Advanced insect robots could be used to deliver medication directly to targeted cells or tissues. This approach promises increased efficacy and reduced side effects compared to systemic delivery methods. For example, micro-robots could transport therapeutic agents to precisely where they are needed within the body, enhancing the overall effectiveness of treatments.

2. Environmental Monitoring and Conservation

In the realm of environmental science, insect robots offer innovative solutions for monitoring and preserving natural ecosystems.

Pollination and Ecosystem Monitoring: Insect robots can mimic the behavior of pollinators, such as bees, to assist in the pollination of crops and plants. This capability is particularly valuable in areas where natural pollinator populations are declining. Additionally, these robots can monitor environmental conditions, track wildlife, and collect data on biodiversity without disrupting the natural balance.

Disaster Response: During environmental disasters, such as oil spills or forest fires, insect robots can be deployed to assess damage, collect samples, and provide real-time data to response teams. Their ability to navigate through hazardous and difficult terrain makes them ideal for situations where human access is limited.

Climate Change Research: By deploying insect robots in various ecosystems, researchers can gather data on climate change impacts. These robots can monitor temperature fluctuations, track shifts in wildlife patterns, and assess the health of vegetation, contributing valuable information to climate models.

3. Military and Security Applications

The military and security sectors stand to benefit significantly from the unique capabilities of insect robots.

Surveillance and Reconnaissance: Insect robots can be used for surveillance and reconnaissance missions, providing detailed information in environments that are difficult for human operatives to access. Their small size allows them to go undetected, making them suitable for gathering intelligence in sensitive or hostile areas.

Search and Rescue Operations: In search and rescue missions, insect robots can navigate through rubble, debris, or collapsed structures to locate survivors. Their agility and ability to operate in tight spaces make them valuable assets in disaster response scenarios.

Explosive Detection: These robots can be equipped with sensors to detect explosives or hazardous materials, improving safety for personnel in high-risk environments. Their ability to move through challenging terrains enhances their effectiveness in locating dangerous substances.

4. Consumer and Everyday Life

The integration of insect robots into everyday life offers intriguing possibilities for consumer applications.

Home Assistance: Small, insect-like robots could assist with household chores, such as cleaning or monitoring home security. Their discreet size and efficient operation could make them useful companions in maintaining a clean and secure home environment.

Entertainment and Education: Insect robots can be used in educational settings to teach students about robotics, biology, and engineering. Their interactive nature makes them engaging tools for learning and experimentation. Additionally, they can be featured in entertainment applications, such as interactive exhibits or robotics competitions.

Personal Health Monitoring: Wearable insect robots or implantable micro-robots could monitor personal health metrics, providing users with real-time feedback on their physiological state. This could lead to proactive health management and personalized medical care.

5. Ethical and Societal Implications

The deployment of insect robots also raises important ethical and societal considerations that must be addressed as the technology evolves.

Privacy Concerns: The use of insect robots for surveillance and monitoring introduces potential privacy issues. Ensuring that these technologies are used responsibly and that appropriate safeguards are in place is crucial to maintaining public trust and ethical standards.

Environmental Impact: While insect robots offer environmental benefits, their production, deployment, and disposal must be managed carefully to minimize ecological impact. Sustainable practices and thorough lifecycle assessments will be necessary to address any unintended consequences.

Social Acceptance: As insect robots become more integrated into various aspects of life, societal acceptance and understanding of these technologies will be essential. Public engagement, transparent communication, and education will play key roles in fostering a positive relationship between technology and society.

Insect robots represent a remarkable fusion of biology and technology, with applications that span healthcare, environmental conservation, military operations, and everyday life. Their ability to operate in challenging environments, perform precise tasks, and contribute to a range of sectors underscores their transformative potential. However, the successful integration of these robots into real-world applications will require continued innovation, thoughtful consideration of ethical implications, and proactive management of their societal impact. As we move forward, it is essential to balance the remarkable benefits of insect robotics with the responsibility of addressing the challenges they present, ensuring that their development and deployment contribute positively to our world.

In the medical field, the potential for insect robots to revolutionize surgery is immense. Their ability to navigate through the human body with minimal invasion opens up new possibilities for procedures that are less traumatic for patients and require shorter recovery times. The precision of these robots, combined with advanced imaging technologies, means that surgeons can perform complex operations with greater accuracy, reducing the risk of complications.

The ethical implications of deploying insect robots in the real world cannot be ignored. As with any new technology, the potential for misuse exists, and it is essential that researchers and policymakers work together to establish guidelines and regulations that ensure these robots are used for the benefit of society. The book has touched on the importance of transparency and accountability in the development and deployment of insect robots, emphasizing the need for a framework that balances innovation with responsibility.

The real-world applications of insect robots are as diverse as the insects themselves. In the natural world, insects play various roles—pollinators, predators, decomposers—and their robotic counterparts are similarly versatile. Current applications include environmental monitoring, where insect robots are used to collect data from hard-to-reach places, such as the canopies of rainforests or the depths of the ocean. Their small size and ability to navigate complex terrains make them ideal for tasks that are beyond the reach of traditional robots.

In the medical field, insect robots are being developed for use in minimally invasive surgeries, where their precision and small size allow them to perform delicate procedures without causing significant damage to surrounding tissues. In disaster response, insect robots are being used to search for survivors in the aftermath of earthquakes or other natural disasters, where their ability to navigate rubble and debris is invaluable.

Looking to the future, the potential applications of insect robots are vast. As technology continues to advance, these robots could play a significant role in agriculture, where they could be used for precision farming, pollination, and pest control. In urban environments, insect robots could be used for infrastructure inspection, maintenance, and even waste management.

Final Thoughts and Broader Significance

As we conclude our journey through the micro world of robotics, it is clear that the field of insect robotics holds immense potential. By drawing inspiration from the natural world, scientists and engineers have developed robotic systems that are more efficient, adaptable, and capable than ever before. These advancements have the potential to transform a wide range of industries and improve our understanding of both robotics and biology. Insect robots, with their unique capabilities and potential applications, represent a significant leap forward in the field of robotics, merging the intricate details of biological systems with innovative technological solutions.

1. Synthesis of Insect Robotics Innovations

Throughout this book, we have delved into the various facets of insect robotics, from their bio-inspired design principles to their diverse applications. The synthesis of biological models with engineering has led to the development of highly efficient, versatile, and adaptable robots. These advancements underscore the potential of cross-disciplinary research in pushing the boundaries of what is possible in robotics.

Insect robots are not merely technological curiosities; they embody a convergence of nature and technology that yields practical solutions to complex problems. By mimicking the behaviors and structures of insects, these robots achieve remarkable feats of agility, efficiency, and miniaturization, offering new capabilities that traditional robots often cannot match.

2. The Impact on Robotics and Engineering

The innovations in insect robotics have profound implications for the broader field of robotics and engineering. The miniaturization techniques and bio-inspired design principles developed for insect robots are paving the way for advancements in various robotic applications. These technologies are likely to influence future robotics research, leading to the creation of even more sophisticated and capable machines.

Furthermore, the interdisciplinary nature of insect robotics research encourages collaboration between biologists, engineers, and computer scientists. This collaborative approach fosters the development of holistic solutions that integrate diverse expertise, leading to breakthroughs that might not be achievable within isolated disciplines.

3. Broader Societal Implications

The societal implications of insect robotics are extensive and multifaceted. As these robots become more integrated into various aspects of life, their impact will be felt across multiple domains:

Healthcare: The ability of insect robots to perform minimally invasive procedures and monitor health conditions could revolutionize medical practices. This transformation has the potential to improve patient outcomes, reduce healthcare costs, and enhance the quality of care.

Environmental Conservation: Insect robots contribute to environmental monitoring and conservation efforts, offering new tools for managing and protecting natural resources. Their application in ecological research and disaster response highlights their role in addressing pressing environmental challenges.

Military and Security: In the military and security sectors, insect robots offer novel solutions for surveillance, reconnaissance, and search-and-rescue operations. Their deployment in these areas enhances operational capabilities and safety, reflecting their strategic value.

Everyday Life: The integration of insect robots into daily life, from home assistance to personal health monitoring, signifies a shift toward more intelligent and responsive technologies. These robots have the potential to enhance convenience, security, and overall quality of life for individuals.

However, with this potential comes responsibility. As we continue to develop and deploy insect robots, it is essential to consider the ethical and environmental implications of our work. By doing so, we can ensure that the benefits of insect robotics are realized in a way that is responsible and sustainable.

Insect robots represent a significant advancement in the field of robotics, but they also have broader implications for the future of robotics as a whole. The development of these robots has challenged traditional notions of what robots can and should be, and has opened up new possibilities for the design and deployment of robotic systems.

One of the key implications of insect robotics is the potential for robots to operate in environments that were previously considered inaccessible. The small size and agility of insect robots allow them to navigate complex terrains and reach places that are beyond the reach of traditional robots. This has significant implications for fields such as disaster response, environmental monitoring, and infrastructure maintenance.

Insect robots also challenge traditional notions of intelligence and autonomy in robotics. These robots are not just tools that follow pre-programmed instructions—they are capable of learning, adapting, and making decisions on their own. This has significant implications for the future of robotics, as it suggests that robots could play a more active role in decision-making processes and could be used in more dynamic and unpredictable environments.

In conclusion, the exploration of insect robotics is far from over. This book has provided a glimpse into the incredible possibilities that lie at the intersection of biology and robotics, but there is still much to learn and discover. As we move forward, the continued collaboration between scientists, engineers, and ethicists will be crucial in shaping the future of insect robotics. The journey has only just begun, and the micro world of robotics promises to be a field of endless innovation and discovery.

In conclusion, the exploration of the micro world of robotics through insect robots has revealed a world of possibilities that were previously unimaginable. These tiny robots represent the cutting edge of technology and science, and they have the potential to revolutionize various industries and fields.

However, the journey of discovery is far from over. The development of insect robots is still in its early stages, and there are many challenges and opportunities that lie ahead. Continued exploration and innovation will be crucial for realizing the full potential of this technology and for ensuring that it is used in ways that benefit society and the environment.

As we look to the future, it is clear that insect robots will play a significant role in shaping the future of robotics. By continuing to explore and develop these tiny robots, we are not just creating tools for the future—we are laying the groundwork for a new era of robotics that is as dynamic, resilient, and adaptive as the natural world itself.

In closing, the exploration of the micro world of robotics through insect-inspired designs reveals a landscape of remarkable innovation and potential. Insect robots stand at the intersection of biology and technology, offering unique solutions to real-world challenges and shaping the future of robotics. Their impact on healthcare, environmental conservation, military operations, and everyday life underscores their significance and promise.

Compilation of References

Abbasi, Q. H., Yang, K., Chopra, N., Jornet, J. M., Abuali, N. A., Qaraqe, K. A., & Alomainy, A. (2016). Nano-communication for biomedical applications: A review on the state-of-the-art from physical layers to novel networking concepts. *IEEE Access : Practical Innovations, Open Solutions*, 4, 3920–3935.

Aggarwal, A., Tam, C. C., Wu, D., Li, X., & Qiao, S. (2023). Artificial intelligence–based chatbots for promoting health behavioral changes: Systematic review. Journal of medical Internet research, 25, e40789. 9.

Aggarwal, M., & Kumar, S. (2022). The use of nanorobotics in the treatment therapy of cancer and its future aspects: A review. *Cureus*, 14(9).

Agrawal, B. (2017). Scalable Data Processing and Analytical Approach for Big Data Cloud Platform (Doctoral dissertation, PhD Thesis).

Agrawal, A., Soni, R., Gupta, D., & Dubey, G. (2024). The role of robotics in medical science: Advancements, applications, and future directions. *Journal of Autonomous Intelligence*, 7(3). Advance online publication. DOI: 10.32629/jai.v7i3.1008

Ahmed, M. S., & Ahmed, N. (2023). A fast and minimal system to identify depression using smartphones: Explainable machine learning–based approach. *JMIR Formative Research*, 7, e28848. DOI: 10.2196/28848 PMID: 37561568

Al Banna, M. H., Taher, K. A., Kaiser, M. S., Mahmud, M., Rahman, M. S., Hosen, A. S., & Cho, G. H. (2020). Application of artificial intelligence in predicting earthquakes: State-of-the-art and future challenges. *IEEE Access : Practical Innovations, Open Solutions*, 8, 192880–192923. DOI: 10.1109/ACCESS.2020.3029859

Alam, F., Naim, M., Aziz, M., & Andyadav, N. (2015). Unique roles of nanotechnology in medicine and cancer-II. *Indian Journal of Cancer*, 52(1), 1–9. DOI: 10.4103/0019-509X.175591 PMID: 26837958

Aleem, S., Huda, N. U., Amin, R., Khalid, S., Alshamrani, S. S., & Alshehri, A. (2022). Machine learning algorithms for depression: Diagnosis, insights, and research directions. *Electronics (Basel)*, 11(7), 1111. DOI: 10.3390/electronics11071111

Algamili, A. S., Mohd, H. M. K., Dennis, J. O., Ahmed, A. Y., Alabsi, S. S., Saeed, S. B. H., & Junaid, M. M. (2021). A review of actuation and sensing mechanisms in MEMS-based sensor devices. *Nanoscale Research Letters*, 16(1), 1–21. DOI: 10.1186/s11671-021-03481-7 PMID: 33496852

Alhanish, A., & Abu Ghalia, M. (2021). Biobased thermoplastic polyurethanes and their capability to biodegradation. Eco-Friendly Adhesives for Wood and Natural Fiber Composites: Characterization, Fabrication and Applications, 85-104.

Aydin, A., & Saranli, U. (2014). Design and implementation of a firefighting robot. Journal of Intelligent & Robotic Systems, 74(3-4), 787-801. https://doi.org/DOI: 10.1007/s10846-013-9931-7

Bai, T., & Tahmasebi, P. (2022). Attention-based LSTM-FCN for earthquake detection and location. *Geophysical Journal International*, 228(3), 1568–1576. DOI: 10.1093/gji/ggab401

Balaji, V., Balaji, M., Chandrasekaran, M., Khan, M. K. A. A., & Elamvazuthi, I. (2015). Optimization of PID control for high-speed line tracking robots. *Procedia Computer Science*, 76, 147–154. Advance online publication. DOI: 10.1016/j.procs.2015.12.329

Barrile, V., Simonetti, S., Citroni, R., Fotia, A., & Bilotta, G. (2022). Experimenting agriculture 4.0 with sensors: A data fusion approach between remote sensing, uavs and self-driving tractors. *Sensors (Basel)*, 22(20), 7910. DOI: 10.3390/s22207910 PMID: 36298261

Bennetts, V. H., Schill, F., & Durand-Petiteville, A. (2019). Autonomous robotic firefighting in GPS-denied environments. IEEE Access, 7, 43735-43751. https://doi.org/DOI: 10.1109/ACCESS.2019.2909527

Bilham, R. (2004). Earthquakes in India and the Himalaya: Tectonics, geodesy and history. *Annals of Geophysics*.

Bonabeau, E., Dorigo, M., & Theraulaz, G. (1999). *Swarm Intelligence: From Natural to Artificial Systems*. Oxford University Press. DOI: 10.1093/oso/9780195131581.001.0001

Breysse, D., Romao, X., Alwash, M., Sbartai, Z. M., & Luprano, V. A. (2020). Risk evaluation on concrete strength assessment with NDT technique and conditional coring approach. *Journal of Building Engineering*, 32, 101541. DOI: 10.1016/j.jobe.2020.101541

Briffault, X. (2018). The Hamilton Scale as an Analyzer for the Epistemological Difficulties in Research on Depression. In *Measuring Mental Disorders* (pp. 55–87). Elsevier. DOI: 10.1016/B978-1-78548-305-9.50002-X

Byeon, H. (2023). Advances in machine learning and explainable artificial intelligence for depression prediction. *International Journal of Advanced Computer Science and Applications*, 14(6). Advance online publication. DOI: 10.14569/IJACSA.2023.0140656

Chaulagain, H., Gautam, D., & Rodrigues, H. (2018). Revisiting major historical earthquakes in Nepal: Overview of 1833, 1934, 1980, 1988, 2011, and 2015 seismic events. *Impacts and insights of the Gorkha earthquake*, 1-17.

Che, C., Zheng, H., Huang, Z., Jiang, W., & Liu, B. (2024). Intelligent robotic control system based on computer vision technology. *arXiv preprint arXiv:2404.01116*.

Chegeni, M. H., Sharbatdar, M. K., Mahjoub, R., & Raftari, M. (2022). New supervised learning classifiers for structural damage diagnosis using time series features from a new feature extraction technique. *Earthquake Engineering and Engineering Vibration*, 21(1), 169–191. DOI: 10.1007/s11803-022-2079-2

Chen, T. Y., Chu, H. T., Tai, Y. M., & Yang, S. N. (2022). Performances of Depression Detection through Deep Learning-based Natural Language Processing to Mandarin Chinese Medical Records: Comparison between Civilian and Military Populations. Taiwanese Journal of Psychiatry, 36(1), 32-38. 12. 12.

Chen, L., & Zhang, X. (2019). Real-Time Seismic Monitoring Using Multi-Sensor Networks. *Sensors (Basel)*.

Chen, M., & Fowler, M. L. (2003, March). The importance of data compression for energy efficiency in sensor networks. In *Conference on Information Sciences and Systems* (p. 13).

Chen, M., & Zhao, L. (2022). Advances in Piezoelectric Sensors for Seismic Monitoring. *Sensors and Actuators*.

Chen, Y., Zhang, H., & Zhao, Y. (2020). A cloud-based detection approach for cryptojacking attacks. *IEEE Access : Practical Innovations, Open Solutions*, 8, 83100–83109.

Chesnitskiy, A. V., Gayduk, A. E., Seleznev, V. A., & Prinz, V. Y. (2022). Bio-inspired micro-and nanorobotics driven by magnetic field. *Materials (Basel)*, 15(21), 7781.

Chilukoti, S. V., Maida, A., & Hei, X. (2022). Diabetic Retinopathy Detection using Transfer Learning from Pre-trained Convolutional Neural Network Models. DOI: 10.36227/techrxiv.18515357

Chiong, R., Budhi, G. S., Dhakal, S., & Chiong, F. (2021). A textual-based featuring approach for depression detection using machine learning classifiers and social media texts. *Computers in Biology and Medicine*, 135, 104499. DOI: 10.1016/j.compbiomed.2021.104499 PMID: 34174760

Chi, Y., Zhao, Y., Hong, Y., Li, Y., & Yin, J. (2024). A perspective on miniature soft robotics: Actuation, fabrication, control, and applications. *Advanced Intelligent Systems*, 6(2), 2300063. DOI: 10.1002/aisy.202300063

Chung, T. S., & Kim, M. S. (2016). Autonomous firefighting mobile robot using embedded controllers. Journal of Mechanical Science and Technology, 30(6), 2597-2604. https://doi.org/DOI: 10.1007/s12206-016-0512-y

Cleland-Huang, J., Chambers, T., Zudaire, S., Chowdhury, M. T., Agrawal, A., & Vierhauser, M. (2024). Human–machine teaming with small unmanned aerial systems in a mapek environment. *ACM Transactions on Autonomous and Adaptive Systems*, 19(1), 1–35. DOI: 10.1145/3618001

Corbi, F., Sandri, L., Bedford, J., Funiciello, F., Brizzi, S., Rosenau, M., & Lallemand, S. (2019). Machine learning can predict the timing and size of analog earthquakes. *Geophysical Research Letters*, 46(3), 1303–1311. DOI: 10.1029/2018GL081251

Costantini, L., Pasquarella, C., Odone, A., Colucci, M. E., Costanza, A., Serafini, G., Aguglia, A., Belvederi Murri, M., Brakoulias, V., Amore, M., Ghaemi, S. N., & Amerio, A. (2021). Screening for depression in primary care with Patient Health Questionnaire-9 (PHQ-9): A systematic review. *Journal of Affective Disorders*, 279, 473–483. DOI: 10.1016/j.jad.2020.09.131 PMID: 33126078

Cuijpers, P., Smits, N., Donker, T., Ten Have, M., & de Graaf, R. (2009). Screening for mood and anxiety disorders with the five-item, the three-item, and the two-item Mental Health Inventory. *Psychiatry Research*, 168(3), 250–255. DOI: 10.1016/j.psychres.2008.05.012 PMID: 19185354

Dahan, F., Alroobaea, R., Alghamdi, W. Y., Mohammed, M. K., Hajjej, F., & Raahemifar, K. (2023). A smart IoMT based architecture for E-healthcare patient monitoring system using artificial intelligence algorithms. *Frontiers in Physiology*, 14, 1125952. DOI: 10.3389/fphys.2023.1125952 PMID: 36793418

De Wolf, T., & Holvoet, T. (2005). *Emergence and Self-Organization: A Statement of Similarities and Differences*. In *Proceedings of the International Workshop on Engineering Self-Organising Applications (ESOA)*, pp. 96-110.

Debnath, P., Chittora, P., Chakrabarti, T., Chakrabarti, P., Leonowicz, Z., Jasinski, M., Gono, R., & Jasińska, E. (2021). Analysis of earthquake forecasting in India using supervised machine learning classifiers. *Sustainability (Basel)*, 13(2), 971. DOI: 10.3390/su13020971

Deepa, V., Kumar, C. S., & Cherian, T. (2022). Ensemble of multi-stage deep convolutional neural networks for automated grading of diabetic retinopathy using image patches. *Journal of King Saud University. Computer and Information Sciences*, 34(8), 6255–6265.

Deshpande, M., & Rao, V. (2017, December). Depression detection using emotion artificial intelligence. In 2017 international conference on intelligent sustainable systems (iciss) (pp. 858-862). IEEE.

Destefano, V., Khan, S., & Tabada, A. (2020). Applications of PLA in modern medicine. [engreg]. *EngRege*, (1), 76–87. DOI: 10. 1016/j PMID: 38620328

Dew Dney, A. K. (1998). Nanotechnology wherein molecular computers control tiny circulatory submarines. *Scientific American*, •••, 100–103.

Ding, X., Yue, X., Zheng, R., Bi, C., Li, D., & Yao, G. (2019). Classifying major depression patients and healthy controls using EEG, eye tracking and galvanic skin response data. *Journal of Affective Disorders*, 251, 156–161. DOI: 10.1016/j.jad.2019.03.058 PMID: 30925266

Dorigo, M., & Şahin, E. (2004). Swarm Robotics: Special Issue Editorial. *Autonomous Robots*, 17(2-3), 111–113. DOI: 10.1023/B:AURO.0000034008.48988.2b

Duan, L., Duan, H., Qiao, Y., Sha, S., Qi, S., Zhang, X., & Wang, C. (2020). Machine learning approaches for MDD detection and emotion decoding using EEG signals. *Frontiers in Human Neuroscience*, 14, 284.

Ducatelle, F., Di Caro, G. A., Pinciroli, C., & Gambardella, L. M. (2011). Self-Organized Cooperation between Robotic Swarms. *Swarm Intelligence*, 5(2), 73–96. DOI: 10.1007/s11721-011-0053-0

Durrah, O., Aldhmour, F. M., El-Maghraby, L., & Chakir, A. (2024). Artificial Intelligence Applications in Healthcare. In *Engineering Applications of Artificial Intelligence* (pp. 175–192). Springer Nature Switzerland. DOI: 10.1007/978-3-031-50300-9_10

Durrant-Whyte, H., & Bailey, T. (2006). Simultaneous Localization and Mapping: Part I. *IEEE Robotics & Automation Magazine*, 13(2), 99–110. DOI: 10.1109/MRA.2006.1638022

Elhoseny, M., Abdelaziz, A., Salama, A. S., Riad, A. M., Muhammad, K., & Sangaiah, A. K. (2018). A hybrid model of internet of things and cloud computing to manage big data in health services applications. *Future Generation Computer Systems*, 86, 1383–1394. DOI: 10.1016/j.future.2018.03.005

Elsharif, A. A. E. F., & Abu-Naser, , S. S. (2022). Retina Diseases Diagnosis Using Deep Learning. *International Journal of Academic Engineering Research*, 6(2).

Ferrer, E. C., Hardjono, T., Pentland, A., & Dorigo, M. (2021). Secure and secret cooperation in robot swarms. Science Robotics, 6(56), eabf1538.

Frässle, S., Marquand, A. F., Schmaal, L., Dinga, R., Veltman, D. J., Van der Wee, N. J., van Tol, M.-J., Schöbi, D., Penninx, B. W. J. H., & Stephan, K. E. (2020). Predicting individual clinical trajectories of depression with generative embedding. *NeuroImage. Clinical*, 26, 102213. DOI: 10.1016/j.nicl.2020.102213 PMID: 32197140

Fujisawa, R., Dobata, S., Kubota, N., Hayashi, N., & Matsuno, F. (2008). Designing Pheromone Communication in Swarm Robotics: Decay and Diffusion Properties of Virtual Pheromones Implemented on an Event-based Simulator. *Swarm Intelligence*, 2(3-4), 185–202.

Fuller, S. B. (2019). Four wings: An insect-sized aerial robot with steering ability and payload capacity for autonomy. *IEEE Robotics and Automation Letters*, 4(2), 570–577. DOI: 10.1109/LRA.2019.2891086

Ganesan, S., Shakya, M., Aqueel, A. F., & Nambiar, L. M. (2011, December). Small disaster relief robots with swarm intelligence routing. In *Proceedings of the 1st International Conference on Wireless Technologies for Humanitarian Relief* (pp. 123-127).

Garg, S., Singh, R., & Jain, R. (2020). Anomaly detection for cryptojacking using behavioral analysis. *2020 2nd International Conference on Advanced Research in Applied Sciences and Engineering (ICARASE)*, 1-6.

Garnier, S., Gautrais, J., & Theraulaz, G. (2007). The Biological Principles of Swarm Intelligence. *Swarm Intelligence*, 1(1), 3–31. DOI: 10.1007/s11721-007-0004-y

Gayathri, S., Gopi, V. P., & Palanisamy, P. (2021). Diabetic retinopathy classification based on multipath CNN and machine learning classifiers. *Physical and Engineering Sciences in Medicine*, 44(3), 639–653. DOI: 10.1007/s13246-021-01012-3 PMID: 34033015

Gupta, A., & Verma, P. (2020). Sensor Technologies for Early Earthquake Detection. Journal of Earthquake Technology.

Gupta, A., Kumar, N., & Agarwal, A. (2021). A hybrid approach to detect cryptojacking attacks on the cloud. *2021 5th International Conference on Intelligent Computing and Control Systems (ICICCS)*, 1035-1040.

Harirchian, E., Jadhav, K., Kumari, V., & Lahmer, T. (2022). ML-EHSAPP: A prototype for machine learning-based earthquake hazard safety assessment of structures by using a smartphone app. *European Journal of Environmental and Civil Engineering*, 26(11), 5279–5299. DOI: 10.1080/19648189.2021.1892829

Harwood, S., & Eaves, S. (2020). Conceptualising technology, its development and future: The six genres of technology. *Technological Forecasting and Social Change*, 160, 120174. DOI: 10.1016/j.techfore.2020.120174 PMID: 32904525

Hemmerling, T. M., & Jeffries, S. D. (2024). Robotic Anesthesia: A Vision for 2050. *Anesthesia and Analgesia*, 138(2), 239–251. DOI: 10.1213/ANE.0000000000006835 PMID: 38215704

Hepp, A. (2020). Artificial companions, social bots and work bots: Communicative robots as research objects of media and communication studies. *Media Culture & Society*, 42(7-8), 1410–1426. DOI: 10.1177/0163443720916412

Hernández, P. D., Ramírez, J. A., & Soto, M. A. (2022). Deep-learning-based earthquake detection for fiber-optic distributed acoustic sensing. *Journal of Lightwave Technology*, 40(8), 2639–2650. DOI: 10.1109/JLT.2021.3138724

Higashino, T., Yamaguchi, H., Hiromori, A., Uchiyama, A., & Yasumoto, K. (2017, June). Edge computing and IoT based research for building safe smart cities resistant to disasters. In *2017 IEEE 37th international conference on distributed computing systems (ICDCS)* (pp. 1729-1737). IEEE.

Hildmann, H., Kovacs, E., Saffre, F., & Isakovic, A. (2019). Nature-inspired drone swarming for real-time aerial data-collection under dynamic operational constraints. *Drones (Basel)*, 3(3), 71. DOI: 10.3390/drones3030071

Holzer, T. L., & Savage, J. C. (2013). Global earthquake fatalities and population. *Earthquake Spectra*, 29(1), 155–175. DOI: 10.1193/1.4000106

Hossen, M. S., Reza, A. A., & Mishu, M. C. (2020, January). An automated model using deep convolutional neural network for retinal image classification to detect diabetic retinopathy. In *Proceedings of the International Conference on Computing Advancements* (pp. 1-8).

Hou, P., Gong, J., & Jiahesilike, A. (2023, October). Disaster Search and Rescue Bionic Insect Robot Based on Edge Computing. In 2023 2nd International Conference on Data Analytics, Computing and Artificial Intelligence (ICDACAI) (pp. 535-541). IEEE.

Housner, G. W. (1955). Properties of strong ground motion earthquakes. *Bulletin of the Seismological Society of America*, 45(3), 197–218. DOI: 10.1785/BSSA0450030197

https://www.google.com/url?

https://www.mayoclinic.org/diseases-conditions/depression/symptoms-causes/syc-20356007

Hu, M., Wei, R., Dai, T., Zou, L., & Li, T. (2008, June). Control strategy for insect-like flapping wing micro air vehicles: Attitude control. In *2008 7th World Congress on Intelligent Control and Automation* (pp. 9043-9048). IEEE.

Huang, C., Song, H., & Li, D. (2018). Detection of cryptojacking malware using machine learning. *2018 IEEE International Conference on Software Quality, Reliability and Security Companion (QRS-C)*, 416-421.

Hu, J., Edsinger, A., Lim, Y. J., Donaldson, N., Solano, M., Solochek, A., & Marchessault, R. (2011). *May. An advanced medical robotic system augmenting healthcare capabilities-robotic nursing assistant. In 2011 IEEE international conference on robotics and automation*. IEEE.

Hu, M., Ge, X., Chen, X., Mao, W., Qian, X., & Yuan, W. E. (2020). Micro/nanorobot: A promising targeted drug delivery system. *Pharmaceutics*, 12(7), 665.

Ibrahim, A. E., Karsiti, M. N., & Elamvazuthi, I. (2017). Fuzzy logic system to control a spherical underwater robot vehicle (URV). International Journal of Simulation Systems Science and Technology. DOI: 10.5013/IJSSST.a.18.01.09

Jagan Mohan, N., Murugan, R., Goel, T., Mirjalili, S., & Roy, P. (2021, December). A novel four-step feature selection technique for diabetic retinopathy grading. *Physical and Engineering Sciences in Medicine*, 44(4), 1351–1366. DOI: 10.1007/s13246-021-01073-4 PMID: 34748191

Jain, K. K. (2019). An overview of drug delivery systems. *Drug Delivery System*. Advance online publication. DOI: 10.1007/978-1-4939-9798-5_1 PMID: 31435914

Jena, S., Mohanty, S., Ojha, M., Subham, K., & Jha, S. (2022). Nanotechnology: An Emerging Field in Protein Aggregation and Cancer Therapeutics. *Bio-Nano Interface: Applications in Food, Healthcare and Sustainability*, 177-207.

Jiang, Y., Zhao, W., Jiang, Y., Sun, K., Huang, X., & Yang, B. (2023). Wireless Multisensors Platform for Distributed Insect Biobot Sensing Network. *IEEE Sensors Journal*, 23(7), 7929–7937. DOI: 10.1109/JSEN.2023.3243916

Jin, C., Wang, K., Oppong-Gyebi, A., & Hu, J. (2020). Application of nanotechnology in cancer diagnosis and therapy - a mini-review. *International Journal of Medical Sciences*, 17(18), 2964–2973. DOI: 10.7150/ijms.49801 PMID: 33173417

Johanson, D. L., Ahn, H. S., & Broadbent, E. (2021). Improving interactions with healthcare robots: A review of communication behaviours in social and healthcare contexts. *International Journal of Social Robotics*, 13(8), 1835–1850. DOI: 10.1007/s12369-020-00719-9

Johnson, T., & White, S. (2021). Innovative Approaches to Earthquake Prediction Using Multi-Sensor Networks. *Earthquake Engineering & Structural Dynamics*.

Jones, M., Smith, T., & Brown, K. (2022). Enhancing cloud security through dynamic blocking of cryptojacking attacks. *Proceedings of the 2022 International Conference on Cloud Computing and Security (ICCCS)*, 82-87.

Jones, S., Milner, E., Sooriyabandara, M., & Hauert, S. (2020). Distributed situational awareness in robot swarms. *Advanced Intelligent Systems*, 2(11), 2000110. DOI: 10.1002/aisy.202000110

Joseph, A., Christian, B., Abiodun, A. A., & Oyawale, F. (2018). A review on humanoid robotics in healthcare. In *MATEC Web of Conferences* (Vol. 153, p. 02004). EDP Sciences. DOI: 10.1051/matecconf/201815302004

Kaiser, M. S., Al Mamun, S., Mahmud, M., & Tania, M. H. (2021). Healthcare robots to combat COVID-19. *COVID-19: Prediction, decision-making, and its impacts*, pp.83-97.

Kalulu, M., Chilikwazi, B., Hu, J., & Fu, G. (2024). Soft Actuators and Actuation: Design, Synthesis and Applications. *Macromolecular Rapid Communications*, •••, 2400282. DOI: 10.1002/marc.202400282 PMID: 38850266

Karki, S. S., & Kulkarni, P. (2021, March). Diabetic retinopathy classification using a combination of efficientnets. In *2021 International Conference on Emerging Smart Computing and Informatics (ESCI)* (pp. 68-72). IEEE.

Kassani, S. H., Kassani, P. H., Khazaeinezhad, R., Wesolowski, M. J., Schneider, K. A., & Deters, R. (2019, December). Diabetic retinopathy classification using a modified xception architecture. In 2019 IEEE international symposium on signal processing and information technology (ISSPIT) (pp. 1-6). IEEE.

Kaur, J. (2024). Revolutionizing Healthcare: Synergizing Cloud Robotics and Artificial Intelligence for Enhanced Patient Care. In *Shaping the Future of Automation With Cloud-Enhanced Robotics* (pp. 272-287). IGI Global.

Kaywan, P., Ahmed, K., Ibaida, A., Miao, Y., & Gu, B. (2023). Early detection of depression using a conversational AI bot: A non-clinical trial. *PLoS One*, 18(2), e0279743. DOI: 10.1371/journal.pone.0279743 PMID: 36735701

Khaled, O., El-Sahhar, M., El-Dine, M. A., Talaat, Y., Hassan, Y. M., & Hamdy, A. (2020, November). Cascaded architecture for classifying the preliminary stages of diabetic retinopathy. In *Proceedings of the 9th International Conference on Software and Information Engineering* (pp. 108-112).

Khalid, M. Y., Arif, Z. U., Tariq, A., Hossain, M., Khan, K. A., & Umer, R. (2024). 3D printing of magneto-active smart materials for advanced actuators and soft robotics applications. *European Polymer Journal*, 205, 112718. DOI: 10.1016/j.eurpolymj.2023.112718

Khang, A., Rath, K. C., Anh, P. T. N., Rath, S. K., & Bhattacharya, S. (2024). Quantum-Based Robotics in the High-Tech Healthcare Industry: Innovations and Applications. In *Medical Robotics and AI-Assisted Diagnostics for a High-Tech Healthcare Industry* (pp. 1-27). IGI Global.

Khang, A. (Ed.). (2023). *AI and IoT-based technologies for precision medicine*. IGI Global.

Khang, A. (Ed.). (2024). *Medical Robotics and AI-Assisted Diagnostics for a High-Tech Healthcare Industry*. IGI Global. DOI: 10.4018/979-8-3693-2105-8

Kim, J. H., Park, H. S., & Kang, J. W. (2017). Development of a fire-detection and extinguishing mobile robot for underground coal mines. Journal of Field Robotics, 34(6), 1152-1169. DOI: 10.1002/rob.21715

Kishore, C., & Bhadra, P. (2021). Targeting Brain Cancer Cells by Nanorobot, a Promising Nanovehicle: New Challenges and Future Perspectives. *CNS & Neurological Disorders - Drug Targets*, 20(6), 531–539. DOI: 10.2174/1871527320666210526154801 PMID: 34042038

Kolpashchikov, D., Gerget, O., & Meshcheryakov, R. (2022). *Robotics in healthcare. Handbook of Artificial Intelligence in Healthcare* (Vol. 2). Practicalities and Prospects.

Kondoyanni, M., Loukatos, D., Maraveas, C., Drosos, C., & Arvanitis, K. G. (2022). Bio-inspired robots and structures toward fostering the modernization of agriculture. *Biomimetics*, 7(2), 69. DOI: 10.3390/biomimetics7020069 PMID: 35735585

Kot, P., Muradov, M., Gkantou, M., Kamaris, G. S., Hashim, K., & Yeboah, D. (2021). Recent advancements in non-destructive testing techniques for structural health monitoring. *Applied Sciences (Basel, Switzerland)*, 11(6), 2750. DOI: 10.3390/app11062750

Krishnababu, K., Kulkarni, G. S., Athmaja Shetty, Y. R., & SN, R. B. (2023). Development of Micro/Nanobots and their Application in Pharmaceutical and Healthcare Industry.

Kubo, H., Naoi, M., & Kano, M. (2024). Recent advances in earthquake seismology using machine learning. *Earth, Planets, and Space*, 76(1), 36. DOI: 10.1186/s40623-024-01982-0

Kumar, A., Choudhary, A., & Jain, P. (2019). Network-based detection of cryptojacking attacks using machine learning. *2019 10th International Conference on Computing, Communication and Networking Technologies (ICCCNT)*, 1-6.

Kumar, N., & Sharma, P. (2020). Enhancing Seismic Prediction through Technological Innovations. *Earthquake Engineering & Structural Dynamics*.

Kumar, S., & Singh, R. (2022). *Multisensor Integration for Real-Time Earthquake Monitoring*. Advances in Earthquake Research.

Kyrarini, M., Lygerakis, F., Rajavenkatanarayanan, A., Sevastopoulos, C., Nambiappan, H. R., Chaitanya, K. K., Babu, A. R., Mathew, J., & Makedon, F. (2021). A survey of robots in healthcare. *Technologies*, 9(1), 8. DOI: 10.3390/technologies9010008

Lahmar, C., & Idri, A. (2022)... *Computer Methods in Biomechanics and Biomedical Engineering. Imaging & Visualization*.

Lee, C. H., & Ke, Y. H. (2021, June). Fundus images classification for diabetic retinopathy using deep learning. In *Proceedings of the 13th International Conference on Computer Modeling and Simulation* (pp. 264-270).

Lee, C., & Kim, H. (2022). Machine learning-based predictive modeling of depression in hypertensive populations. *PLoS One*, 17(7), e0272330. DOI: 10.1371/journal.pone.0272330 PMID: 35905087

Lee, D., & Kim, S. (2021). Integration of Infrared Sensors in Seismic Monitoring Systems. *Journal of Seismology*.

Lee, D., Park, S., & Kim, J. (2020). Development of a Novel Seismic Sensor System Mimicking Snake Vibration Detection Mechanisms. *Journal of Seismology*.

Lee, D., & Yoon, S. N. (2021). Application of artificial intelligence-based technologies in the healthcare industry: Opportunities and challenges. *International Journal of Environmental Research and Public Health*, 18(1), 271. DOI: 10.3390/ijerph18010271 PMID: 33401373

Lee, J., & Park, S. (2022). Comprehensive Early Warning Systems for Earthquakes Using Multi-Sensor Integration. *Journal of Seismology*.

Lee, O. E., Lee, H., Park, A., & Choi, N. G. (2024). My precious friend: Human-robot interactions in home care for socially isolated older adults. *Clinical Gerontologist*, 47(1), 161–170. DOI: 10.1080/07317115.2022.2156829 PMID: 36502295

Li, J., Zhou, X., & Zhou, X. (2019). A signature-based detection approach for cryptojacking malware. *2019 IEEE 11th International Conference on Advanced Infocomm Technology (ICAIT)*, 1-5.

Li, L., & Chew, Z. J. (2018). Microactuators: Design and technology. In Smart sensors and MEMS (pp. 313-354). Woodhead Publishing.

Li, S., & Liu, L. (2018). Multi-sensor fusion for robotic fire detection and suppression. Sensors, 18(2), 564. DOI: 10.3390/s18020564

Li, Y., & Wu, Q. (2015). Research on the control system of an autonomous firefighting robot. Procedia Engineering, 99, 876-885. DOI: 10.1016/j.proeng.2014.12.611

Lim, J., Tewolde, G., Kwon, J., & Choi, S. (2019). Design and implementation of a network robotic framework using a smartphone-based platform. IEEE Access, 7, 1-1. DOI: 10.1109/ACCESS.2019.2916464

Liu, K. (2018). Security analysis of mobile device-to-device network applications. IEEE Internet of Things Journal, (c), 1. https://doi.org/DOI: 10.1109/JIOT.2018.2854318

Liu, H., Wang, F., Wu, W., Dong, X., & Sang, L. (2023). 4D printing of mechanically robust PLA/TPU/Fe3O4 magneto-responsive shape memory polymers for smart structures. *Composites. Part B, Engineering*, 248, 110382.

Li, W., Hu, D., & Yang, L. (2023). Actuation mechanisms and applications for soft robots: A comprehensive review. *Applied Sciences (Basel, Switzerland)*, 13(16), 9255. DOI: 10.3390/app13169255

Lydon, D., Kromanis, R., Lydon, M., Early, J., & Taylor, S. (2022). Use of a roving computer vision system to compare anomaly detection techniques for health monitoring of bridges. *Journal of Civil Structural Health Monitoring*, 12(6), 1299–1316. DOI: 10.1007/s13349-022-00617-w

Macrorie, R., Marvin, S., & While, A. (2021). Robotics and automation in the city: A research agenda. *Urban Geography*, 42(2), 197–217. DOI: 10.1080/02723638.2019.1698868

Mahato, S., & Paul, S. (2020). Classification of depression patients and normal subjects based on electroencephalogram (EEG) signal using alpha power and theta asymmetry. *Journal of Medical Systems*, 44(1), 1–8. DOI: 10.1007/s10916-019-1486-z PMID: 31834531

Maheswari, R., Gomathy, V., & Sharmila, P. (2018). Cancer detecting nanobot using positron emission tomography. *Procedia Computer Science*, 133, 315–322. DOI: 10.1016/j.procs.2018.07.039

Majcherczyk, N., & Pinciroli, C. (2020, May). SwarmMesh: A distributed data structure for cooperative multi-robot applications. In *2020 IEEE International Conference on Robotics and Automation (ICRA)* (pp. 4059-4065). IEEE.

Malik, A., Bashir, M., Lodhi, F. S., Jadoon, Z. G., Tauqir, A., & Khan, M. A. (2024). Depression, Anxiety and Stress using Depression, Anxiety, and Stress Scoring System (DASS-21) Among the Students of Women Medical and Dental College Abbottabad, Pakistan. [JIIMC]. *Journal of Islamic International Medical College*, 19(2), 103–107.

Manickam, P., Mariappan, S. A., Murugesan, S. M., Hansda, S., Kaushik, A., Shinde, R., & Thipperudraswamy, S. P. (2022). Artificial intelligence (AI) and internet of medical things (IoMT) assisted biomedical systems for intelligent healthcare. *Biosensors (Basel)*, 12(8), 562. DOI: 10.3390/bios12080562 PMID: 35892459

Manjunath, A., & Kishore, V. (2014). The promising future in medicine: Nanorobots. *Biomedical Science and Engineering*, 2(2), 42–47.

Manoonpong, P., Patanè, L., Xiong, X., Brodoline, I., Dupeyroux, J., Viollet, S., Arena, P., & Serres, J. R. (2021). Insect-inspired robots: Bridging biological and artificial systems. *Sensors (Basel)*, 21(22), 7609. DOI: 10.3390/s21227609 PMID: 34833685

Marcelloni, F., & Vecchio, M. (2008). A simple algorithm for data compression in wireless sensor networks. *IEEE Communications Letters*, 12(6), 411–413. DOI: 10.1109/LCOMM.2008.080300

Marcus, H. J., Ramirez, P. T., Khan, D. Z., Layard Horsfall, H., Hanrahan, J. G., Williams, S. C., Beard, D. J., Bhat, R., Catchpole, K., Cook, A., Hutchison, K., Martin, J., Melvin, T., Stoyanov, D., Rovers, M., Raison, N., Dasgupta, P., Noonan, D., Stocken, D., & Paez, A. (2024). The IDEAL framework for surgical robotics: Development, comparative evaluation and long-term monitoring. *Nature Medicine*, 30(1), 61–75. DOI: 10.1038/s41591-023-02732-7 PMID: 38242979

Marriwala, N., & Chaudhary, D. (2023). A hybrid model for depression detection using deep learning. *Measurement. Sensors*, 25, 100587.

Martinez-De-Dios, J. R., & Ollero, A. (2017). Integrated perception and control for a firefighting robot team. Robotics and Autonomous Systems, 90, 104-115. DOI: 10.1016/j.robot.2016.12.005

Martin, S., & Szeliga, W. (2010). A catalog of felt intensity data for 570 earthquakes in India from 1636 to 2009. *Bulletin of the Seismological Society of America*, 100(2), 562–569. DOI: 10.1785/0120080328

Matsushima, T., Yoshikawa, Y., Matsuo, K., Kurahara, K., Uehara, Y., Nakao, T., Ishiguro, H., Kumazaki, H., & Kato, T. A. (2024). Development of depression assessment tools using humanoid robots- Can tele-operated robots talk with depressive persons like humans? *Journal of Psychiatric Research*, 170, 187–194. DOI: 10.1016/j.jpsychires.2023.12.014 PMID: 38154335

McLurkin, J., & Yamins, D. (2005, June). Dynamic Task Assignment in Robot Swarms. In Robotics: Science and Systems (Vol. 8, No. 2005).

Militano, L., Arteaga, A., Toffetti, G., & Mitton, N. (2023). The cloud-to-edge- to-iot continuum as an enabler for search and rescue operations. *Future Internet*, 15(2), 55. DOI: 10.3390/fi15020055

Miller, A., & Barber, J. (2018). Autonomous robotic fire response: A review. Robotics, 7(3), 41. DOI: 10.3390/robotics7030041

Mirats-Tur, J. M., & Corominas Murtra, A. (2009). *A Survey on SLAM Techniques.* In *Proceedings of the International Conference on Intelligent Robots and Systems (IROS)*, pp. 2072-2077.

Mohapatra, A. K., & Mohanty, W. K. (2010, December). An overview of seismic zonation studies in India. In *Proc. Indian Geotechnical Conference, GEOtrendz, December* (pp. 16-18).

Mousavi, S. M., & Beroza, G. C. (2020). A machine-learning approach for earthquake magnitude estimation. Geophysical Research Letters, 47(1), e2019GL085976.

Mou, Y. (2020). Seismic Prediction Using Animal Behavior and Advanced Sensor Technologies. *International Journal of Geophysics*.

Mou, Y. (2021). Advanced Sensor Networks for Earthquake Prediction: A Review. *Seismological Research Letters*.

Mudaser, W., Padungweang, P., Mongkolnam, P., & Lavangnananda, P. (2021, December). Diabetic retinopathy classification with pre-trained image enhancement model. In 2021 IEEE 12th Annual Ubiquitous Computing, Electronics & Mobile Communication Conference (UEMCON) (pp. 0629-0632). IEEE.

Mukherjee, A., & Rakshit, M. (2016). Design and development of an intelligent firefighting robot. Procedia Computer Science, 92, 395-400. DOI: 10.1016/j.procs.2016.07.391

Nakata, T., Otsuki, K., & Khan, S. H. (1990). Active faults, stress field, and plate motion along the Indo-Eurasian plate boundary. *Tectonophysics*, 181(1-4), 83–95. DOI: 10.1016/0040-1951(90)90009-W

Nance, E., Pun, S. H., Saigal, R., & Sellers, D. L. (2021). Drug delivery to the central nervous system. *Nature Reviews. Materials*, 7(4), 314–331. DOI: 10.1038/s41578-021-00394-w PMID: 38464996

Narziev, N., Goh, H., Toshnazarov, K., Lee, S. A., Chung, K. M., & Noh, Y. (2020). STDD: Short-term depression detection with passive sensing. *Sensors (Basel)*, 20(5), 1396. DOI: 10.3390/s20051396 PMID: 32143358

Nasir, N., Oswald, P., Alshaltone, O., Barneih, F., Al Shabi, M., & Al-Shammaa, A. (2022, February). Deep DR: detection of diabetic retinopathy using a convolutional neural network. In 2022 Advances in Science and Engineering Technology International Conferences (ASET) (pp. 1-5). IEEE.

Nasr, M., Islam, M. M., Shehata, S., Karray, F., & Quintana, Y. (2021). Smart healthcare in the age of AI: Recent advances, challenges, and future prospects. *IEEE Access : Practical Innovations, Open Solutions*, 9, 145248–145270. DOI: 10.1109/ACCESS.2021.3118960

Nazir, T., Nawaz, M., Rashid, J., Mahum, R., Masood, M., Mehmood, A., Ali, F., Kim, J., Kwon, H. Y., & Hussain, A. (2021). Detection of diabetic eye disease from retinal images using a deep learning based CenterNet model. *Sensors (Basel)*, 21(16), 5283. DOI: 10.3390/s21165283 PMID: 34450729

Negenborn, R. R., & van de Wouw, N. (2015). Multi-agent systems and sensor integration for firefighting robotics. Control Engineering Practice, 37, 74-88. DOI: 10.1016/j.conengprac.2014.12.003

Ngo, T. Q. L., Wang, Y. R., & Chiang, D. L. (2021). Applying artificial intelligence to improve on-site non-destructive concrete compressive strength tests. *Crystals*, 11(10), 1157. DOI: 10.3390/cryst11101157

Nguyen, Q. H., Muthuraman, R., Singh, L., Sen, G., Tran, A. C., Nguyen, B. P., & Chua, M. (2020, January). Diabetic retinopathy detection using deep learning. In *Proceedings of the 4th international conference on machine learning and soft computing* (pp. 103-107).

Niku, S. B. (2020). *Introduction to robotics: analysis, control, applications.* John Wiley & Sons.

Nouyan, S., Groß, R., Bonani, M., Mondada, F., & Dorigo, M. (2009). Teamwork in Self-Organized Robot Colonies. *IEEE Transactions on Evolutionary Computation*, 13(4), 695–711. DOI: 10.1109/TEVC.2008.2011746

O'Hara, K., Roalter, L., & Simmel, D. (2008). *Swarm Intelligence Approaches to Robotic Surveillance.* In *Proceedings of the International Conference on Unmanned Aircraft Systems (ICUAS)*, pp. 1-10.

Olawade, D. B., David-Olawade, A. C., Wada, O. Z., Asaolu, A. J., Adereni, T., & Ling, J. (2024). Artificial intelligence in healthcare delivery: Prospects and pitfalls. *Journal of Medicine, Surgery, and Public Health*, 3, 100108. DOI: 10.1016/j.glmedi.2024.100108

Othman, N., & Khan, M. A. A. (2017). A hybrid fuzzy-logic and neural-network-based control system for firefighting robots. Journal of Intelligent & Fuzzy Systems, 32(2), 1477-1488. DOI: 10.3233/JIFS-162248

Padhan, S., Mohapatra, A., Ramasamy, S. K., & Agrawal, S. (2023). Artificial intelligence (AI) and Robotics in elderly healthcare: Enabling independence and quality of life. *Cureus*, 15(8). Advance online publication. DOI: 10.7759/cureus.42905 PMID: 37664381

Pamadi, A. M., Ravishankar, A., Nithya, P. A., Jahnavi, G., & Kathavate, S. (2022, March). Diabetic retinopathy detection using MobileNetV2 architecture. In *2022 International Conference on Smart Technologies and Systems for Next Generation Computing (ICSTSN)* (pp. 1-5). IEEE.

Panesar, D. K., & Shindman, B. (2012). The effect of segregation on transport and durability properties of self consolidating concrete. *Cement and Concrete Research*, 42(2), 252–264. DOI: 10.1016/j.cemconres.2011.09.011

Park, S., Kim, D., & Lee, S. (2021). Blockchain-based detection and mitigation of cryptojacking attacks in cloud environments. *2021 6th International Conference on Advanced Information Systems and Engineering (ICAISE)*, 1-5.

Park, J., & Lee, H. (2021). *Combining Ground Vibration Sensors and Electromagnetic Field Sensors for Earthquake Prediction*. Seismic Engineering Journal.

Parvathi, R., & Vignesh, U. (2023). Diabetic Retinopathy Detection Using Transfer Learning. In AI and IoT-Based Technologies for Precision Medicine (pp. 177-204). IGI Global.

Parvathi, R., Vignesh, U. Diabetic retinopathy detection using transfer learning

Paskal, A., & Stefan, M. (2019). Real-time navigation and control of a firefighting robot. IEEE Robotics and Automation Letters, 4(4), 3245-3252. https://doi.org/DOI: 10.1109/LRA.2019.2918237

Patel, M., Reddy, P., & Sharma, A. (2021). Artificial Neural Networks in Seismic Data Analysis for Earthquake Prediction. *Applied Geophysics*.

Patel, R., & Raut, R. (2023). Prediction Model Using Artificial Neural. *IEEE Network*.

Patel, R., Shah, D., & Patel, N. (2021). Integration of file integrity monitoring and endpoint detection and response for cryptojacking mitigation. *2021 International Conference on Computational Science and Computational Intelligence (CSCI)*, 276-280. IEEE.

Patel, S., & Kumar, R. (2022). The Role of Interdisciplinary Approaches in Earthquake Prediction. *Journal of Geophysical Research*.

Patil, R., Patil, S., Todakari, N., Devkar, A., & Raut, R. (2023, May). Earthquake Depth & Magnitude Prediction Model Using Artificial Neural Network. In 2023 4th International Conference for Emerging Technology (INCET) (pp. 1-5). IEEE. DOI: 10.1109/INCET57972.2023.10170413

Peng, T., & Zhou, X. (2016). An intelligent firefighting robot with an integrated multi-sensor system. Sensors, 16(12), 2100. DOI: 10.3390/s16122100

Podder, I., Fischl, T., & Bub, U. (2023, March). Artificial intelligence applications for MEMS-based sensors and manufacturing process optimization. In Telecom (Vol. 4, No. 1, pp. 165-197). MDPI.

Priya, A., Garg, S., & Tigga, N. P. (2020). Predicting anxiety, depression and stress in modern life using machine learning algorithms. *Procedia Computer Science*, 167, 1258–1267. DOI: 10.1016/j.procs.2020.03.442

Priya, A., Garg, S., & Tigga, N. P. (2020). Predicting anxiety, depression and stress in modern life using machine learning algorithms. Procedia Computer Science, 167, 1258-1267. 27. Islam, M. R., Kabir, 27. M. A., Ahmed, A., Kamal, A. R. M., Wang, H., & Ulhaq, A. (2018). Depression detection from social network data using machine learning techniques. *Health Information Science and Systems*, 6(1). Advance online publication. DOI: 10.1007/s13755-018-0046-0

Queralta, J. P., Taipalmaa, J., Pullinen, B. C., Sarker, V. K., Gia, T. N., Tenhunen, H., & Westerlund, T. (2020). Collaborative multi-robot search and rescue: Planning, coordination, perception, and active vision. *IEEE Access : Practical Innovations, Open Solutions*, 8, 191617–191643.

Ragno, L., Borboni, A., Vannetti, F., Amici, C., & Cusano, N. (2023). Application of social robots in healthcare: Review on characteristics, requirements, technical solutions. *Sensors (Basel)*, 23(15), 6820. DOI: 10.3390/s23156820 PMID: 37571603

Ratnakumar, R., & Vignesh, U. (2024). Machine Learning-Based Environmental, Social, and Scientific Studies Using Satellite Images: A Case Series. In AI and Blockchain Optimization Techniques in Aerospace Engineering (pp. 149-163). IGI Global.

Ratnakumar, R., & Vignesh, U. (2024). *Machine learning-based environmental, social, and scientific studies using satellite images: A case series*. AI and Blockchain Optimization Techniques in Aerospace Engineering. DOI: 10.4018/979-8-3693-1491-3.ch007

Reddy, S., Fox, J., & Purohit, M. P. (2019). Artificial intelligence-enabled healthcare delivery. *Journal of the Royal Society of Medicine*, 112(1), 22–28. DOI: 10.1177/0141076818815510 PMID: 30507284

Reddy, S., & Kumar, P. (2021). Real-Time Data Analysis for Earthquake Magnitude Prediction Using Neural Networks. *IEEE Transactions on Geoscience and Remote Sensing*.

Ribeiro, J., Lima, R., Eckhardt, T., & Paiva, S. (2021). Robotic process automation and artificial intelligence in industry 4.0–a literature review. *Procedia Computer Science*, 181, 51–58. DOI: 10.1016/j.procs.2021.01.104

Robert, A. F. J. (2005). 2005 Current Status of Nanomedicine and Medical Nanorobotics. *Journal of Computational and Theoretical Nanoscience*, 2, 1–25.

Robert, A. F. J. (2005). Microbivores: Artificial Mechanical Phagocytes using Digest and Discharge Protocol. *Journal of Evolution and Technology / WTA*, 14, 1–52.

Rouet-Leduc, B., Hulbert, C., Lubbers, N., Barros, K., Humphreys, C. J., & Johnson, P. A. (2017). Machine learning predicts laboratory earthquakes. *Geophysical Research Letters*, 44(18), 9276–9282. DOI: 10.1002/2017GL074677

Rubio, F., Valero, F., & Llopis-Albert, C. (2019). A review of mobile robots: Concepts, methods, theoretical framework, and applications. *International Journal of Advanced Robotic Systems*, 16(2), 1729881419839596. DOI: 10.1177/1729881419839596

Saha, H. N., Das, N. K., Pal, S. K., Basu, S., Auddy, S., Dey, R., . . . Maity, T. (2018, January). A cloud based autonomous multipurpose system with self-communicating bots and swarm of drones. In 2018 IEEE 8th annual computing and communication workshop and conference (CCWC) (pp. 649-653). IEEE.

Samad, T., Iqbal, S., Malik, A. W., Arif, O., & Bloodsworth, P. (2018). A multi-agent framework for cloud-based management of collaborative robots. *International Journal of Advanced Robotic Systems*, 15(4), 1729881418785073. DOI: 10.1177/1729881418785073

Sanjana, S., Shadin, N. S., & Farzana, M. (2021, November). Automated diabetic retinopathy detection using transfer learning models. In 2021 5th International Conference on Electrical Engineering and Information Communication Technology (ICEEICT) (pp. 1-6). IEEE.

Saranya, P., Devi, S. K., & Bharanidharan, B. (2022, March). Detection of diabetic retinopathy in retinal fundus images using densenet based deep learning model. In 2022 international mobile and embedded technology conference (MECON) (pp. 268-272). IEEE.

Sarkar, K., Shiuly, A., & Dhal, K. G. (2024). Revolutionizing concrete analysis: An in-depth survey of AI-powered insights with image-centric approaches on comprehensive quality control, advanced crack detection and concrete property exploration. *Construction & Building Materials*, 411, 134212. DOI: 10.1016/j.conbuildmat.2023.134212

Sasikumar, S., Santhakumar, S., Jayapal, R., & Thanigaivelan, R. (2024). Actuators in Medical Devices. In Robotics and Automation in Healthcare (pp. 137-149). Apple Academic Press.

Scaglione, A., & Servetto, S. D. (2002, September). On the interdependence of routing and data compression in multi-hop sensor networks. In *Proceedings of the 8th annual international conference on Mobile computing and networking* (pp. 140-147).

Scheper, K. Y., Karásek, M., De Wagter, C., Remes, B. D., & De Croon, G. C. (2018, May). First autonomous multi-room exploration with an insect-inspired flapping wing vehicle. In *2018 IEEE International Conference on Robotics and Automation (ICRA)* (pp. 5546-5552). IEEE. DOI: 10.1109/ICRA.2018.8460702

Schranz, M., Di Caro, G. A., Schmickl, T., Elmenreich, W., Arvin, F., Şekercioğlu, A., & Sende, M. (2021). Swarm intelligence and cyber-physical systems: Concepts, challenges and future trends. *Swarm and Evolutionary Computation*, 60, 100762. DOI: 10.1016/j.swevo.2020.100762

Sebastian, A., Elharrouss, O., Al-Maadeed, S., & Almaadeed, N. (2023, January 18). A Survey on Deep-Learning-Based Diabetic Retinopathy Classification. *Diagnostics (Basel)*, 13(3), 345. DOI: 10.3390/diagnostics13030345 PMID: 36766451

Selvi, R. T., Elakya, R., & Vignesh, U. (2024). Securing Sensitive Patient Data in Healthcare Settings Using Blockchain Technology. In Blockchain and IoT Approaches for Secure Electronic Health Records (EHR) (pp. 73-88). IGI Global.

Selvi, R. T., Elakya, R., & Vignesh, U. (2024). *Securing sensitive patient data in healthcare settings using blockchain technology, Blockchain and IoT Approaches for Secure Electronic Health Records.* EHR.

Shah, F. M., Ahmed, F., Joy, S. K. S., Ahmed, S., Sadek, S., Shil, R., & Kabir, M. H. (2020, June). Early depression detection from social network using deep learning techniques. In 2020 IEEE region 10 symposium (TENSYMP) (pp. 823-826). IEEE.

Shankar, K., Sait, A. R. W., Gupta, D., Lakshmanaprabu, S., Khanna, A., & Pandey, H. M. (2020). Automated detection and classification of fundus diabetic retinopathy images using synergic deep learning model. *Pattern Recognition Letters*, 133, 210–216. DOI: 10.1016/j.patrec.2020.02.026

Sharma, S., Yadav, R., & Tanwar, S. (2020). Cloud-based machine learning framework for detection of cryptojacking attacks. *2020 International Conference on Artificial Intelligence in Information and Communication (ICAIIC)*, 1-6.

Sharpe, T., & Webb, B. (1999). *Simulated and Situated Models of Chemical Trail Following in Ants.* In *Proceedings of the Fifth European Conference on Artificial Life (ECAL)*, pp. 317-324.

Sheth, H. C., Ray, J. S., Bhutani, R., Kumar, A., & Smitha, R. S. (2009). Volcanology and eruptive styles of Barren Island: An active mafic stratovolcano in the Andaman Sea, NE Indian Ocean. *Bulletin of Volcanology*, 71(9), 1021–1039. DOI: 10.1007/s00445-009-0280-z

Siderska, J. (2020). Robotic Process Automation—A driver of digital transformation? *Engineering Management in Production and Services*, 12(2), 21–31. DOI: 10.2478/emj-2020-0009

Singh, A., Vohra, V., & Bhatia, V. (2021). Detection of cryptojacking attacks using signature-based approach. *2021 6th International Conference on Intelligent Computing and Control Systems (ICICCS)*, 868-871.

Singh, M., & Gupta, R. (2020). Advances in Seismic Sensor Technology. *Sensors (Basel)*.

Smith, A., & Johnson, B. (2021). Integrating Biosensors and Machine Learning for Early Earthquake Detection. *Journal of Earthquake Engineering*.

Smith, L., & Green, J. (2022). Integrating Biosensors with Machine Learning Models for Earthquake Prediction. *Computational Geosciences*.

Soljacic, F., Law, T., Chita-Tegmark, M., & Scheutz, M. (2024). Robots in healthcare as envisioned by care professionals. *Intelligent Service Robotics*, 17(3), 1–17. DOI: 10.1007/s11370-024-00523-8

Spencer, B. F.Jr, Hoskere, V., & Narazaki, Y. (2019). Advances in computer vision-based civil infrastructure inspection and monitoring. *Engineering (Beijing)*, 5(2), 199–222. DOI: 10.1016/j.eng.2018.11.030

Spielberger, C. D. (1983). State-trait anxiety inventory for adults.

Srimadhur, N. S., & Lalitha, S. (2020). An end-to-end model for detection and assessment of depression levels using speech. *Procedia Computer Science*, 171, 12–21. DOI: 10.1016/j.procs.2020.04.003

Stern, A. F. (2014). The hospital anxiety and depression scale. *Occupational Medicine*, 64(5), 393–394. DOI: 10.1093/occmed/kqu024 PMID: 25005549

Sudarmadji, P. W., Pakan, P. D., & Dillak, R. Y. (2020, November). Diabetic retinopathy stages classification using improved deep learning. In *2020 International Conference on Informatics, Multimedia, Cyber and Information System (ICIMCIS)* (pp. 104-109). IEEE.

Suhail, M., Khan, A., Rahim, M. A., Naeem, A., Fahad, M., Badshah, S. F., & Janakiraman, A. K. (2022). Micro and nanorobot-based drug delivery: An overview. *Journal of Drug Targeting*, 30(4), 349–358.

Taufiqurrahman, S., Handayani, A., Hermanto, B. R., & Mengko, T. L. E. R. (2020, November). Diabetic retinopathy classification using a hybrid and efficient MobileNetV2-SVM model. In 2020 IEEE Region 10 Conference (Tencon) (pp. 235-240). IEEE.

Thoduparambil, P. P., Dominic, A., & Varghese, S. M. (2020). EEG-based deep learning model for the automatic detection of clinical depression. *Physical and Engineering Sciences in Medicine*, 43(4), 1349–1360. DOI: 10.1007/s13246-020-00938-4 PMID: 33090373

Thomalla, F., & Larsen, R. K. (2010). Resilience in the context of tsunami early warning systems and community disaster preparedness in the Indian Ocean region. *Environmental Hazards*, 9(3), 249–265. DOI: 10.3763/ehaz.2010.0051

Tiberti, G., Minelli, F., & Plizzari, G. (2015). Cracking behavior in reinforced concrete members with steel fibers: A comprehensive experimental study. *Cement and Concrete Research*, 68, 24–34. DOI: 10.1016/j.cemconres.2014.10.011

Tran, M.-N., & Kim, Y. (2021). Named data networking based disaster response support system over edge computing infrastructure. *Electronics (Basel)*, 10(3), 335. DOI: 10.3390/electronics10030335

Tripathi, R., Kumar, A., & Kumar, A. (2020). Architecture and application of nanorobots in medicine. In *Control Systems Design of Bio-Robotics and Bio-mechatronics with Advanced Applications* (pp. 445–464). Academic Press.

Types of depression. (n.d.). Beyond Blue.(2023) https://www.beyondblue.org.au/mental-health/depression/types-of-depression

Uddin, M. Z., Dysthe, K. K., Følstad, A., & Brandtzaeg, P. B. (2022). Deep learning for prediction of depressive symptoms in a large textual dataset. *Neural Computing & Applications*, 34(1), 721–744. DOI: 10.1007/s00521-021-06426-4

Varadharajan, V. S., St-Onge, D., Adams, B., & Beltrame, G. (2020). Soul: Data sharing for robot swarms. *Autonomous Robots*, 44(3), 377–394. DOI: 10.1007/s10514-019-09855-2

Verma, B., Yudheksha, G. K., & Sanjana Reddy, P. (2023). Parvathi, R., Vignesh, U. An Intelligent Flood Automation System Using IoT and Machine Learning. *Advances in Transdisciplinary Engineering*, 32, 444–449.

Verma, B., Yudheksha, G. K., Sanjana Reddy, P., & Vignesh, U. (2023). An Intelligent Flood Automation System Using IoT and Machine Learning. In *Recent Developments in Electronics and Communication Systems* (pp. 444–449). IOS Press.

Vignesh, P., Maheswari, R., Vijaya, P., & Vignesh, U. (2024). Machine Learning for Aerospace Object Categorization. In AI and Blockchain Optimization Techniques in Aerospace Engineering (pp. 164-180). IGI Global.

Vignesh, U., & Ratnakumar, R. (2024). An Empirical Review on Clustering Algorithms for Image Segmentation of Satellite Images. AI and Blockchain Optimization Techniques in Aerospace Engineering, 33-52.

Vignesh, U., Parvathi, R., & Goncalves, R. (2023). *"Structural and Functional Data Processing in Bio-Computing and Deep Learning." Structural and Functional Aspects of Biocomputing Systems for Data Processing 2023*. IGI Global.

Vignesh, U., Parvathi, R., & Goncalves, R. (2023). *Structural and functional aspects of biocomputing systems for data processing*. Structural and Functional Aspects of Biocomputing Systems for Data Processing. DOI: 10.4018/978-1-6684-6523-3

Vignesh, U., & Ratnakumar, R. (2024). *An empirical review on clustering algorithms for image segmentation of satellite images*. AI and Blockchain Optimization Techniques in Aerospace Engineering. DOI: 10.4018/979-8-3693-1491-3.ch002

Vignesh, U., Ratnakumar, R., & Al-Obaidi, A. S. M. (2024). *AI and blockchain optimization techniques in aerospace engineering*. AI and Blockchain Optimization Techniques in Aerospace Engineering. DOI: 10.4018/979-8-3693-1491-3

Wamba, S. F., Queiroz, M. M., & Hamzi, L. (2023). A bibliometric and multidisciplinary quasi-systematic analysis of social robots: Past, future, and insights of human-robot interaction. *Technological Forecasting and Social Change*, 197, 122912. DOI: 10.1016/j.techfore.2023.122912

Wang, G., Kong, Y., Sun, T., & Shui, Z. (2013). Effect of water–binder ratio and fly ash on the homogeneity of concrete. *Construction & Building Materials*, 38, 1129–1134. DOI: 10.1016/j.conbuildmat.2012.09.027

Wang, J., Bai, Y., & Xia, B. (2020). Simultaneous diagnosis of severity and features of diabetic retinopathy in fundus photography using deep learning. *IEEE Journal of Biomedical and Health Informatics*, 24(12), 3397–3407. DOI: 10.1109/JBHI.2020.3012547 PMID: 32750975

Wang, T., & Li, X. (2021). Continuous Monitoring of Electromagnetic Anomalies for Earthquake Prediction. *Geophysical Journal International*.

Wang, Y., Wang, Y., Mushtaq, R. T., & Wei, Q. (2024). Advancements in Soft Robotics: A Comprehensive Review on Actuation Methods, Materials, and Applications. *Polymers*, 16(8), 1087. DOI: 10.3390/polym16081087 PMID: 38675005

Wang, Y., Zhang, L., & Wu, X. (2022). Cloud-based behavioral analysis framework for detection of cryptojacking attacks. *Proceedings of the 2022 International Conference on Cyber Security and Cloud Computing (CSCC)*, 67-72.

Wang, Z., Li, X., Yao, M., Li, J., Jiang, Q., & Yan, B. (2022). A new detection model of microaneurysms based on improved FC-DenseNet. *Scientific Reports*, 12(1), 1–9. DOI: 10.1038/s41598-021-04750-2 PMID: 35046432

Weisong, J. (n.d.). The Sensory Capabilities of Snakes in Earthquake Prediction. (Details of publication not provided).

Werger, B. B., & Mataric, M. J. (1996). Robotic "Food" Chains: Externalization of State and Program for Minimal-Agent Foraging. In *From Animals to Animats 4: Proceedings of the Fourth International Conference on Simulation of Adaptive Behavior* (pp. 625-634). MIT Press.

Williams, J., & Brown, H. (2021). *Enhancing Earthquake Prediction Accuracy with Biosensors*. Earthquake Science.

Wodrich, M., & Bilchev, G. (1997). Cooperative Distributed Search: The Ants' Way. *Control and Cybernetics*, 26(3), 413–446.

Wowk, B. (1988). *Cell repair technology* (Vol. 21-30). Cryonics.

Wu, S., Chen, T., & Guo, X. (2020). Secure container deployment for cryptojacking prevention. *2020 International Conference on Intelligent Transportation, Big Data & Smart City (ICITBS)*, 1-

Yang, G., Pang, Z., Deen, M. J., Dong, M., Zhang, Y. T., Lovell, N., & Rahmani, A. M. (2020). Homecare robotic systems for healthcare 4.0: Visions and enabling technologies. *IEEE Journal of Biomedical and Health Informatics*, 24(9), 2535–2549. DOI: 10.1109/JBHI.2020.2990529 PMID: 32340971

Zhang, D., & Wang, D. (2019, April). Heterogeneous social sensing edge computing system for deep learning based disaster response: demo abstract. In *Proceedings of the International Conference on Internet of Things Design and Implementation* (pp. 269-270).

Zhang, Y., & Liu, H. (2021). The Use of Magnetometers in Seismic Anomaly Detection. *Journal of Applied Geophysics*.

Zhang, Y., Liu, H., & Wang, X. (2020). Utilization of Induction Coil Magnetometers for Seismic Anomaly Detection. *Geophysical Research Letters*.

Zhu, C., Wang, H., Liu, X., Shu, L., Yang, L. T., & Leung, V. C. (2014). A novel sensory data processing framework to integrate sensor networks with mobile cloud. *IEEE Systems Journal*, 10(3), 1125–1136. DOI: 10.1109/JSYST.2014.2300535

Zhu, J., Wang, Z., Gong, T., Zeng, S., Li, X., Hu, B., Li, J., Sun, S., & Zhang, L. (2020). An improved classification model for depression detection using EEG and eye tracking data. *IEEE Transactions on Nanobioscience*, 19(3), 527–537. DOI: 10.1109/TNB.2020.2990690 PMID: 32340958

Zhu, T., Liu, X., Wang, J., Kou, R., Hu, Y., Yuan, M., Yuan, C., Luo, L., & Zhang, W. (2023). Explainable machine-learning algorithms to differentiate bipolar disorder from major depressive disorder using self-reported symptoms, vital signs, and blood-based markers. *Computer Methods and Programs in Biomedicine*, 240, 107723. DOI: 10.1016/j.cmpb.2023.107723 PMID: 37480646

Zou, Y., Zhang, W., & Zhang, Z. (2016). Liftoff of an electromagnetically driven insect-inspired flapping-wing robot. *IEEE Transactions on Robotics*, 32(5), 1285–1289. DOI: 10.1109/TRO.2016.2593449

About the Contributors

U. Vignesh, is currently an Assistant Professor Senior Grade 2 in School of Computer Science and Engineering, Vellore Institute of Technology (VIT) - Chennai. Prior to his recent appointment at the VIT, he was a Post-Doctoral Fellow in National Institute of Technology (NIT), Trichy – India. Dr. Vignesh received his undergraduate degree in B.Tech (IT) as well as his M.Tech (IT) degree from Anna University - Chennai, and his PhD in Computer Science and Engineering from VIT University - Chennai. Dr. Vignesh published several papers in preferred Journals, patents and chapters in books, and participated in a range of forums on computer science, social science, etc. He also presented various academic as well as research-based papers at several national and international conferences. His research activities are currently twofold: while the first research activity is set to explore the developmental role that society needs with technology such as, Artificial Intelligence; the second major research theme that he is pursuing is focused on the bioinformatics and data mining.

Annavarapu Chandra Sekhara Rao is currently an Associate Professor in Department of Computer Science and Engineering, Indian Institute of Technlogy (ISM), Dhanbad, Jharkhand.

A. Saleem Raja is an Associate Professor and researcher with 16+ years of experience teaching courses in both undergraduate and postgraduate levels. Published over 26 articles in peer-reviewed journals and conferences. Trained project students in bachelor and master level in field of web application development, mobile application development, IoT application development, cisco networks and python data science. Organized practical workshops for students and teachers in reputed colleges. Supervised more than 5 applied research projects in Oman.

***

Sharmistha Dey is currently working as Assistant Professor in the School of Computing Applications and Technology. She has a total experience of more than 15 years of teaching and one year administrative experience in IMT Ghaziabad. She has been associated with Galgotias University since November 2023. She is a firm believer in productivity and efficiency at work. She Exhibits an honest work ethic and the ability to excel in a fast-paced, time-sensitive environment. Being a passionate teacher, she believes that teaching is not merely restricted to making the students understand the underlying concepts of a course but also to developing critical thinking and evaluating alternate approaches for problem-solving. She always puts his efforts toward the overall development of her students.

S. Geetha received the B.E., from the Madurai Kamaraj University, M.E., and Ph.D. degrees in Computer Science and Engineering from Anna University, Chennai, in 2000, 2004 and 2011 respectively. She has 14+ years of teaching experience. Currently, she is a professor at School of Computing Science and Engineering at VIT-University, Chennai Campus. She has published more than 50 papers in reputed IEEE International Conferences and refereed Journals. She joins the review committee for IEEE Transactions on Information Forensics and Security and IEEE Transactions on Image Processing, Springer Multimedia Tools and Security, Elsevier – Information Sciences. She was an editor for the Indian Conference proceedings of ICCIIS 2007 and RISES-2013. Her research interests include multimedia security, intrusion detection systems, machine learning paradigms and information forensics. She is a recipient of University Rank and Academic Topper Award in B.E. and M.E. in 2000 and 2004 respectively. She is also a pride recipient of the "Best Academic Researcher Award 2013" of ASDF Global Awards.

Monica Gupta, a distinguished scholar in Electronics and Communication Engineering, earned her Ph.D. from Delhi Technological University in 2022, highlighting a relentless commitment to academic excellence. With over 17 years of teaching experience, she currently serves as an Associate Professor at Bharati Vidyapeeth's College of Engineering, New Delhi. A prolific researcher, Dr. Gupta has made substantial contributions in Low Power Memory Design, Image and Video Processing, VLSI, IoT, Artificial Intelligence and Machine Learning, evident in her numerous publications in reputable journals and conferences. Beyond academia, she has initiated the IOSC-BVP student club, in collaboration with Intel OneAPI Student ambassador, addressing the skill gap between graduates and industry expectations. Inaugurated on November 7, 2022, the club focuses on skill enhancement programs, organizing national-level events to deepen students' understanding of crucial areas like Artificial Intelligence, Internet of Things, Machine Learning, and Website Development. Dr. Monica Gupta's impact extends beyond the classroom, shaping the future of engineering by bridging academic knowledge with industry needs. Her dynamic leadership and commitment to excellence make her a trailblazer in the landscape of Electronics and Communication Engineering education.

P. Jeevanasree is a student in the Department of Computer Science and Engineering. Vel Tech Rangarajan Dr. Sagunthala R&D Institute of Science and Technology. Avadi, Chennai, India.She is currently pursuing her B.E degree in Computer Science and Engineering in Vel Tech University.

S.P. Gayathri finished Ph.D. in the Department of Computer Science and Applications in Gandhigram Rural Institute (DU), Dindigul, TN, India and currently working as a Guest Teacher in the same department. She has 14 years of teaching experience in the field of computer science. She has published many research articles in reputed journals and contributed book chapters. Her research interest is Digital and Medical Image Processing.

Karthigai Selvi S. is currently working as an assistant professor in the School of Computing Applications and Technology. She has a total experience of more than 12 years of teaching. She has been associated with Galgotias University since 2023. She served as a project consultant at Mother Teresa's Women's University, India. Her research work focuses on human brain image processing and sustainable development. She bagged a best presenter award at the 4th World Environment Summit in 2023 and a national award at the Student Research Convention held in 2013. She has published more than thirty papers in reputed journals (SCI, Scopus, and Web of Science) and at international and national conferences. She has served as a resource person at universities and colleges. She is passionate about teaching and shows much interest in promoting students to acquire a good career. She is actively involved in motivating the students to participate in many competitions.

Krishan veer Singh is currently working as an Assistant Professor in Galgotias University. He has completed his MCA, M Tech and PhD from Jawaharlal Neheru University, New Delhi. He has more than 15 years of teaching experience in various reputed organizations of Delhi University. His research interest is cloud computing, grid computing,computational intelligence. He has a good number of publications in reputed journal and conferences